AF388997

Decarbonization and Circular Economy in the Sustainable Development and Renovation of Buildings and Neighborhoods

Decarbonization and Circular Economy in the Sustainable Development and Renovation of Buildings and Neighborhoods

Editors

Pilar Mercader-Moyano
Paula M. Esquivias

MDPI • Basel • Beijing • Wuhan • Barcelona • Belgrade • Manchester • Tokyo • Cluj • Tianjin

Editors
Pilar Mercader-Moyano
Universidad de Sevilla
Spain

Paula M. Esquivias
Universidad de Valladolid
Spain

Editorial Office
MDPI
St. Alban-Anlage 66
4052 Basel, Switzerland

This is a reprint of articles from the Special Issue published online in the open access journal *Sustainability* (ISSN 2071-1050) (available at: https://www.mdpi.com/journal/sustainability/special_issues/Decarbonization_Renovation_Buildings_Neighborhoods).

For citation purposes, cite each article independently as indicated on the article page online and as indicated below:

LastName, A.A.; LastName, B.B.; LastName, C.C. Article Title. *Journal Name* **Year**, *Article Number, Page Range.*

ISBN 978-3-03943-479-4 (Hbk)
ISBN 978-3-03943-480-0 (PDF)

Contents

About the Editors

Pilar Mercader-Moyano (Dr) has a PhD in Architecture, Master's in Architecture and Historical Heritage and in Expert Assessment and Repair of Buildings. She is a Senior Lecturer at the Department of Architectural Constructions I, High Technical School of Architecture, University of Seville. She has been focused on continuing professional development enriched by research and professional experience in Architecture for the last twenty years. She combines her research and teaching activities with her professional activities, working as an architect in public administration for eight years and with her own architecture studio where she has developed refurbishment, expert analysis, and reparation projects linked to research projects in the field of sustainable construction design and environmental impact of buildings.

Paula M. Esquivias (Dr) is a building scientist with a background in Architecture and Architectural Engineering and Master's in Architecture and Historical Heritage working in the field of sustainable building design and environmental modelling. She is as Postdoctoral Researcher at the University of Valladolid (UVa), Spain, where she is nowadays researching about the optimization of building thermal systems. Previously, she has participated in several research projects at the University of Seville (US), Spain, covering aspects such as daylighting and visual comfort, thermal conditioning, thermal systems, renewable systems, and passive strategies for indoor environmental comfort and quality. Her main research line is the effect of solar radiation on buildings and human comfort, covering mainly visible and infrared radiation and their interconnection. This interest has been continuously reflected in her research, teaching, and practised activities over the last thirteen years.

 *sustainability*

Editorial

Decarbonization and Circular Economy in the Sustainable Development and Renovation of Buildings and Neighbourhoods

Pilar Mercader-Moyano [1,*] **and Paula M. Esquivias** [2,*]

[1]   Departamento de Construcciones Arquitectónicas I, Escuela Técnica Superior de Arquitectura, Universidad de Sevilla, Avenida de Reina Mercedes n°2, 41012 Seville, Spain
[2]   GIR Termotecnia, Departamento de Ingeniería Energética y Fluidomecánica, Universidad de Valladolid, Paseo del Cauce n° 59, 47011 Valladolid, Spain
*   Correspondence: pmm@us.es (P.M.-M.); paula.esquivias@uva.es (P.M.E.)

Received: 21 September 2020; Accepted: 22 September 2020; Published: 24 September 2020

Abstract: In recent years, the building sector has been turning towards intervening in the existing city building stock. In fact, it is generally accepted that the refurbishment of buildings and the urban regeneration based on sustainability must form the axis of reformulation of the building sector. Nowadays, achieving sustainable urban development inevitably involves improving existing buildings, thereby preventing the need for city growth, and for the emptying of established neighbourhoods. Furthermore, considering the whole life cycle, it is well known the great amount of greenhouse emissions derived from the construction sector, so in order to reach a decarbonized society it is important to provide eco-efficient construction materials and solutions, adding the principles of circular economy and resource efficiency. The articles of this special issue show different aspects to be considered in order to reach a decarbonized and circular building stock.

Keywords: circular economy; decarbonization of construction sector; refurbishment of buildings; neighbourhood regeneration; eco-efficient construction solutions; construction and waste management; indoor environmental quality; resource efficiency

---

1. Special Issue "Decarbonization and Circular Economy in the Sustainable Development and Renovation of Buildings and Neighbourhoods"

In recent years, the building sector has been turning towards intervening in the existing city building stock. In fact, it is generally accepted that the refurbishment of buildings and urban regeneration based on sustainability must form the axis of reformulation of the building sector. At present, achieving sustainable urban development inevitably involves improving existing buildings, thereby preventing the need for city growth, and for the emptying of established neighbourhoods. Furthermore, considering the whole life cycle, the great amount of greenhouse emissions derived from the construction sector is well known; thus, in order to reach a decarbonized society, it is important to provide eco-efficient construction materials and solutions, adding the principles of circular economy and resource efficiency.

Therefore, the theme of this Special Issue is "Decarbonization and Circular Economy in the Sustainable Development and Renovation of Buildings and Neighbourhoods" in response to the objectives not only raised at Horizon 2020 but by all the people who seek a more sustainable world. This Special Issue of *Sustainability* focuses on, but is not limited to:

- Obtaining an overview of the environmental problems that arise from construction activity, focusing on refurbishment as an alternative to the current crisis in the construction sector, as well as on actions designed to minimize environmental effects on the environment;

- Searching for new alternatives to conventional construction solutions that minimize the environmental impact of the construction activity, improve indoor environmental quality of buildings, build or refurbish, always from a rentable and optimal cost in time, and implement a circular economy and an efficient resource and waste management;
- Minimizing the consumption of material resources, energy consumption and CO_2 emissions in construction and looking for the proper management of construction and demolition waste and the opportunities for their recycling and reuse;
- Sustainable planning and urban development, for an ordered and sustainable growth.

2. Background and Contents

The European Union (EU) is committed to establish a sustainable, competitive, secure and decarbonised energy system [1]. Furthermore, the European Green Deal [2] boosts the efficient use of resources in order to get a clean and circular economy.

Buildings have a capital importance in both communications as the building sector in the European Union represents around 36% of greenhouse emissions, 40% of the final energy consumption, 50% of extracted materials, 30–50% material resource use, 21% of total water abstracted and is responsible for over 35% of the EU's total waste generation with very significant life cycle impacts, particularly associated with extraction and processing stages [1–6].

Decarbonization means a decrease in the specific amount of carbon (or CO_2) emitted per unit of primary energy consumed. In the building sector, decarbonization can be achieved through the energy efficiency measures focused on reducing the energy demand, reducing the energy consumption and increasing the use of low-carbon technologies, such as renewable energy sources. Decarbonization in the building sector involves the materiality of the built environment throughout their whole life cycle, users' energy habits and the performance and efficiency of the building and neighbourhood systems [4].

A circular economy is one "that is restorative and regenerative by design and aims to keep products, components and materials at their highest utility and value at all times" [7]. In a circular economy, waste and resource use are minimised, and when a product reaches the end of its life it is used again to create further value [8]. In the building sector, the circular economy is applied by closing water, energy and material loops, so design for disassembly has a central role and promotes buildings, construction materials and products to be intentionally designed for material recovery, value retention and meaningful next use [9,10].

In this context, the new Circular Economy Action Plan [5] promotes circularity principles throughout the lifecycle of buildings in order to increase the material efficiency and reduce the climate impacts. These goals are aligned to the 'Renovation Wave' initiative [11] as around 60% of the current EU's buildings were built under limited or non-existent energy efficiency requirements [12,13] so they require a partial or total refurbishment in order to fulfil the EU's Energy Efficiency objectives for 2050 [12,14]. Furthermore, the current inefficient building stock is at the base of the energy poverty that around 50 million Europeans suffer [2], affecting their comfort, sanitation and living conditions.

Energy renovation strategies of social housing are addressed in the article written by Belinda López-Mesa, Marta Monzón-Chavarrías and Almudena Espinosa-Fernández [15]. The two studied strategies are compared based on the regulation compliance, the energy use, the emissions reduction, the thermal comfort and the cost-effectiveness of the measures. The results show that renovating the whole building with efficient solutions, including external insulation for existing brick-facing walls, and drawing up coordination plans in order to guarantee a coordinated new image of the residential estates is the best strategy, achieving a 69%–79% reduction of CO_2 emissions and a 75%–97% reduction of heating use, as well as being the only one capable of transforming these buildings into nearly zero energy buildings (nZEBs).

One example on reducing energy consumption by implementing low-carbon technologies is exposed by Eloy Velasco-Gómez, Ana Tejero-González, Javier Jorge-Rico and F. Javier Rey-Martínez [16].

Evaporative cooling systems are energy-efficient, cost-effective and environmentally friendly solutions for cooling the indoor environment by evaporating water in dry air. It is a technology that replicates a natural process with a low energy consumption, so it can substitute, improve the performance of, or complement conventional mechanical compression cooling systems. The system depends on the wet surface in contact with the air, so the authors have designed a cooling pad that enables maximum wetted surface with minimum pressure drop so it improves the currently high efficiency of the system.

However, decarbonising the building stock must look beyond the energy consumption during the use phase of the buildings. The energy used in the manufacture of construction product and during the construction process of the building also has an important role. Manufacturing construction products represents 5–10% of total energy consumption in the EU [17] and the greenhouse emissions related to material extraction and manufacturing of construction products are estimated at 5–12% of total national greenhouse gas (GHG) emissions in the EU [5]. Furthermore, the material requirement of buildings currently represents one of the greatest resource use challenges in terms of mass of resources used. Concrete, aggregate materials (sand, gravel and crushed stone) and bricks make up to 90% (by weight) of all materials used [6].

In this context, looking for alternatives that minimize the extraction of raw materials, incorporate recycled materials and reduce the energy consumption related to the manufacture of construction products is another way to move towards decarbonization and a circular building sector. This approach is addressed by the article written by Cătălina Mihaela Grădinaru, Radu Muntean, Adrian Alexandru Șerbănoiu, Vasilică Ciocan and Andrei Burlacu [18]. Focused on the manufacture of concrete, they present how to rationalise its use of natural mineral aggregates and reduce its energy consumption by replacing mineral resources with vegetal waste for non-structural concretes. The exposed ecological concretes substitute 50% of their mineral aggregate volume, thus reducing the extraction of raw materials, and have mechanical properties used for concrete screeds or for modular elements for buildings closures.

The article written by Carolina Piña Ramírez, Alejandra Vidales Barriguete, Julián García Muñoz, Mercedes del Río Merino and Patricia del Solar Serrano [19] is also focused in the replacement of raw materials for sustainable alternatives. In this case, recycled materials are used as reinforcement of cement mortars used for external continuous coating as a mean to seek new uses for construction and demolition waste from buildings. The explored options were characterized obtaining similar values to those cement mortars reinforced with commercial fibres, so it is a viable way to reduce raw material extraction and energy consumption from manufacture, and reuse construction and demolition waste.

Pilar Mercader-Moyano and Jesús Roldán-Porras [20] go beyond manufacture of materials and explore the greenhouse gas emissions related to the phases related to transportation of materials to the worksite and the construction process, applied to foundations and structures. The variables related to those phases are not contemplated in the sustainability assessment in the Spanish legislation and the results they present show the importance of these variables to be considered when choosing the lower contaminant option.

Apart from the contribution of certain phases of the life cycle, it is important to consider the environmental impact of the whole life cycle of a building. This is the focus of the article written by Patricia González-Vallejo, Radu Muntean, Jaime Solís-Guzmán and Madelyn Marrero [21]. The authors compare the environmental impact of housing projects in Spain and Romania using the OERCO2 tool considering the difference in the constructive solutions for structures in both countries. A total of 24 projects were assessed, differentiated by the material of their structure and according to the type of their foundation. They found that buildings with metallic structures were more economical but more contaminant than reinforced concrete, and they ranked the impact of the materials being metal, concrete, cement and ceramic products those with greater environmental impact.

One of the circular economy principles regarding materials is related to their durability. This concept is also related to the economic investment for the replacement of the material or construction product. These aspects should be carefully taken into account especially when working on social housing, as their inhabitants are economically, and frequently energy, vulnerable. The article written by Antonio Dominguez-Delgado, Helena Domínguez-Torres and Carlos-Antonio Domínguez-Torres [22] studies not only the energy impact but also the economic one of cool roof constructive solutions through their life cycle in order to evaluate their performance over time as roofs are elements continuously exposed to outdoor climate conditions affecting to the durability of the solution.

Ernesto Antonini, Andrea Boeri, Massimo Lauria and Francesca Giglio also studied durability as an indicator for circular technologies, added to the reversibility of such technologies as design for disassembly is one of the principles of the circular design approach [23]. As the authors indicate "no clear definitions are yet available to link these concepts to indicators suitable for measuring Circular Economy and, specifically, Circular Building Technologies", so they discuss whether reversibility and durability may represent suitable indicators in order to promote the circular assessment of construction technologies, providing useful assessment tools to designers to assist their choices early.

A circular building sector must optimise the use of resources and result in zero waste to landfill. The measures are based on incorporating recycled content and reducing resource waste in the manufacturing process of construction products; minimise the use of new materials, select environmental-friendly construction products, and establish an adequate construction and demolition waste (CDW) management in order to recover quality materials, as CDW is the largest waste stream in EU [17,24–26]. In this context, design for disassembly, one pillar of the circular economy, includes the reduction of CDW as it promotes that buildings and products are designed intentionally for material recovery, value retention and meaningful next use [10].

In this sense, Eduardo Vázquez-López, Federico Garzia, Roberta Pernetti, Jaime Solís-Guzmán and Madelyn Marrero [27] analyse the costs that imply the selective dismantling and waste management which also have a repercussion on the environmental performance of a demolition project. They develop a method to calculate the end-of-life cost through the quantification of demolished materials considering demolition, load, transport and waste management costs. They found that demolition cost was the weight factor and it was mostly due to foundation and structure demolition. Compared to the life cycle cost, the end-of-life cost represents approximately 5% of the total.

CDW also has to be considered when executing refurbishments in order to update existing building to current legislative requirements. This is the focus of the article written by Begoña Blandón, Luís Palmero and Giacomo di Ruocco [28]. The authors studied the CDW management impact on interventions destinated to repopulate the abandoned rural areas as these areas do not count with the facilities for waste management of the cities, and repopulating the rural areas is another way to move towards decarbonization and circular economy as it reuses and revalue existing built environments. The authors analysed interventions carried out in recent years, reviewed the existing European management protocols specifying their application to Mediterranean popular housing and presented a representative case in order to be conscious about the existing reality regarding the destiny of the CDW in the rural environment.

However, users are also participants on the sustainability of the building stock and the built environment, and they have an important role so tools are needed to transmit, in an understandable manner, how their decisions related to buildings can be sustainable. The article written by Milagrosa Borrallo-Jiménez, Maria Lopez de Asiain, Rafael Herrera-Limones and María Lumbreras Arcos [29] defines a method for the development of tools able to transmit to society scientific knowledge related to sustainability, evaluating the environmental impact of their actions in order to actively involve the actors towards a more sustainable built environment. They present a specific tool experimentally developed for the context of Seville prioritizing communicative actions in order to reach citizens and professionals, but also researchers in the building sector.

The participatory dimension not only has an influence on the built properties of the persons; it can go beyond and be applied to the urban regeneration actions and projects, as shown in the article written by Maria Lopez De Asiaín and Vicente Díaz-García [30]. The authors reviewed the participatory dimension of different urban regeneration actions carried out in Spain and the impact of this dimension on the results obtained by environmental, economic and social urban improvements in order to improve the resilience of the cities. They analysed several cases in order to assess their effectiveness and relevance of the actions on the urban development and resilience. From their study, success and fail reasons can be obtained, identifying weaknesses for future participatory actions.

The final reflection of this editorial is that although several cases shown before are thorough and applied to refurbishment, which is important as is a mean to reuse and revalue the existing buildings, the principles of decarbonization and circular economy must be incorporated at the very early stages of building and urban design. In this context, environmental assessment schemes are playing an important role moving the building sector towards its sustainability. The article written by María M. Serrano-Baena, Paula Triviño-Tarradas, Carlos Ruiz-Díaz and Rafael E. Hidalgo Fernández [31] analyses the implications of the BREEAM scheme on the design of hotels for the Spanish context. They present seven case studies and the measures implemented in order to achieve the scheme targets and conclude that although considering the requirements for getting a good environmental assessment may limit the hotel design, it is easier implemented at the early design stage so these limitations can be reduced.

Author Contributions: Conceptualization and methodology, P.M.-M.; investigation and writing—original draft preparation, P.M.E.; writing—review and editing and supervision, P.M.-M. All authors have read and agreed to the published version of the manuscript.

Funding: This research received no external funding.

Acknowledgments: We are grateful to the authors of the contributions to this special issue for their valuable research towards a more decarbonized and circular building sector.

Conflicts of Interest: The authors declare no conflict of interest.

References

1. European Commission. *A Clean Planet for All. A European Strategic Long-Term Vision for a Prosperous, Modern, Competitive and Climate Neutral Economy*; Communication from the Commission COM (2018) 773 final; European Commission: Brussels, Belgium, 2018.
2. European Commission. *The European Green Deal*; Communication from the Commission COM (2019) 640 final; European Commission: Brussels, Belgium, 2019.
3. European Commission. *Clean Energy for All Europeans*; Communication from the Commission COM (2016) 860 final; European Commission: Brussels, Belgium, 2016.
4. European Commission. *Sustainable Products in a Circular Economy-towards an EU Product Policy Framework contributing to the Circular Economy*; Commission Staff Working Document SWD (2019) 91 final; European Commission: Brussels, Belgium, 2019.
5. European Commission. *A New Circular Economy Action Plan: For a Cleaner and More Competitive Europe*; COM (2020) 98 final; European Commission: Brussels, Belgium, 2020.
6. Herczeg, M.; McKinnon, D.; Milios, L.; Bakas, I.; Klaassens, E.; Svatikova, K.; Widerberg, O. *Resource Efficiency in the Building Sector*; DG Environment FEA91117; Ecorys: Rotterdam, The Netherlands, 2014.
7. MacArthur Foundation. Circular Economy Reports & Publications from the Ellen MacArthur Foundation. Available online: https://www.ellenmacarthurfoundation.org/assets/downloads/publications/EllenMacArthurFoundation_Growth-Within_July15.pdf (accessed on 15 July 2020).
8. European Commission. *Closing the Loop-An EU Action Plan for the Circular Economy*; Communication from the Commission to the European Parliament, the Council, the European Economic and Social Committee and the Committee of the Regions COM(2015) 614 final; European Commission: Brussels, Belgium, 2015.
9. Pomponi, F.; Moncaster, A. Circular economy for the built environment: A research framework. *J. Clean. Prod.* **2017**, *143*, 710–718. [CrossRef]
10. Faculty of Architecture and the Built Environment, Delft University of Technology. Circular Built Environment. 2020. Available online: https://www.tudelft.nl/en/architecture-and-the-built-environment/research/research-themes/circular-built-environment/ (accessed on 10 June 2020).
11. European Commission. Renovation Wave. Energy. Available online: https://ec.europa.eu/energy/topics/energy-efficiency/energy-efficient-buildings/renovation-wave_en (accessed on 8 June 2020).

12. European Commission. *An EU Strategy on Heating and Cooling*; Communication from the European Commission COM (2016) 51 final; European Commission: Brussles, Belgium, 2016.

13. European Commission. *Impact Assessment Accompanying the Document Proposal for a Directive of the European Parliament and of the Council*; Commission Staff Working Document SWD (2016) 414 final; European Commission: Brussels, Belgium, 2016.

14. European Commission. Commission Recommendation (EU) 2019/1019 of 7 June 2019 on building modernisation. *Off. J. Eur. Union* **2019**, *165*, 70–128.

15. López-Mesa, B.; Monzón-Chavarrías, M.; Espinosa-Fernández, A. Energy Retrofit of Social Housing with Cultural Value in Spain: Analysis of Strategies Conserving the Original Image vs. Coordinating Its Modification. *Sustainability* **2020**, *12*, 5579. [CrossRef]

16. Velasco-Gómez, E.; Tejero-González, A.; Jorge-Rico, J.; Rey-Martínez, F.J. Experimental Investigation of the Potential of a New Fabric-Based Evaporative Cooling Pad. *Sustainability* **2020**, *12*, 7070. [CrossRef]

17. European Commission. *Resource Efficiency Opportunities in the Building Sector*; Communication from the Commission COM (2014) 445 final; European Commission: Brussels, Belgium, 2014.

18. Grădinaru, C.M.; Muntean, R.; Șerbănoiu, A.A.; Ciocan, V.; Burlacu, A. Sustainable Development of Human Society in Terms of Natural Depleting Resources Preservation Using Natural Renewable Raw Materials in a Novel Ecological Material Production. *Sustainability* **2020**, *12*, 2651. [CrossRef]

19. Ramírez, C.P.; Barriguete, A.V.; Muñoz, J.G.; Merino, M.d.; Serrano, P.d. Ecofibers for the Reinforcement of Cement Mortars for Coating Promoting the Circular Economy. *Sustainability* **2020**, *12*, 2835. [CrossRef]

20. Mercader-Moyano, P.; Roldán-Porras, J. Evaluating Environmental Impact in Foundations and Structures through Disaggregated Models: Towards the Decarbonisation of the Construction Sector. *Sustainability* **2020**, *12*, 5150. [CrossRef]

21. González-Vallejo, P.; Muntean, R.; Solís-Guzmán, J.; Marrero, M. Carbon Footprint of Dwelling Construction in Romania and Spain. A Comparative Analysis with the OERCO2 Tool. *Sustainability* **2020**, *12*, 6745. [CrossRef]

22. Dominguez-Delgado, A.; Dominguez-Torres, H.; Dominguez-Torres, C. Energy and Economic Life Cycle Assessment of Cool Roofs Applied to the Refurbishment of Social Housing in Southern Spain. *Sustainability* **2020**, *12*, 5602. [CrossRef]

23. Antonini, E.; Boeri, A.; Lauria, M.; Giglio, F. Reversibility and durability as potential indicators for Circular building Technologies. *Sustainability* **2020**, *12*, 7659. [CrossRef]

24. European Union. Level(s)-A common EU framework of core sustainability indicators for office and residential buildings. Parts 1 and 2: Introduction to Level(s) and how it works. *JCR Tech. Rep.* **2017**. [CrossRef]

25. European Commission. Circular Economy Principles for Buildings Design. 2020. Available online: https://ec.europa.eu/docsroom/documents/39984 (accessed on 15 August 2020).

26. Hertwich, E.; Lifset, R.; Pauliuk, S.; Heeren, N. *Resource Efficiency and Climate Change: Material Efficiency Strategies for a Low-Carbon Future*; Report of the International Resource Panel 978-92-807-3771-4; United Nations Environment Programme: Nairobi, Kenya, 2020.

27. Vázquez-López, E.; Garzia, F.; Pernetti, R.; Solís-Guzmán, J.; Marrero, M. Assessment Model of End-of-Life Costs and Waste Quantification in Selective Demolitions: Case Studies of Nearly Zero-Energy Buildings. *Sustainability* **2020**, *12*, 6255. [CrossRef]

28. Blandón, B.; Palmero, L.; di Ruocco, G. The Revaluation of Uninhabited Popular Patrimony under Environmental and Sustainability Parameters. *Sustainability* **2020**, *12*, 5629. [CrossRef]

29. Borrallo-Jiménez, M.; LopezdeAsiaín, M.; Herrera-Limones, R.; Arcos, M.L. Towards a Circular Economy for the City of Seville: The Method for Developing a Guide for a More Sustainable Architecture and Urbanism (GAUS). *Sustainability* **2020**, *12*, 7421. [CrossRef]

30. LopezDeAsiain, M.; Díaz-García, V. The Importance of the Participatory Dimension in Urban Resilience Improvement Processes. *Sustainability* **2020**, *12*, 7305. [CrossRef]

31. Serrano-Baena, M.M.; Triviño-Tarradas, P.; Ruíz-Díaz, C.; Rafael, E.; Fernández, H. Implications of BREEAM Sustainability Assessment on the Design of Hotels. *Sustainability* **2020**, *12*, 6550. [CrossRef]

 *sustainability* 

Article

Energy Retrofit of Social Housing with Cultural Value in Spain: Analysis of Strategies Conserving the Original Image vs. Coordinating Its Modification

Belinda López-Mesa *, Marta Monzón-Chavarrías * and Almudena Espinosa-Fernández *

Department of Architecture, University of Zaragoza, c/María de Luna 3, 50018 Zaragoza, Spain
* Correspondence: belinda@unizar.es (B.L.-M.); monzonch@unizar.es (M.M.-C.); almudenaef@unizar.es (A.E.-F.)

Received: 2 June 2020; Accepted: 7 July 2020; Published: 10 July 2020

Abstract: Two renovation strategies were considered and compared in this paper for the energy renovation of social housing in condo buildings built in 1945–1969, situated in residential estates that have been declared urban areas of interest. One of the strategies, named here the image conservation approach, consists of renovating the building with very efficient solutions, except for the brick-facing walls, which will only have insulation in the existing cavities. The other strategy, the image coordination approach, consists of renovating the whole building with efficient solutions, including external insulation for existing brick-facing walls, and drawing up coordination plans for each of the residential estates in order to guarantee a coordinated new image of the residential estates. By means of three case studies in the city of Zaragoza and four scenarios of improvement, the two strategies are compared regarding the regulation compliance, the energy use, the emissions reduction, the thermal comfort, and the cost-effectiveness of the measures. The results show that the image coordination approach is the best strategy in regard to the studied aspects, achieving a 69%–79% reduction of CO_2 emissions and a 75%–97% reduction of heating use, as well as being the only one capable of transforming these buildings into nearly zero energy buildings (nZEBs).

Keywords: social housing; major renovation; nZEB renovation; cultural value; condominium

---

1. Introduction

The European Union (EU) has set targets for reducing its greenhouse gas emissions progressively up to 2050 by means of the 2020 climate and energy package, the 2030 climate and energy framework, and the 2050 long-term strategy [1]. Sustainable construction is expected to play a big part in meeting the European Union's climate and energy goals since the built environment is estimated to account for approximately 40% of energy consumption and 36% of CO_2 emissions in the EU [2]. Sustainable, climate-proofed buildings are needed on a massive scale to meet the objectives for 2020 and 2030, and to achieve a climate-neutral Europe by 2050 [2]. Now, the EU is urging countries to bring the building renovation wave into postpandemic recovery plans.

The majority of the building floor area in Spain is composed of residential buildings (82.67% in 2013) [3]. Since the existing residential buildings at present will represent the major part of the housing stocks of Spanish cities in 2050 [4,5], energy retrofit has become one of the most important strategies for the decarbonization of the residential sector. The European Directive 2010/31/EU on Energy Efficiency in Buildings introduced the need to transform buildings into nearly zero energy (nZEB) by 2050, and the European Directive 2018/844/EU amended the existing Directive 2010/31/EU in order to accelerate the decarbonization of the parks in the EU.

The energy retrofit of multifamily residential buildings built in 1945–1969 is of great importance in Spain because it is one of the countries in Europe with the highest share of multifamily dwellings (70.78% in 2010) and due to the fact that this is the interval of years with the highest share of dwellings (18.62%

in 2014) in Spain among the thermally significant periods identified by the European Commission [3]. This stock of buildings was built after the wars and before the first thermal standard was introduced in Spain in 1979, and therefore it is highly energy inefficient. In this period, the share of social housing in Spain (understood as that which has been built with the support of the State) was significantly higher than the mean value for the country (26.6% for the years 1952–2016 [6]). Specifically, the share of social housing was 30%–50% between 1945–1958, 70%–90% between 1959–1965, and 30%–40% between 1966–1969 [6], i.e., 50% as mean value for the period 1945–1969.

The public administrations in Spain started to promote an energy retrofit of social housing estates built in the mentioned period about fifteen years ago. This is the case, for example, of the city of Zaragoza, where 19 multifamily social housing estates (including 7981 dwellings [7]) started being studied in 2004 (together with another two residential estates of single-family houses) [8] and demonstrative building renovation projects were developed in four of these estates in 2008–2011 [9]. These estates were designed in the period of 1942–1964 and built in the period of 1945–1965, except for one of them (Balsas de Ebro Viejo), whose construction took place between 1964 and 1975. They had been declared urban areas of interest (UAI, *Conjuntos Urbanos de Interés* in Spanish) in the General Plan of Urban Planning of Zaragoza (PGOUZ) of 2001 and 2007 due to their character of unity, bringing together representative environmental values of types of residential building and urban areas characteristic of its time (Figure 1). These estates are characterized by buildings of poor quality in the regimes of condominiums with walk-up free-market private apartments, public places with poor quality environments, and selective population loss. It must be noted that social rental housing parks are virtually nonexistent in Spain (1.5%) [6] and social private housing parks are freed from 10 to 30 years after their construction, depending on the autonomous community. For this reason, the interventions on the common areas of the property in multifamily houses must be agreed by the community of neighbors of each condominium. The pilot interventions included the improvement of the envelope and systems, the installation of elevators, and, in given cases, the improvement of the surrounding areas.

These pilot building renovation projects in the residential estates had important qualitative effects for Zaragoza in terms of capacity building and visualization of the model. This positively contributed to reach a deep renovation rate in the whole city of about 0.1% (300 dwellings per year [10] out of a total of 277,411 main regular dwellings [11]) during the subsequent years, which the Zaragoza Municipal Housing Company (ZMHC) aims to increase up to 1% by 2030.

At this moment, the ZMHC, with whom one of the authors has a close contact by means of a collaboration agreement that has been running for eight years [12], is facing an important challenge regarding its renovation strategies in the UAI, which has to do with the loss of character of unity of these residential estates as a consequence of the major renovations. No specific restrictions had been defined in the PGOUZ for the building renovations in UAI, and since the dwellings' usable floor areas are notably small (more of the 70% of dwellings have usable floor areas ranging between 34 m^2 to 56 m^2), the façades have been externally insulated in the interventions conducted so far, changing the external appearance of the buildings (Figure 2). This can bring loss of the character of unity (Figure 2a) when different interventions propose different visual solutions. In other occasions, the visual solution has been carefully designed to remain very close to the original image (Figure 2b), and it is even difficult to appreciate which buildings are externally renovated and which are original from a certain distance (Figure 2(b2)). The character of unity loss is more likely to affect those renovated buildings whose façades are composed of load-bearing walls of facing bricks. Sixteen out of the nineteen residential estates have buildings with facing brick façades. The share of facing brick façades is about 74% for the whole of the UAI [13].

Figure 1. Example of social housing estate declared as urban area of interest (UAI) in Zaragoza: San Jorge.

Figure 2. (**a**) Example of unity character loss after renovation with external insulation in *Girón* residential estate: (**a0**) original building image, (**a1**) example 1 of renovated building, (**a2**) example 2 of renovated building; (**b**) example of unity character continuation after renovation with external insulation in *Alférez Rojas* residential estate: (**b0**) original building image, (**b1**) example of renovated building, (**b2**) renovated building in context.

The tensions and conflicts between the sustainability goals and the preservation of the modern built heritage have broadly been addressed in the literature [14–21]. Some authors took a stand against the demolition of old housing buildings, arguing that important synergies can be found between sustainability and conservation, e.g., [4,5,18,21]. Other authors focused on the design challenge of finding an architectural balance between existing and newly added qualities when designing the renovation of housing blocks with specific architectural heritage values, e.g., [14,15,19,20]. Oorschot et al. [20] considered the careful analysis of the architectural and cultural value of the original building by recording which characteristics must be preserved and which could be modified to be of great importance. In Amsterdam, buildings with historic and visual values were observed to offer fewer possibilities for interventions on the outside, especially the façades and the roofs, whereas inside insulation creates more inconvenience, especially if tenants stay in their homes during the intervention [14]. Dowson et al. [16] estimated that in UK there are restrictions on external wall insulation for a further 1.2 million dwellings in conservation areas and 300,000 listed dwellings [22] because of the change of appearance. Ben and Steemers [17] argued that some measures are generally not considered appropriate for listed buildings in UK, such as external wall insulation and double (or triple) glazing, and proposed several viable improvement strategies, including cavity wall insulation, roof insulation, secondary glazing, and draughtproofing, for the Brunswick Centre in London.

In the case of the UAI of Zaragoza, the main cons of inside insulation strategies discussed by the technicians of the municipality and the researchers were the following:

- It reduces the already small usable floor areas. This disadvantage is very difficult to avoid by merging different dwellings in one, as in the case of Amsterdam [14] because of the condominium regime of the buildings, or by expanding the area of the dwellings because the floor additions would affect the image as much as, or even more than, the insulation.
- It obliges occupants to temporarily leave their homes during the intervention. This is an inconvenience that could be partially mitigated by carefully planning the stages with more impact on the occupants' daily life and reducing their duration as much as possible.

For these reasons, the internal insulation strategies were discarded in Zaragoza and this position was accepted in this paper as a starting point. The disadvantage of external insulation is, as already mentioned, the potential loss of the residential estates' character of unity, and two possible options were considered in this study and have been discussed with the ZMHC to overcome the problem:

- One of the options is to use image conservation-oriented energy retrofit strategies that do not compromise the façades image, as in the case of the Brunswick Centre in London [17], but without using inside insulation so that the usable floor area remains equal after the intervention. This implies energy improvement of roofs, windows, plastered walls, and existing wall cavities. Plastered walls are externally insulated, and brick-facing walls are improved by pumping insulation into the existing wall cavities, which are 3 cm thick [13].
- The other option is to assume the preponderance of environmental matters and mitigate their negative impact on cultural aspects as much as possible by drawing up coordination plans for each of these areas in order that the individual interventions in each of the buildings follow some common rules that ensure a final image of unity. This implies that the whole envelope is deeply renovated. All façades, including brick-facing walls, are externally insulated.

This paper focused on comparing different energy retrofit strategies for the case of the UAI in Zaragoza that do not reduce the usable floor area, with the aim to know if the image conservation approach and the image coordination one can ensure comparable levels of performance regarding the regulation compliance, the energy use, the emissions reduction, the thermal comfort, and the cost-effectiveness of the measures, in order to support with data the decisions of the local administration. The scientific contribution of this paper is in the field of design research [23,24], and focuses on the improvement of the observed phenomenon in architectural design of dealing with the tensions between the sustainability goals and the preservation of the modern built heritage.

2. Materials and Methods

In order to compare the different energy retrofit strategies, three case studies were used of real social housing buildings located in three of the 19 multifamily UAI of Zaragoza. The research methodology is composed of the next steps, which we develop in the following sections:

1. Selection of cases studies
2. Scenarios definition
3. Definition of comparison criteria and simulation methods
4. Results analysis and discussion
5. Conclusions.

2.1. Selection of Cases Studies

In previous research works, we had cataloged the 19 multifamily residential estates of Zaragoza declared as UAI [25], identified the construction characteristics of the buildings [13,26], done an energy and acoustic characterization of their envelopes [27,28], and simulated the buildings energy use through a simplified methodology [7,26,27]. Important conclusions from these previous studies have been useful to define the criteria to select our case studies.

On the one hand, we tried to select cases studies representative of the most common energy efficiency performances found among the buildings of the 19 residential estates. The energy performance rating for the indicator of nonrenewable primary energy was found to be an E for 88.5% of the buildings, F for 5.2%, and G for 3.1%, with the remaining 3.2% being renovated buildings in 2016 [29]. Regarding the energy performance rating for the indicator CO_2 emissions, the share was 80.9% of E and 15.9% of D [29]. As can be seen, by far the majority of buildings had a rating of E for nonrenewable primary energy and a rating of E for CO_2 emissions. Our three case studies had an E estimation for nonrenewable primary energy rating and an E estimation for CO_2 emissions rating, except for one of them (the *Alférez Rojas* case), which had an F estimation for nonrenewable primary energy rating (Table 1).

Table 1. Construction characteristics of the case studies (the façade coding is the same as in [13]).

Case Study	Social Housing Block in Residential Estate *Girón*	Social Housing Block in Residential Estate *Alférez Rojas*	Social Housing Block in Residential Estate *Balsas de Ebro Viejo*
Year of Construction	1957	1960	1969
Nonrenewable Primary Energy Rating [29]	E	F	E
CO_2 Emissions Rating [29]	E	E	E
Ground Floor Façade	Façade type F1: 36 cm facing solid brick + 1 cm mortar cement coating + 3 cm nonventilated air space + 4 cm hollow brick + 1.5 cm plastering		Main solution (façade type F9): 12 cm facing solid brick + 1 cm mortar cement coating + 3 cm air space + 4 cm hollow brick + 1.5 cm plastering Solution below windows (façade type F10): 1.5 cm mortar cement exterior coating + 9 cm hollow brick + 1 cm mortar cement coating + 3 cm nonventilated air space + 4 cm hollow brick + 1.5 cm plastering
1st to 3rd Floor Façade	Façade type F2: 24 cm facing solid brick + 1 cm mortar cement coating + 3 cm nonventilated air space + 4 cm hollow brick + 1.5 cm plastering		
4th Floor Façade	Façade type F3: 1.5 cm mortar cement exterior coating + 24 cm hollow brick + 1 cm mortar cement coating + 3 cm nonventilated air space + 4 cm hollow brick + 1.5 cm plastering		
Roof	Pitched roof with ventilated chamber		
Ground	Concrete slab-on-ground	Suspended concrete floor	Slab in contact with uninhabitable room
Windows	Wood frame with single glass 3 mm. Dark shutter		

On the other hand, we selected cases studies with different construction solutions observed in the 19 residential estates in order to check whether these different solutions can make an important difference or not regarding the energy performance of the renovated building:

- Most of the buildings presented façades with thick brick-bearing walls without thermal insulation, and only some of the buildings, built at the end of the studied period, presented structures of pillars with the corresponding thinning of the façade without thermal insulation [13]. Two of our case studies (the *Alférez Rojas* and *Girón* cases) presented thick brick-bearing walls, and one of them presented thinner walls (the *Balsas de Ebro Viejo* case), as can be seen in Table 1.

- With regard to the grounds, most of the buildings (65% of them) presented concrete slab-on-grounds, some of them (26%) had suspended reinforced concrete one-way floors, and the remaining part (9%) had reinforced concrete one-way slabs in contact with uninhabitable rooms, such as storage rooms [13]. Each of our case studies presented one of these three solutions (Table 1).

The general image and urban layout of each of these case studies can be observed in Figure 3. The specific *U*-value of the different construction solutions for the envelope can be found in Appendix A. Since the thermal mass is important in the energy retrofit of buildings from this period [30], the internal areal heat capacity and periodic thermal transmittance for the different façades and grounds are provided in Appendix A to evaluate their thermal inertia, as suggested by [31,32]. The calculations were made according to EN ISO 13786:2017 [33].

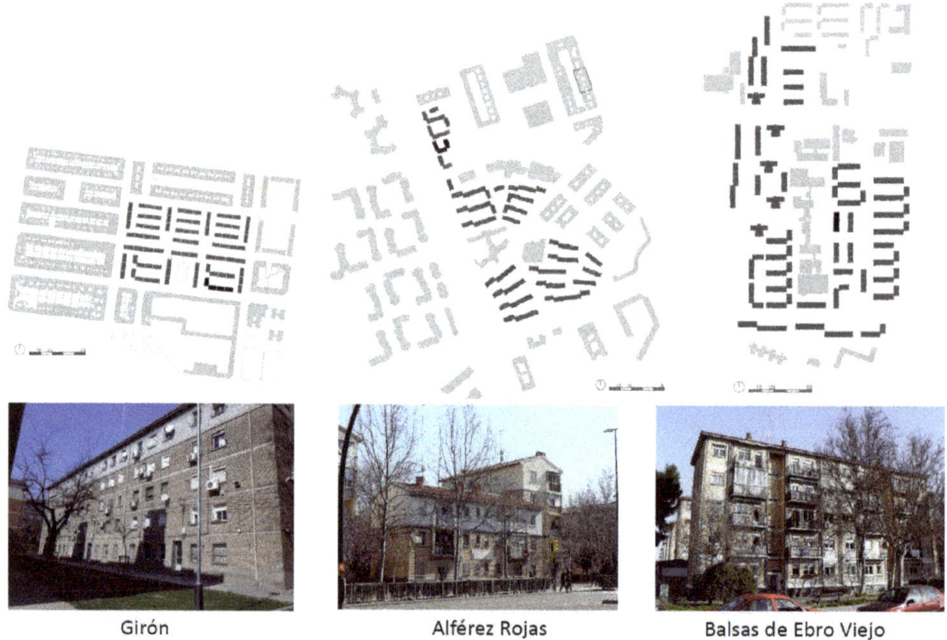

Figure 3. General image and urban layout of the three case studies (plans legend: highlighted black are the buildings selected as case studies, highlighted dark grey are the buildings of the same residential estate, highlighted light grey are other buildings in the surroundings).

2.2. Scenarios Definition

Five scenarios were considered for each of the three case studies (Figure 4):

- Scenario 0: original state. The envelope is the original one, described in Table 1 and thermally characterized in Appendix A. With regard to the heating and domestic hot water (DHW) systems, all of them are individual, and so are the cooling systems when they exist. For the sake of comparison, we considered, for the heating and DHW systems, an individual regular gas boiler in each of the dwellings, since it is the most extensive system in Spain [34]. For the cooling system, an electrical single zone system with a seasonal energy efficiency of 2.52 was considered, as recommended by the user guide for the official tool for energy rating in Spain when the system is not known or nonexistent [35].

- Scenario 1: the envelope is improved, avoiding the external insulation of the façades made out of facing bricks. Efficient solutions are considered for the rest of elements of the envelope, even when installing very thick insulations is technically difficult, such as in the case of the external insulation of plastered façades and grounds, in order to compensate the thin insulation of brick facing walls. This scenario follows the image conservation approach. Specifically, the energy improvement of the envelope is made by means of:

  - Insulation of roofs by means of 25 cm thick mineral wool blankets installed in the air chamber. According to the literature, efficient solutions for roofs have a U-value between 0.09 and 0.2 W/m^2·K [36,37]. As can be observed in Appendix A, the U-value of our renovated roofs is 0.12 W/m^2·K.
  - Secondary glazing. Since in Spain the original windows are aligned with the inside surface of the façade, the additional glazing is external. For the calculations, the inside glazing was considered not to contribute to insulation, since many of them present serious problems of permeability. According to the literature, very efficient solutions for windows have a U-value between 0.80 and 1.26 W/m^2·K [36,37]. Our windows have a U-value of 1.04 W/m^2·K (Appendix A). The permeability of the new window is 3 m^3/h·m^2 at 100 Pa (class 4, according to EN 12207:2016).
  - Grounds are improved for the case of suspended floors with 20 cm thick external Polyurethane (PUR) insulation, and for the case of slabs in contact with uninhabitable rooms with 20 cm thick external PUR insulation. According to the literature, efficient solutions for grounds have a U-value between 0.2 and 0.25 W/m^2·K [37,38]. In our case studies, the U-value of these two types of renovated grounds are considerably lower, specifically 0.16 and 0.15 W/m^2·K, respectively (Appendix A). For the case of the slab-on-ground, perimeter insulation is installed because insulating the whole slab would imply important interferences with occupants.
  - External insulation of 16 cm thick for plastered walls. According to the literature, efficient solutions for façades have a U-value between 0.15 and 0.3 W/m^2·K [36–38]. As can be observed in Appendix A, the U-value of our renovated plastered façades is quite low, specifically 0.15 W/m^2·K.
  - Cavity wall insulation of 3 cm thick for the brick facing walls. These walls have a U-value that ranges from 0.60 to 0.73 W/m^2·K (Appendix A).

- Scenario 2: the envelope is improved, insulating externally the whole façade. Efficient solutions are used for the elements of the envelope that they are technically easy to install (roofs and windows), whereas for more problematic elements of the envelope (grounds and external insulation of walls) more technically feasible solutions are considered. This scenario is thought to follow the image coordination approach. Specifically, the improvement is made by means of:

  - Same solutions for roofs and windows as in scenario 1.
  - Grounds are improved for the case of suspended floors with 10 cm thick external PUR insulation, and for the case of slabs in contact with uninhabitable rooms with 10 cm thick external PUR insulation. The U-value of these two types of renovated grounds is 0.28 W/m^2·K (Appendix A). For the slab-on-ground, perimeter insulation is used.
  - External insulation of 10 cm thick for façades. The U-values of our renovated façades in this scenario range from 0.21 to 0.23 W/m^2·K (Appendix A).

- Scenario 3: the envelope improvement is the same as in scenario 1, and additionally:

  - The heating and DHW system is substituted by gas condensing boilers.
  - Solar panels are installed in order to guarantee that at least 60% of the energy for DHW is solar, as prescribed in the Spanish Technical Code for this location if the system is changed.

- Scenario 4: the envelope improvement is the same as in scenario 2, and as in scenario 3 the heating system is substituted by gas condensing boilers and 60% of the energy for DHW is solar.

The values for the parameters of the envelope and the heating, cooling, and DHW systems that define their contribution to the buildings energy efficiency can be found in Appendix A.

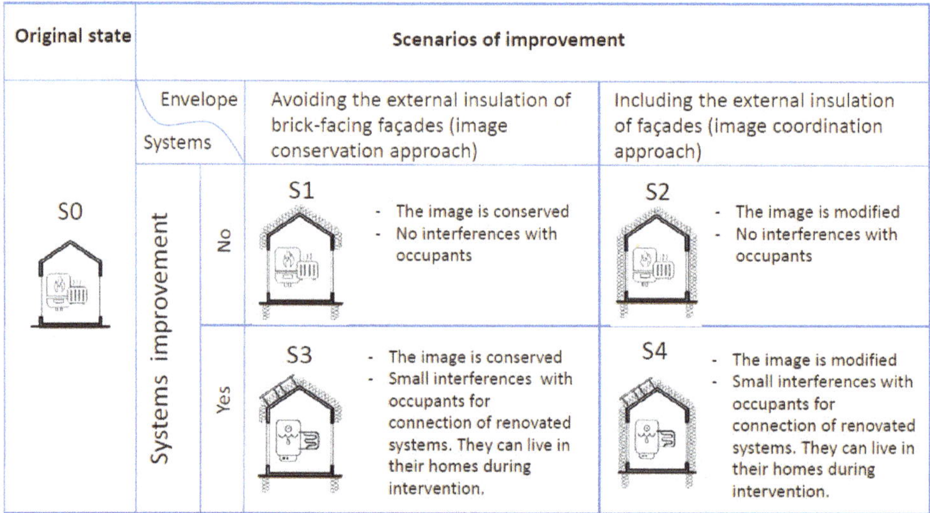

Figure 4. Studied scenarios for each of the three case studies.

2.3. Definition of Comparison Criteria and Simulation Methods

The five scenarios of the three case studies were compared in terms of: energy regulation compliance; energy use, CO_2 emissions, and thermal comfort performance; and cost-effectiveness of the measures.

2.3.1. Methodology to Check the Energy Regulation Compliance

In December 2019, a royal decree modifying the Basic Document on Energy Saving (DB-HE) of the Buildings Technical Code was approved in Spain. This recently updated regulation sets, on the one hand, more restrictive requirements for major renovations—which are compulsory when more than 25% of the envelope is improved, as is our case— and, on the other hand, defines for the first time specific requirements for the transformation of existing buildings into nZEBs—which are not mandatory, but must be met if an nZEB renovation is aimed for.

In this paper, we checked the compliance or not of the five scenarios with the two set of requirements: the compulsory major renovation requirements and the optional nZEB renovation requirements.

For both levels of renovation (major renovation and nZEB), the main parameters that are regulated are:

- The envelope global heat transfer coefficient K-value ($W/m^2 \cdot K$) must be less than a limit value Klimit. The K-value depends on the U-values of the envelope elements and thermal bridges and it is the heat transfer coefficient of ISO 14683:2017, divided by the exchange area [39]. The maximum allowable Klimit depends on the climate zone and the compactness of the building and is the same for major renovation and nZEB renovation.
- The value of the solar control of windows will be lower than 2 $kW \cdot h/m^2 \cdot month$ for the month of July for households, for both major renovation and nZEB renovation. The solar control is the

ratio between the solar gains in July of all windows of the envelope with their solar protections activated and the useful floor area of the building.

- Nonrenewable primary energy (kW·h/m^2·year) and total primary energy (kW·h/m^2·year) use are limited by a fixed value depending on the climate zone. The values for nZEB renovation are more restrictive than for major renovation. Specifically, for our climate zone D, the limit values for nonrenewable primary energy are 70 kW·h/m^2·year for major renovation and 38 kW·h/m^2·year for nZEB renovation, and the limit values for total primary energy are 105 kW·h/m^2·year for major renovation and 76 kW·h/m^2·year for nZEB renovation.

Additionally, the regulation sets other requirements, such as:

- A maximum U-value is set for each element of the envelope that is added, replaced, or modified (Table 2). Our four scenarios of improvement fulfill these values except for the brick facing walls in scenarios 1 and 3.

Table 2. Maximum U-value according to the Spanish Buildings Technical Code for climate zone D.

Part of the Envelope	U-value lim (W/m^2·K)
Façade—Exterior Wall	0.41
Roof	0.35
Ground	0.65
Windows	1.80

- In our climate zone, at least 60% of the energy demand for DHW has to be obtained from renewable sources, with local origin or nearby the building. This is applicable to those renovations that change the building installations. We made scenarios 3 and 4 fulfil this requirement.
- The accepted permeability of the windows depends on the climate zone. In our location, class 3 or higher, as defined in EN 12207:2017, is demanded. We chose class 4 with lower permeability and therefore better energy performance.

We made our four scenarios of improvement fulfil the additional requirements (except for the U-value of the brick-facing walls in scenarios 1 and 3), and afterwards we checked whether the main requirements were met or not for major renovation and nZEB renovation. To check the compliance of these main requirements, the official tool for energy certification of buildings in Spain was used, named HULC version 2.0.1960.1156 [39].

2.3.2. Methodology to Estimate Energy Use, CO$_2$ Emissions, and Thermal Comfort

To estimate energy use and CO$_2$ emissions, we used the HULC version 2.0.1960.1156 software [37], and to characterize the thermal comfort the Design Builder version 6 software [40]. This implies a total of 30 simulations.

The same building models were introduced in the two software tools regarding the shapes and orientations of the buildings, their construction solutions (Table 1 and Appendix A), and thermal bridges (Appendix B). The linear thermal transmittances of thermal bridges were calculated using the catalogue of thermal bridges of the HULC version 2.0.1960.1156 software [39], and the same values were used in the Design Builder model. The use and climate conditions established by default in the HULC software were used in the Design Builder software and can be found in Appendix C. The renewal air flow was estimated according to the calculation method proposed in the Spanish Technical Code. The result, 33 L/s, was the same for the five scenarios (scenarios 0–4), since it depends on the household layout, which was kept the same in the improved scenarios. It includes the infiltrations plus the ventilation through the use of the windows. Additionally, we considered the following values for the infiltration air flow rate per internal volume through the envelope (n50), both in HULC and Design

Builder: 5.72 h^{-1} for scenario 0 (original building), as reported by Fernández-Agüera et al. [41] for multifamily social housing in Mediterranean Europe, and 4.91 h^{-1} for the improved scenarios (1 to 4), since according to Almeida et al. [42], the average window contribution for the room permeability is 15% and we changed windows with air permeability of 50 m^3/h·m^2 to 3 m^3/h·m^2 (a reduction of 94%). Night ventilation of four renovations per hour during summer was considered in the two software tools, since all the studied buildings allow for crossed ventilation.

With regard to the solar control of windows, we defined two parameters:

- The total transmittance of solar energy of windows when all the solar devices are activated: according to the support document of the Spanish Technical Code for calculation of envelope parameters [43], it is of 0.15 for scenario 0 (characterized by a dark exterior shutter with single glass) and 0.05 for the improved scenarios 1 to 4 (characterized by a light shutter and a window with triple glass).
- The activation of the solar protection: the activation values by default in HULC, for 30% for the summer period, were considered in both software tools. The way this parameter is introduced in Design Builder can be seen in Appendix C.

To estimate the thermal comfort, we used the adaptative comfort approach according to the ASHRAE55-2017 standard included in the Design Builder software for naturally ventilated buildings. This method has been proven to be useful for mixed-mode buildings such as ours (i.e., a combination of operable windows and mechanical conditioning) [44]. An annual simulation was run at intervals of one hour. Once the simulation was done, the data (air temperature, mean radiant temperature, operative temperature, exterior air temperature, renovations per hour by means of mechanical ventilation, natural ventilation, and infiltrations) in csv format were exported to the software CONFADAPT-ASH55 version 1.0 [45], which allowed us to obtain the hours inside and outside the 80% and 90% acceptability limits, indicating the percentage of occupants expected to be comfortable at the indicated indoor. We used the 90% acceptability limit to do the estimations with a higher standard of thermal of comfort.

2.3.3. Methodology to Estimate the Cost-Effectiveness of the Scenarios

We first estimated the total investment renovation costs of the measures for each of the four improvement scenarios of the three case studies by means of the CYPE software and database [46], and afterwards we divided it by the total surface to obtain the renovation investment per m^2 for each case. To estimate the cost-effectiveness of each scenario of improvement, the costs-to-effectiveness ratio [47] that we established divided the total investment costs by the saved kW·h of nonrenewable primary energy for the five scenarios of the three case studies.

3. Discussion and Results

3.1. Energy Regulation Compliance

Only one of the scenarios complied with the nZEB renovation requirements, scenario 4, and two of the scenarios, scenarios 2 and 4, complied with the major renovation requirements (Table 3). For the case study of Balsas de Ebro Viejo, it was necessary to increase the percentage of solar energy for domestic hot water (DHW) up to 65% in order to make it comply with the nZEB renovation requirements (we name this scenario 4* in Tables 3 and 4 and Tables A1–A3). This building presented a worse energy performance in all the scenarios (with 27%–59% more nonrenewable primary energy and total primary energy use), which could be explained by the fact that it is the only one with thinner façades and a corresponding lower thermal mass, as shown by the values of the periodic thermal transmittance in Appendix A.

This implies that if social housing in Zaragoza is renovated following the image conservation approach (scenarios 1 and 3), the energy regulation for major renovation, which is compulsory, could not be fulfilled, unless additional measures were taken on the top of those that we have studied here. Only the image coordination approach provides solutions that would comply with current regulations (scenarios 2 and 4) and could even transform existing social housing buildings into nZEBs when improving both the envelope and the systems (scenario 4). This was the case for the two case studies with massive load-bearing façades (*Girón* and *Alférez Rojas*). For the case of the building with thinner façades and a structure of pillars (*Balsas de Ebro Viejo*), the only scenario fulfilling the major renovation requirements was scenario 4. In order for the image conservation approach (scenarios 1 and 3) to fulfil the major renovation requirements, it would be necessary to reduce the K-value (Table 3). This would require, for example, adding internal insulation, but as already discussed this is not considered a feasible solution due to the small size of households.

Table 3. Results of compliance with the Spanish Buildings Technical Code.

| Scenario | K-Value (W/m^2·K) | | Qsol, jul (kW·h/m^2·month) | | Nonrenewable Primary Energy (kW·h/m^2·year) | | | Total Primary Energy (kW·h/m^2·year) | | | Compliance with Requirements | |
	Calculated	Limit	Calculated	Limit	Calculated	MR Limit	n ZEB Limit	calculated	MR Limit	n ZEB Limit	MR	nZEB
*Girón* case study												
0	1.59	0.65	1.59	2	103.5	70	38	105.5	105	76	NO	NO
1	0.94	0.65	0.35	2	80.0	70	38	82.0	105	76	NO	NO
2	0.53	0.65	0.35	2	64.1	70	38	66.1	105	76	YES	NO
3	0.94	0.65	0.35	2	46.9	70	38	69.3	105	76	NO	NO
4	0.53	0.65	0.35	2	33.7	70	38	56.0	105	76	YES	YES
Alférez Rojas case study												
0	1.93	0.65	1.93	2	113.0	70	38	115.8	105	76	NO	NO
1	1.01	0.65	0.48	2	83.9	70	38	86.9	105	76	NO	NO
2	0.33	0.65	0.48	2	51.6	70	38	54.8	105	76	YES	NO
3	1.01	0.65	0.48	2	54.2	70	38	73.9	105	76	NO	NO
4	0.33	0.65	0.48	2	27.6	70	38	47.5	105	76	YES	YES
Balsas de Ebro Viejo case study												
0	1.83	0.66	1.79	2	165.0	70	38	168.3	105	76	NO	NO
1	0.96	0.66	0.66	2	101.5	70	38	105.3	105	76	NO	NO
2	0.36	0.66	0.66	2	72.3	70	38	76.2	105	76	NO	NO
3	0.96	0.66	0.66	2	61.2	70	38	87.9	105	76	NO	NO
4	0.36	0.66	0.66	2	38.9	70	38	65.8	105	76	YES	NO
4 *	0.36	0.66	0.66	2	37.2	70	38	65.8	105	76	YES	YES

Note: MR: major renovation; nZEB: nearly zero energy buildings; highlighted grey are the unfulfilled limits; * Additional 5% of DHW (in total 65%).

Table 4. CO$_2$ emissions of the three case studies for the five studied scenarios.

Case Study	*Girón*	*Alférez Rojas*	*Balsas de Ebro Viejo*
Scenario	CO$_2$ Emissions (kg CO$_2$/m^2·year)	CO$_2$ Emissions (kg CO$_2$/m^2·year)	CO$_2$ Emissions (kg CO$_2$/m^2·year)
0	21.6	23.46	34.4
1	16.6	17.25	20.82
2	13.23	10.35	14.52
3	9.59	10.95	12.28
4	6.78	5.26	7.49
4*	-	-	7.14

Note: *Additional 5% of DHW (in total 65%).

3.2. Energy Use, CO_2 Emissions, and Thermal Comfort

When looking at the CO_2 emissions of the three case studies for the five scenarios considered (Table 4), we observed that:

- The energy retrofit of the envelope following the image conservation approach (scenario 1) could be responsible for the reduction of 23%–39% of CO_2 emissions, depending on the case study, whereas the improvement of the envelope following the image coordination approach (scenario 2) could entail a reduction of 39%–58% of CO_2 emissions, according to simulations.
- If, additionally, the systems are improved, the total reduction of CO_2 emissions following the image conservation approach (scenario 3) could be up to 53%–64%, whereas following the image conservation approach (scenario 4) it could be up to 69%–79%, according to simulations.

Therefore, the image coordination approach could bring about an additional 13%–24% reduction of CO_2 emissions, as compared to the image conservation approach, which is quite a significant figure considering the objectives of the EU to reduce emissions by 80%–95% for 2050.

When the CO_2 emissions are broken down into heating, cooling, and DHW emissions for the three case studies and five scenarios (Figures 5–7), we observed that:

- The envelope energy retrofit would significantly contribute to the reduction of emissions caused by the use of heating systems: a 45%–64% reduction in heating CO_2 emissions with the image conservation approach, and 75%–96% with the image coordination approach. On the other hand, it would increase the emissions caused by the use of cooling systems: a 9%–27% increase in cooling CO_2 emissions with the image conservation approach and 13%–39% with the image coordination approach. However, these increases are not so significant because CO_2 emissions due to cooling only represent a 6%–8% of total emissions in the original buildings (scenario 0) for this city and these case studies. The reason for this low need of cooling systems may be due to the combination of high internal thermal inertia, night ventilation, and solar protection, which together would significantly reduce the cooling demand in the five scenarios [48].
- The proposed systems improvement would reduce 5%–20% of CO_2 emissions of heating systems and 69% of DWH systems.

Therefore, the most significant CO_2 emissions reductions would be caused by the deep energy retrofit of the envelope in the image coordination approach and the improvement of the DHW production systems.

When analyzing the energy use (broken down into heating, cooling, and DHW, as well as into renewable and nonrenewable) and thermal comfort for the three case studies and five scenarios (Figures 8–10), we observed that:

- The reason why the emissions get so significantly reduced by the deep renovation of the envelope is the significant saving in heating energy use: a 45%–64% saving with the image conservation approach (scenario 1) and 75%–97% for the image coordination one (scenario 2), according to simulations. This reduction in heating energy use could be achieved without compromising the comfort. In fact, it would actually increase the percentage of hours of 90% acceptability, with an additional 48%–54% of hours for the image conservation approach (scenario 1) and 45%–55% for the image coordination approach (scenario 2). The comfort achieved would be quite similar for both approaches.
- The reason why the emissions get so significantly reduced by the improvement of the DHW production systems is mainly the installation of renewable sources. As can be seen in Figures 8–10, the energy use for DHW would only be reduced by 21% in all the cases, and the rest of the emissions reduction (48%) would be achieved by the use of sun as energy source.

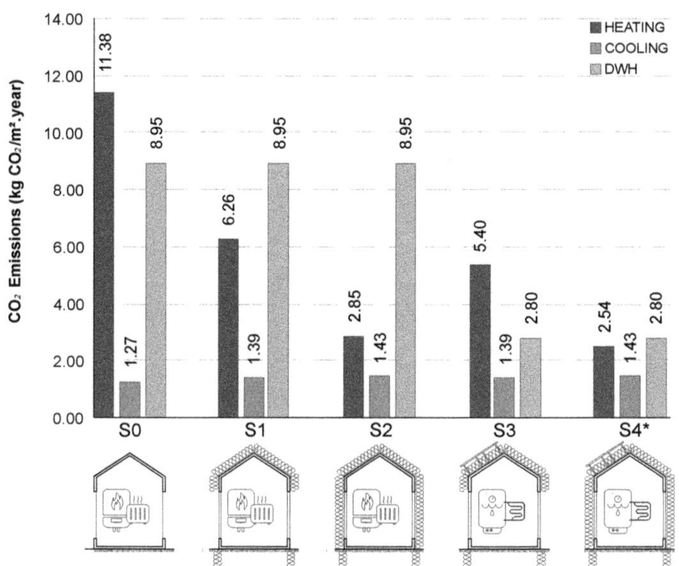

Figure 5. CO_2 emissions for the case study of *Girón* (S#: scenario #).

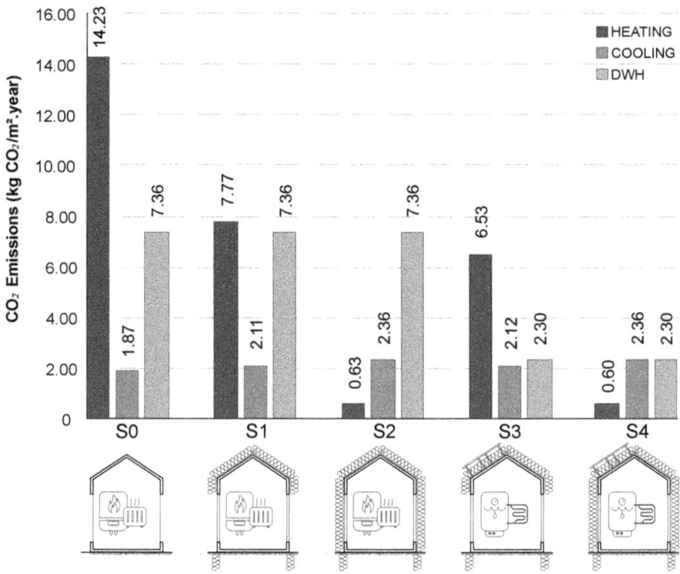

Figure 6. CO_2 emissions for the case study of *Alférez Rojas* (S#: scenario #).

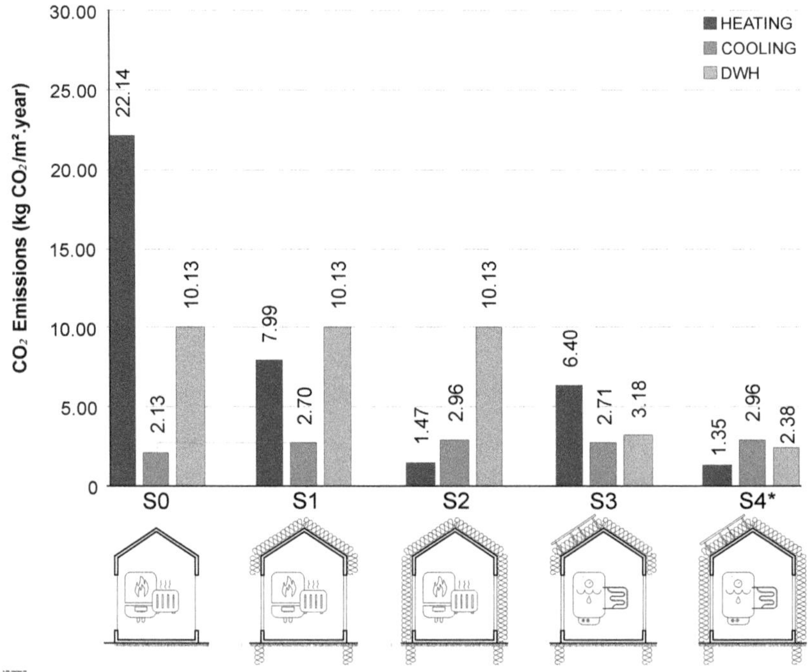

Figure 7. CO$_2$ emissions for the case study of *Balsas de Ebro Viejo* (S#: scenario #).

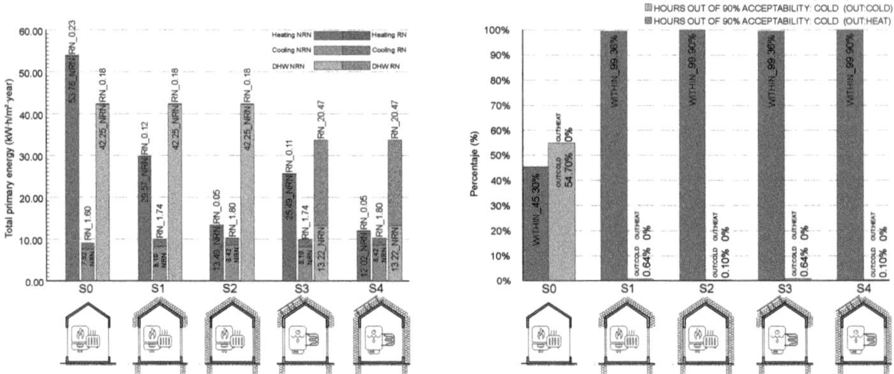

Figure 8. Primary energy use (**left**) and percentage of hours within the 90% of acceptability comfort range (**right**) in *Girón* case study (S#: scenario #; NRN: nonrenewable; RN: renewable).

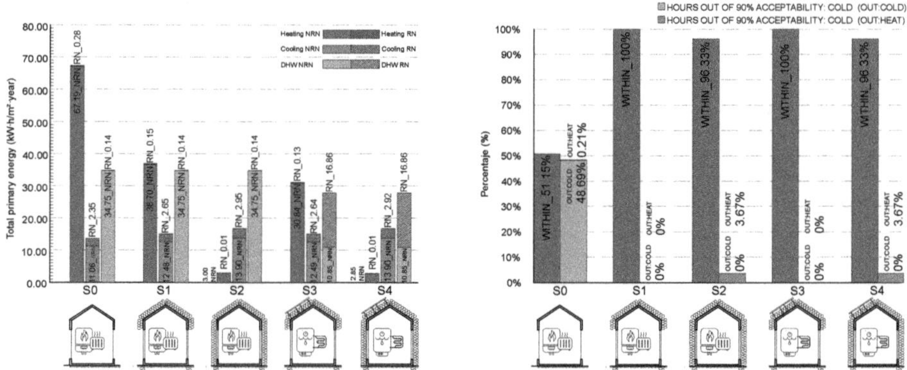

Figure 9. Primary energy use (**left**) and percentage of hours within the 90% of acceptability comfort range (**right**) in *Alférez Rojas* case study (S#: scenario #; NRN: nonrenewable; RN: renewable).

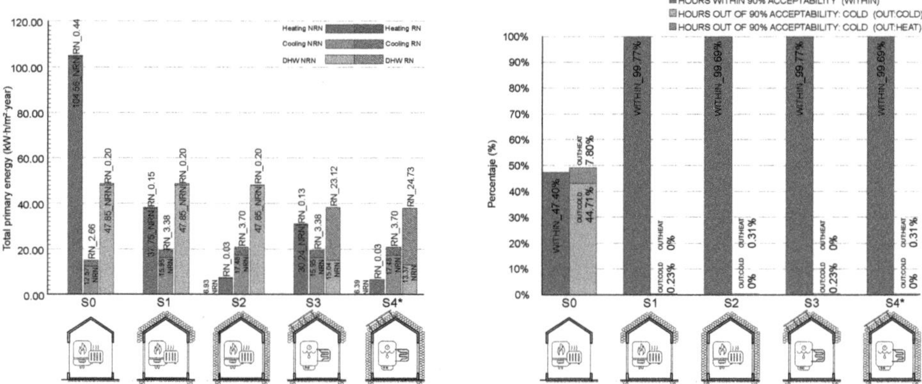

Figure 10. Primary energy use (**left**) and percentage of hours within the 90% of acceptability comfort range (**right**) in *Balsas de Ebro Viejo* case study (S#: scenario #; NRN: nonrenewable; RN: renewable).

3.3. Cost-Effectiveness of the Scenarios

To complete the analysis, we performed a cost-effectiveness analysis of the scenarios for the three case studies. As can be seen in Table 5, the most cost-effective scenario is number 2, corresponding to the deep renovation of the envelope following the image coordination approach, with a cost of 0.69€–2.02€ per nonrenewable kW·h saved, depending on the case study. The next most cost-effective scenario is the one improving the envelope following the image conservation approach (scenario 1), followed by the scenarios that also improve systems (scenarios 4 and 3). Again, the image coordination approach was proven to be more advantageous from the sustainability point of view than the image conservation one.

Table 5. Costs analysis (TC: total renovation investment costs; RC/S: renovation investment costs per usable surface; C-E: cost-effectiveness: NRN: nonrenewable).

Scenario	*Girón* Case Study			*Alférez Rojas* Case Study			*Balsas de Ebro Viejo* Case Study		
	TRC (€)	RC/S (€/m^2)	C-E (€/NRN kW·h Saved)	TRC (€)	RC/S (€/m^2)	C-E (€/NRN kW·h Saved)	TRC (€)	RC/S (€/m^2)	C-E (€/NRN kW·h Saved)
1	139,402	70.07	2.97	89,850	57.16	1.96	51,221	55.08	0.87
2	158,975	79.90	2.02	96,411	61.33	0.99	59,350	63.82	0.69
3	289,299	145.41	5.26	210,805	134.10	3.84	141,590	152.25	2.15
4	308,872	155.25	3.80	217,367	138.27	2.25	148,700	159.89	1.71

4. Conclusions

How to conduct the energy renovation of social housing of the period 1945–1969 in Spain, without compromising the character of unity of the residential estates to which they belong, is a challenge that some city councils are facing.

In Spain, the energy retrofit of social housing with cultural value presents different challenges from those identified in other countries, such as England or the Netherlands [14–22], since inside insulation is not considered a feasible option because of the small size of the households and the difficulty of merging dwellings in condo buildings.

Two possible approaches were considered in this paper to deal with the tensions between the sustainability goals and the preservation of the modern built heritage. One of them, named here the image conservation approach, consists of renovating the building with very efficient solutions, except for the brick-facing walls, which will only have insulation in their existing 3 cm thick cavities. The other strategy, the image coordination approach, consists of renovating the whole building with efficient solutions, including external insulation for existing brick-facing walls, and drawing up coordination plans for each of the residential estates in order to guarantee a coordinated new image of the residential estates.

By means of three case studies in the city of Zaragoza and four scenarios of improvement, the two strategies were compared regarding the regulation compliance, the energy use, the emissions reduction, the thermal comfort, and the cost-effectiveness of the measures.

The results show that from the points of view of environment (energy use and emissions) and cost-effectiveness, the image coordination approach is the best strategy because:

- It would bring about an additional 13%–24% reduction of CO_2 emissions as compared to the image conservation approach. In total, it could contribute to up to 69%–79% reduction of CO_2 emissions in comparison with the original buildings, which is quite close to the objectives of the EU for 2050 to reduce emissions in the building sector by 80%–95%.
- It implies an additional reduction of heating energy use of 29%–51% as compared to the image conservation approach. The heating energy use can be reduced in total up to 75%–97% in comparison with the original buildings with this approach, according to simulations.
- Even if it implies an additional increase of the cooling energy use of 4%–13% as compared to the image conservation approach and a total increase of 13%–39% with regard to the original buildings (scenario 0), this is not so significant when looking at the whole picture because cooling only represents a 6%–8% of total emissions in scenario 0 thanks to the high thermal inertia, night ventilation, and solar protection of these types of buildings, which are maintained or improved in the subsequent scenarios.

Regarding thermal comfort, both strategies provide high levels of comfort, reaching 96%–100% of hours within the 90% acceptability.

The analysis also shows that the image coordination strategy is the only one that can ensure the fulfilment of the compulsory major renovation requirements recently modified in Spain, and can even transform existing social housing buildings into nZEBs by improving both the envelope and the systems.

For these reasons, the image coordination approach is the only strategy recommended in this paper, unless the boundary conditions were changed, for example by defining innovative solutions to facilitate merging households in condo buildings and using inside insulation solutions.

The buildings from this period with thick brick-bearing walls show a better energy performance before and after renovation than buildings with structures of pillars and the corresponding thinning of the façade, due to the higher thermal inertia of their façades making it easier and cheaper (up to 14%–62%) to achieve the major renovation and nZEB standards.

Author Contributions: Conceptualization, B.L.-M.; data curation, M.M.-C. and A.E.-F.; formal analysis, B.L.-M., M.M.-C. and A.E.-F.; funding acquisition, B.L.-M.; investigation, B.L.-M., M.M.-C. and A.E.-F.; methodology, B.L.-M., M.M.-C. and A.E.-F.; project administration, B.L.-M.; resources, B.L.-M.; software, M.M.-C. and A.E.-F.; validation, B.L.-M., M.M.-C. and A.E.-F.; visualization, A.E.-F.; writing—original draft, B.L.-M.; writing—review and editing, B.L.-M., M.M.-C. and A.E.-F. All authors have read and agreed to the published version of the manuscript.

Funding: This research was funded by the Ministry of Economy and Competitiveness of Spain, project "BIA2013-44001-R Integrated design protocol for social housing retrofit and urban regeneration"; and by FEDER 2014–2020, project "Construyendo Europa desde Aragón", research group T37_17R "Grupo de Investigación en Arquitectura".

Acknowledgments: The authors would like to express their gratitude to "Sociedad Municipal Zaragoza Vivienda" and to Juan Rubio del Val (director of the Observatory *Ciudad* 3-R) for providing information about the case studies and the exchange of ideas.

Conflicts of Interest: The authors declare no conflict of interest.

Appendix A

Table A1. Thermal characterization of the elements of the envelope and systems of the simulation models (Y_{12}: periodic thermal transmittance; k_1: internal areal thermal capacity).

Scenarios and Cases		Façade	Roof	Ground	Windows	Systems
Scenario 0	G irón	F1 U-value = 1.05 W/m²·K Y_{12} = 0.11 W/m²·K K_1 = 41 kJ/m²·K	U-value = 2.25 W/m²·K	U-value = 3.81 W/m²·K	Old wood frame U-value = 2.0 W/m²·K; Single glass U-value = 5.6 W/m²·K; g = 0.85; Dark shutter Window U-value = 5.08 W/m²·K [13]	Regular gas boiler Performance 81% [34]
	Alférez Rojas	F2 U-value = 1.23 W/m²·K Y_{12} = 0.30 W/m²·K K_1 = 45 kJ/m²·K F3 U-value = 0.78 W/m²·K Y_{12} = 0.13 W/m²·K K_1 = 43 kJ/m²·K		U-value = 2.21 W/m²·K		
	Balsas de Ebro Viejo	F9 U-value = 1.49 W/m²·K Y_{12} = 0.89 W/m²·K K_1 = 48 kJ/m²·K F10 U-value = 1.20 W/m²·K Y_{12} = 0.75 W/m²·K K_1 = 49 kJ/m²·K	U-value = 2.31 W/m²·K	U-value = 1.92 W/m²·K		
Scenario 1	Girón	F1 U-value = 0.60 W/m²·K Y_{12} = 0.04 W/m²·K K_1 = 41 kJ/m²·K	U-value = 0.12 W/m²·K	U-value = 3.81 W/m²·K; Perimeter insulation: 1 m, R = 3.0 m²·K/W;	PVC frame U-value = 1.80 W/m²·K Triple glass U-value = 0.85 W/m²·K; g = 0.70 Light shutter Window U-value = 1.04 W/m²·K	Regular gas boiler Performance 81% [34]
	Alférez Rojas	F2 U-value = 0.66 W/m²·K Y_{12} = 0.11 W/m²·K K_1 = 42 kJ/m²·K F3 U-value = 0.15 W/m²·K Y_{12} = 0.00 W/m²·K K_1 = 41 kJ/m²·K		U-value = 0.16 W/m²·K		
	Balsas de Ebro Viejo	F9 U-value = 0.73 W/m²·K Y_{12} = 0.33 W/m²·K K_1 = 45 kJ/m²·K F10 U-value = 0.15 W/m²·K Y_{12} = 0.03 W/m²·K K_1 = 42 kJ/m²·K	U-value = 0.12 W/m²·K	U-value = 0.15 W/m²·K		

Table A1. *Cont.*

Scenarios and Cases		Façade	Roof	Ground	Windows	Systems
Scenario 2	Girón	F1 U-value = 0.23 W/m^2·K Y_{12} = 0.00 W/m^2·K K_1 = 41 kJ/m^2·K	U-value =0.12 W/m^2·K	U-value = 3.81 W/m^2·K; Perimeter insulation: 1 m, R = 3.0 m^2·K/W;	PVC frame U-value = 1.80 W/m^2·K Triple glass U-value = 0.85 W/m^2·K; g = 0.70 Light shutter Window U-value = 1.04 W/m^2·K	Regular gas boiler Performance 81% [34]
	Alférez Rojas	F2 U-value = 0.23 W/m^2·K Y_{12} = 0.01 W/m^2·K K_1 = 41 kJ/m^2·K F3 U-value = 0.21 W/m^2·K Y_{12} = 0.01 W/m^2·K K_1 = 41 kJ/m^2·K		U-value = 0.28 W/m^2·K		
	Balsas de Ebro Viejo	F9 U-value = 0.22 W/m^2·K Y_{12} = 0.04 W/m^2·K K_1 = 42 kJ/m^2·K F10 U-value = 0.23 W/m^2·K Y_{12} = 0.05 W/m^2·K K_1 = 43 kJ/m^2·K	U-value = 0.12 W/m^2·K	U-value = 0.28 W/m^2·K		
Scenario 3	Girón	F1 U-value = 0.60 W/m^2·K Y_{12} = 0.04 W/m^2·K K_1 = 41 kJ/m^2·K	U-value = 0.12 W/m^2·K	U-value = 3.81 W/m^2·K; Perimeter insulation: 1 m, R = 3.0 m^2·K/W;	PVC frame U-value = 1.80 W/m^2·K Triple glazing U-value = 0.85 W/m^2·K; g = 0.70 Light shutter U-value window = 1.04 W/m^2·K	Gas condensing boiler Performance 100% 60% solar energy DHW
	Alférez Rojas	F2 U-value = 0.66 W/m^2·K Y_{12} = 0.11 W/m^2·K K_1 = 42 kJ/m^2·K F3 U-value = 0.15 W/m^2·K Y_{12} = 0.00 W/m^2·K K_1 = 41 kJ/m^2·K		U-value = 0.16 W/m^2·K		
	Balsas de Ebro Viejo	F9 U-value = 0.73 W/m^2·K Y_{12} = 0.33 W/m^2·K K_1 = 45 kJ/m^2·K F10 U-value = 0.15 W/m^2·K Y_{12} = 0.03 W/m^2·K K_1 = 42 kJ/m^2·K	U-value = 0.12 W/m^2·K	U-value = 0.15 W/m^2·K		

Table A1. *Cont.*

Scenarios and Cases		Façade	Roof	Ground	Windows	Systems
Scenarios 4 and 4*	*Girón*	$\underline{F1}U$-value = 0.23 W/m²·K Y_{12} = 0.00 W/m²·K K_1 = 41 kJ/m²·K	U-value = 0.12 W/m²·K	U-value = 3.81 W/m²·K; Perimeter insulation: 1 m, R = 3.0 m²·K/W;	PVC frame U-value = 1.80 W/m²·K Triple glass U-value = 0.85 W/m²·K; g = 0.70 Light shutter U-value window =1.04 W/m²·K	Gas condensing boiler Performance 100% 60% solar energy DHW (scenario 4) 65% solar energy DHW (scenario 4*)
	Alférez Rojas	$\underline{F2}$ U-value = 0.23 W/m²·K Y_{12} = 0.01 W/m²·K K_1 = 41 kJ/m²·K $\underline{F3}$ U-value = 0.21 W/m²·K Y_{12} = 0.01 W/m²·K K_1 = 41 kJ/m²·K		U-value = 0.28 W/m²·K		
	Balsas de Ebro Viejo	$\underline{F9}U$-value = 0.22 W/m²·K Y_{12} = 0.04 W/m²·K K_1 = 42 kJ/m²·K $\underline{F10}$ U-value = 0.23 W/m²·K Y_{12} = 0.05 W/m²·K K_1 = 43 kJ/m²·K	U-value = 0.12 W/m²·K	U-value = 0.28 W/m²·K		

Note: * Additional 5% of DHW (in total 65%)

Appendix B

Table A2. Linear thermal transmittance of thermal bridges (W/mK).

Scenario	Floor-To-Wall Junctions	Wall-To-Wall Junctions: Exterior Corners	Wall-To-Wall Junctions: Interior Corners	Window-To-Wall Junctions: Ledge	Window-To-Wall Junctions: Lintel	Window-To-Wall Junctions: Jambs	Slab-On-Ground	Suspended Concrete Floors and Slabs in Contact with Uninhabitable Rooms
				Girón case study				
0	0.99	0.15	−0.22	0.10	0.58	0.37	3.51	-
1, 3	0.67	0.08	−0.12	0.12	0.55	0.34	3.76	-
2, 4	0.07	0.04	−0.06	0.08	0.09	0.03	3.88	-
				Alférez Rojas case study				
0	0.98	0.15	−0.22	0.11	0.59	0.37	-	−2.67
1, 3	0.18	0.08	−0.19	0.15	0.70	0.43	-	0.94
2, 4	0.08	0.05	−0.07	0.08	0.09	0.04	-	0.19
				Balsas de Ebro Viejo case study				
0	1.06	0.19	-	0.10	0.56	0.36	-	−6.11
1, 3	0.61	0.09	-	0.12	0.51	0.32	-	0.9
2, 4, 4 *	0.06	0.04		0.08	0.09	0.04	-	0.81

Note: * Additional 5% of DHW (in total 65%).

Appendix C

Table A3. Use and climate conditions in the simulation models.

Parameters	Values
Occupancy density (3 people)	0.05 people/m^2
Hours of operation and occupancy: Monday to Friday: 7 am to 3 pm	25% occupation
Hours of operation and occupancy: Monday to Friday: 3 pm to 8 pm	50% occupation
Hours of operation and occupancy: Monday to Friday: 8 pm to 7 am	100% occupation
Hours of operation, activity and occupancy: Saturday and Sunday	100% occupation
Climate equipment operating hours (heating)	7 a.m. to 10 p.m.
Climate equipment operating hours (cooling)	9 a.m. to 8 p.m.
Running of the air conditioning system from Monday to Sunday	7 days/week
Summer period	1st June–30th September
Winter period	1st October–31st May
Metabolic factor: "Standing/walking" option	1
Clothing values (CLO)	Summer 0.5 Clo; winter: 1.0 Clo
Load due to general lighting	4.4 W/m^2–100lux (10% (0–7 h, 30% (7–18 h), 50% (18–19 h), 100% (19–23 h), 50% (23-24 h)
High set point temperature (June-September)	27 °C (0:00–7:00, 23:00–23:59) 25 °C (15:00–22:59)
Low set point temperature (January-May, October-December)	17 °C (0:00–7:00, 23:00–23:59) 20 °C (7:00–22:59)
Relative humidity of the indoor air	50%
Infiltration air flow rate per internal volume through the envelope (n50)	5.72 h^{-1} for scenario 0 [41], and 4.91 h^{-1} for the improved scenarios (1–4 and 4 *)
Renewal air flow	33 L/s
Night ventilation	4 h^{-1} (June–September; 1–8 h)
Activation of solar control in summer	Window shadingType: blind with low reflectivity slats (scenario 0), blind with high reflectivity slats (scenarios 1–4 and 4*) Position: outside Slat angle control type: fix; vertical Operation schedule: residential CTE shading: On: June–September; 30% 0–24 h Off: January–May; October–December

Note: * Additional 5% of DHW (in total 65%).

References

1. European Commission. Climate Strategies & Targets—European Commission. Available online: https://ec.europa.eu/clima/policies/strategies_en (accessed on 9 April 2020).
2. European Commission. Sustainable Buildings for Europe's Climate-Neutral Future | EASME. Available online: https://ec.europa.eu/easme/en/news/sustainable-buildings-europe-s-climate-neutral-future (accessed on 9 April 2020).
3. European Commission. EU Buildings Factsheets. Available online: https://ec.europa.eu/energy/eu-buildings-factsheets_en (accessed on 9 April 2020).
4. Palacios-Munoz, B.; López-Mesa, B.; Gracia-Villa, L. Influence of refurbishment and service life of reinforced concrete buildings structures on the estimation of environmental impact. *Int. J. Life Cycle Assess.* **2019**, *24*, 1913–1924. [CrossRef]
5. Palacios-Munoz, B.; Peuportier, B.; Gracia-Villa, L.; López-Mesa, B. Sustainability assessment of refurbishment vs. new constructions by means of LCA and durability-based estimations of buildings lifespans: A new approach. *Build. Environ.* **2019**, *160*, 106203. [CrossRef]
6. Trilla Bellart, C.; Bosch Meda, J. *El Parque Público y Protegido de Viviendas en España: Un Análisis Desde el Contexto Europeo*; Fundación Alternativas: Madrid, Spain, 2018; ISBN 978-84-15860-87-7.
7. Monzón, M.; López-Mesa, B. Buildings performance indicators to prioritise multi-family housing renovations. *Sustain. Cities Soc.* **2018**, *38*, 109–122. [CrossRef]
8. Ruiz Palomeque, L.G.; Rubio del Val, J. *Nuevas Propuestas de Rehabilitación Urbana en Zaragoza. Estudio de Conjuntos Urbanos de Interés*; Sociedad Municipal de Rehabilitación Urbana de Zaragoza: Zaragoza, Spain, 2006.
9. Rubio del Val, J. Veinte años rehabilitando en Zaragoza: Primeras actuaciones de rehabilitación edificatoria sobre Conjuntos Urbanos de Interés de vivienda social, impulsados por la Sociedad Municipal Zaragoza Vivienda (2004–2017). In *Nuevos Enfoques en la Rehabilitación Energética de la Vivienda Hacia la Convergencia Europea La Vivienda Social en Zaragoza, 1939–1979*; López-Mesa, B., Ed.; Prensas de la Universidad de Zaragoza: Zaragoza, Spain, 2018; pp. 55–74. ISBN 978-84-17358-41-9.
10. Rubio del Val, J. Presentación del Informe GTR 2018. Los Municipios, Pieza Clave en un Marco de Cooperación Institucional. Available online: https://catedrazaragozavivienda.wordpress.com/2019/07/01/videos-de-la-vi-jornada-de-la-catedra-zaragoza-vivienda-sobre-estrategias-locales-de-regeneracion-urbana/ (accessed on 10 April 2020).
11. INE. Censos de Población y Viviendas. 2011. Available online: https://www.ine.es/censos2011_datos/cen11_datos_resultados.htm# (accessed on 10 April 2020).
12. López-Mesa, B.; Tejedor Bielsa, J. The Cátedra Zaragoza Vivienda, in search of interdisciplinary solutions to the problem of obsolescence of housing and city. *Inf. Constr.* **2015**, *67*, ed009.
13. Kurtz, F.; Monzón, M.; López-Mesa, B. Energy and acoustics related obsolescence of social housing of Spain's postwar in less favoured urban areas: The case of Zaragoza. *Inf. Constr.* **2015**, *67*.
14. Van Hal, A.; Dulski, B.; Postel, A.M. Reduction of CO_2 emissions in houses of historic and visual importance. *Sustainability* **2010**, *2*, 443–460. [CrossRef]
15. Moorhouse, J.; Littlewood, J. The Low-Carbon Retrofit of a UK Conservation Area Terrace: Introducing a Pattern Book of Energy-Saving Details. In *Sustainability in Energy and Buildings*; Springer: Berlin/Heidelberg, Germany, 2012; Volume 12, ISBN 9783642275081.
16. Dowson, M.; Poole, A.; Harrison, D.; Susman, G. Domestic UK retrofit challenge: Barriers, incentives and current performance leading into the Green Deal. *Energy Policy* **2012**, *50*, 294–305. [CrossRef]
17. Ben, H.; Steemers, K. Energy retrofit and occupant behaviour in protected housing: A case study of the Brunswick Centre in London. *Energy Build.* **2014**, *80*, 120–130. [CrossRef]
18. Van Krugten, L.T.F.; Hermans, L.M.C.; Havinga, L.C.; Pereira Roders, A.R.; Schellen, H.L. Raising the energy performance of historical dwellings. *Manag. Environ. Qual. Int. J.* **2016**, *27*, 740–755. [CrossRef]
19. Oorschot, L. New ways in retrofitting postwar Dutch walk-up apartment buildings: Carbon neutrality, cultural value and user preferences. In Proceedings of the 15th International Docomomo Conference—Metamorphosis: The Continuity of Change, Cankarjev Dom, Ljubljana, Slovenia, 28–31 August 2018; IDC: Framingham, MA, USA, 2018; pp. 442–448.

20. Oorschot, L.; Spoormans, L.; El Messlaki, S.; Konstantinou, T.; de Jonge, T.; van Oel, C.; Asselbergs, T.; Gruis, V.; de Jonge, W. Flagships of the dutch welfare state in transformation: A transformation framework for balancing sustainability and cultural values in energy-efficient renovation of postwar walk-up apartment buildings. *Sustainability* **2018**, *10*, 2562. [CrossRef]

21. Karoglou, M.; Kyvelou, S.S.; Boukouvalas, C.; Theofani, C.; Bakolas, A.; Krokida, M.; Moropoulou, A. Towards a preservation-sustainability nexus: Applying LCA to reduce the environmental footprint of modern built heritage. *Sustainability* **2019**, *11*, 6147. [CrossRef]

22. Bowman, J. *Home Truths: A Low-Carbon Strategy to Reduce UK Housing Emissions by 80% by 2050*; University of Oxford Environmental Change Institute: Oxford, UK, 2007.

23. Blessing, L.T.M.; Chakrabarti, A. *DRM, A Design Research Methodology*; Springer: Dordrecht, The Netherlands; Heidelberg, Germany; London, UK; New York, NY, USA, 2009; ISBN 9781848825864.

24. Chakrabarti, A.; Blessing, L. A review of theories and models of design. *J. Indian Inst. Sci.* **2015**, *95*, 325–340.

25. Monzón-Chavarrías, M.; López-Mesa, B. Identificación y catalogación de los casos de estudio: Los Conjuntos Urbanos de Interés de Zaragoza. In *Nuevos Enfoques en la Rehabilitación Energética de la Vivienda Hacia la Convergencia Europea. La Vivienda Social en Zaragoza, 1939–1979*; López-Mesa, B., Ed.; Prensas de la Universidad de Zaragoza: Zaragoza, Spain, 2018; pp. 95–123. ISBN 978-84-17358-41-9.

26. Kurtz, F.; López-Mesa, B. Definición constructiva de los casos de estudio. In *Nuevos Enfoques en la Rehabilitación Energética de la Vivienda Hacia la Convergencia Europea. La Vivienda Social en Zaragoza, 1939–1979*; López-Mesa, B., Ed.; Prensas de la Universidad de Zaragoza: Zaragoza, Spain, 2018; pp. 125–167. ISBN 978-84-17358-41-9.

27. Monzón, M.; López-Mesa, B. Simplified model to determine the energy demand of existing buildings. Case study of social housing in Zaragoza, Spain. *Energy Build.* **2017**, *149*, 483–493.

28. Monzón-Chavarrías, M.; López-Mesa, B. Caracterización energética y acústica de las soluciones constructivas de la envolvente. In *Nuevos Enfoques en la Rehabilitación Energética de la Vivienda Hacia la Convergencia Europea. La Vivienda Social en Zaragoza, 1939–1979*; López-Mesa, B., Ed.; Prensas de la Universidad de Zaragoza: Zaragoza, Spain, 2018; pp. 169–187. ISBN 978-84-17358-41-9.

29. López-Mesa, B.; Monzón-Chavarrías, M. El desarrollo de indicadores de obsolescencia física más precisos para la definición de estrategias de rehabilitación de vivienda social. In *Nuevos Enfoques en la Rehabilitación Energética de la Vivienda Hacia la Convergencia Europea. La Vivienda Social en Zaragoza, 1939–1979*; López-Mesa, B., Ed.; Prensas de la Universidad de Zaragoza: Zaragoza, Spain, 2018; pp. 243–265. ISBN 978-84-17358-41-9.

30. Domínguez-Amarillo, S.; Fernández-Agüera, J.; Sendra, J.J.; Roaf, S. The performance of Mediterranean low-income housing in scenarios involving climate change. *Energy Build.* **2019**, *202*, 109374. [CrossRef]

31. Rossi, M.; Rocco, V.M. External walls design: The role of periodic thermal transmittance and internal areal heat capacity. *Energy Build.* **2014**, *68*, 732–740. [CrossRef]

32. Bienvenido-Huertas, D.; Rubio-Bellido, C.; Pulido-Arcas, J.A.; Pérez-Fargallo, A. Towards the implementation of periodic thermal transmittance in Spanish building energy regulation. *J. Build. Eng.* **2020**, *31*, 101402. [CrossRef]

33. CEN/TC 89. *EN ISO 13786:2017 Thermal Performance of Building Components—Dynamic Thermal Characteristics—Calculation Methods (ISO 13786:2017, Corrected Version 2018-03)*; European Committee for Standardization: Brussels, Belgium, 2018.

34. AICIA (Universidad de Sevilla). *Escala de Calificación Energética. Edificios Existentes*; IDAE, Ed.; Gobierno de España: Madrid, Spain, 2011.

35. AICIA (Universidad de Sevilla). *CALENER-VYP: Viviendas y Edificios Terciarios Pequeños y Medianos. Manual de Usuario*; IDAE, Ed.; Gobierno de España: Madrid, Spain, 2009.

36. Rose, J.; Thomsen, K.E.; Mørck, O.C.; Gutierrez, M.S.M.; Jensen, S.Ø. Refurbishing blocks of flats to very low or nearly zero energy level–technical and financial results plus co-benefits. *Energy Build.* **2019**, *184*, 1–7. [CrossRef]

37. Harkouss, F.; Fardoun, F.; Biwole, P.H. Passive design optimization of low energy buildings in different climates. *Energy* **2018**, *165*, 591–613. [CrossRef]

38. Aguacil, S.; Lufkin, S.; Rey, E.; Cuchi, A. Application of the cost-optimal methodology to urban renewal projects at the territorial scale based on statistical data—A case study in Spain. *Energy Build.* **2017**, *144*, 42–60. [CrossRef]

39. AICIA (Universidad de Sevilla) HULC. Available online: https://www.codigotecnico.org/index.php/menu-recursos/menu-aplicaciones/282-herramienta-unificada-lider-calener.html (accessed on 28 February 2020).

40. Design Builder Software Design Builder. Available online: https://designbuilder.co.uk/ (accessed on 28 February 2020).

41. Fernández-Agüera, J.; Domínguez-Amarillo, S.; Sendra, J.J.; Suárez, R. An approach to modelling envelope airtightness in multi-family social housing in Mediterranean Europe based on the situation in Spain. *Energy Build.* **2016**, *128*, 236–253. [CrossRef]

42. Almeida, R.M.S.F.; Ramos, N.M.M.; Pereira, P.F. A contribution for the quantification of the influence of windows on the airtightness of Southern European buildings. *Energy Build.* **2017**, *139*, 174–185. [CrossRef]

43. Ministerio de Transporte, M.y.A.U. DA DB-HE/1 Cálculo de parámetros característicos de la envolvente. Available online: https://www.codigotecnico.org/index.php/menu-ahorro-energia.html (accessed on 8 May 2020).

44. Parkinson, T.; de Dear, R.; Brager, G. Nudging the adaptive thermal comfort model. *Energy Build.* **2020**, *206*, 109559. [CrossRef]

45. Sol-Arq CONFADAPT-ASH55. Available online: https://www.seiscubos.com/recursos/confadapt-ash55 (accessed on 8 May 2020).

46. CYPE Ingenieros CYPE Generador de Precios. España. Available online: http://www.generadordeprecios.info/ (accessed on 22 June 2020).

47. Capacity4dev Cost-Effectiveness Analysis. Available online: https://europa.eu/capacity4dev/evaluation_guidelines/wiki/cost-effectiveness-analysis-0 (accessed on 23 April 2020).

48. Imessad, K.; Derradji, L.; Messaoudene, N.A.; Mokhtari, F.; Chenak, A.; Kharchi, R. Impact of passive cooling techniques on energy demand for residential buildings in a Mediterranean climate. *Renew. Energy* **2014**, *71*, 589–597. [CrossRef]

Article

Experimental Investigation of the Potential of a New Fabric-Based Evaporative Cooling Pad

Eloy Velasco-Gómez, Ana Tejero-González *, Javier Jorge-Rico and F. Javier Rey-Martínez

Research Group in Thermal Engineering, Department of Energy and Fluidmechanics, School of Engineering, Universidad de Valladolid, 47011 Valladolid, Spain; eloy@eii.uva.es (E.V.-G.); javiejorgerico@gmail.com (J.J.-R.); rey@eii.uva.es (F.J.R.-M.)
* Correspondence: anatej@eii.uva.es

Received: 6 August 2020; Accepted: 27 August 2020; Published: 30 August 2020

Abstract: Direct evaporative coolers are energy-efficient, economic solutions to supplying cooling demand for space conditioning. Since their potential strongly depends on air hygrothermal conditions, they are traditionally used in dry and hot climates, though they can be used in many applications and climates. This work proposes a new direct evaporative cooling system with a fabric-based pad. Its design enables maximum wetted surface with minimum pressure drop. Its performance has been experimentally characterized in terms of saturation efficiency, air humidification, pressure drop, and level of particles, based on a full factorial Design of Experiments. Factors studied are air dry bulb temperature, specific humidity, and airflow. Saturation efficiencies obtained for a 25 cm pad are above the values achieved by other alternative evaporative cooling (EC) pads proposed in the literature, with lower pressure drops.

Keywords: direct evaporative cooling; new pad materials; wet fabric; saturation efficiency; pressure drop

---

1. Introduction

Evaporative cooling is a phenomenon that occurs in nature when water comes into contact with unsaturated air, hence its occurrence near water bodies like waterfalls or the sea, after a storm, or due to sweat evaporation from skin. Its applicability for human comfort dates back to ancient Egypt [1]. The consequent simplicity of evaporative cooling systems makes them cost-effective, energy-efficient, and environmentally friendly solutions for a wide range applications and climates [2]. They permit achieving appropriate indoor conditions in otherwise non-conditioned indoor spaces such as greenhouses, storehouses, factories, and farms, where ventilation during the cooling season is insufficient, and mechanical cooling is unfeasible. However, their use for indoor comfort is also crucial. On the path to Net Zero Energy Buildings, they can either substitute, improve the performance of, or complement conventional mechanical compression cooling systems.

Direct Evaporative Coolers (DEC) enable air cooling as it passes through wetted pads or sprayed water. As water evaporates in air, heat and mass transfer transforms the sensible heat of air into latent heat in an adiabatic process, lowering the air Dry Bulb Temperature (DBT) of treated air towards its Wet Bulb Temperature (WBT), while its specific humidity increases. Materials of commercialized pads range from aspen wood or absorbent plastic foams to "rigid-media" pads made of corrugated materials including cellulose, plastic, and fiberglass [3].

Cellulose commercialized pads are extensively studied in the literature, both experimentally and theoretically, through mathematical models. Sheng & Nanna [4] characterize the saturation efficiency and humidity increase in the treated air of a cellulose corrugated panel in terms of inlet air DBT, air velocity and pad thickness. Besides saturation efficiency and humidification increase, Malli et al. [5], Franco et al. [6] and Warke & Deshmukh [7] characterize the pressure drop generated.

Barzegar et al. [8] compare the performance of cellulose pads from two different manufacturers, though only in terms of air velocity. Al-Badri & Al-Waaly [9] compare the saturation efficiency of three different cellulose pads in terms of inlet air DTB, air WBT and water temperature and ratio. He et al. [10] compare the performance of a Munters CELdek®cellulose pad to a PVC fibre pad, while Franco et al. [11] compares a cellulose pad to a plastic mesh-based pad. The latter also approaches the study of heat and mass coefficients, and so do Al-Badri and AlWaaly [9]. To the knowledge of the authors, only He et al. [10] provide detailed observations on water entrainment.

There exists wide research on the use of alternative materials for DEC media. Most works focus on pads made of locally available vegetable fibres such as coconut [12–16], jute [12,17,18], luffa [17,19] and palash [13,14], as well as wood fibres [17,20].

Some research also considers the use of porous stones [20,21] and ceramics [22] as DEC media. More unusual EC medias are made of bulk charcoal [18,23], sliced wood [24], perforated bamboo [25], composite material with rice byproducts [26] or shredded latex foam, palm fruit fibre, etc. [18].

However, existing literature barely approaches the use of textiles for evaporative cooling, despite the advantage of their wicking ability, as studied by Xu et al. [27] for indirect evaporative cooling systems. They analysed six different fabrics compared to Kraft paper, with favourable results, but nonetheless observed that the use of fabrics may cause mechanical problems, as wet surfaces distort if adhered to a support sheet.

Liao et al. [28] studied nonwoven fabric and coir as alternatives to EC pads, providing correlations for saturation efficiency and pressure drop in terms of pad thickness. Alam et al. [12] also studied coir, together with sackcloth, with the former performing better due to a maximum temperature drop of 4 °C. Both studies highlight the energy efficiency and low cost of the solution proposed. Liao and Chiu [29] experimentally studied a new pad made of coarse and fine fabric PVC sponge mesh, obtaining efficiencies of up to 86.32% and 85.5% for 15 cm thickness pads, respectively.

Fabrics have occasionally been considered in combination with passive cooling solutions for buildings. Ghosal et al. [30] studied the cooling effect of a wetted shade fabric installed in a green house, achieving up to 6 °C temperature drop. Esparza et al. [31] designed a fabric-based evaporative cooling system with a 3 mm thick wool and polyester membrane as an alternative to water roof ponds. Their solution achieved the lowest temperatures under the three climate conditions studied: hot sub-humid, warm sub-humid and hot humid.

Together with capillary diffusion and wicking ability, the main assets of the use of fabrics relies on their low cost, it being possible to reuse discarded clothes. Microbial growth is the main concern shared by all evaporative cooling equipment but specially noticed for pads based on organic materials, both conventional and alternative. Proper maintenance based on bleedoff, makeup water quality and complete emptying of the sump during non-operating periods would avoid this issue [32,33]. As usually found in commercial systems, fabric-based EC can also include ozonisers or UV lamps for pathogen inactivation in working water. Moreover, simple pad configurations made with fabrics can be easily disassembled and washed, facilitating the maintenance of the media.

Organic fibres may also be disregarded due to concerns about system durability. For instance, Jute, Luffa, Comm and Palm show cooling efficiency degradation with time. Deterioration is mainly due to bio-degradation, and among the previous materials, jute fibres show the worst performance [17]. Nonetheless, replacement of media in non-elaborated pads made of organic, widely available materials would be cost-effective.

The aim of this work is to design, build and experimentally characterize a new Evaporative Cooling Pad made of cotton fabric as an alternative to conventional pads, which is different from the proposals already studied in the literature. Saturation efficiency, air humidification, pressure drop and particles are presented and contrasted with results obtained in existing research.

2. Materials and Methods

This section describes the design, construction and experimental characterization of the target evaporative cooling system.

2.1. Design and Construction of the Fabric EC System

A fabric-based evaporative cooling (EC) pad has been designed and built with a 25 x 1600 cm^2 piece of cotton fabric. It was obtained from a 200 x 200 cm^2 cloth cut into eight pieces, then sewed. The cloth has been arranged and stretched on a wire structure, creating a 25 cm thick hexagonal pad with homogeneous air paths separated with the aid of spacers (Figure 1). The cross section is 0.18 m^2. This design aims at avoiding any mechanical issues related to problems observed in the literature [27]. Since the cloth does not excessively tighten, in case of slight deterioration of the cotton fibres the system would not dismantle.

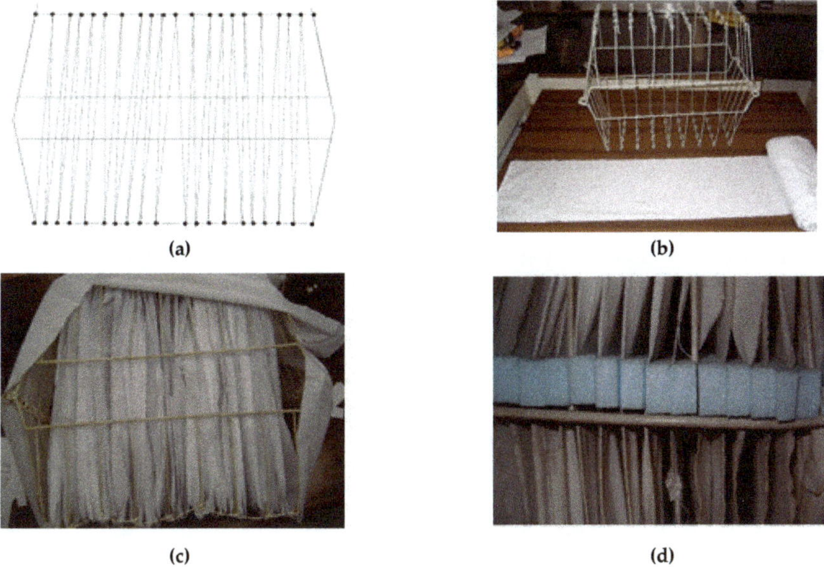

Figure 1. Construction of the fabric evaporative cooling (EC) pad: (**a**) scheme, (**b**) wire structure and cotton cloth, (**c**) cloth arrangement within the structure, and (**d**) detail of the spacers.

The pad is installed within a methacrylate case 45 cm wide, 45 cm high and 25 cm deep. Since the pad is hexagonal, non-useful areas are blocked with additional methacrylate plates to avoid air bypass. Water is pumped from a lower deposit to an upper distributor. The water pump provides 9.5 l/min and requires 9 W. The whole system is connected to air inlet and outlet plenums. These plenums are pyramidal shaped and 60 cm long to favour uniform airflow through the pad. (Figure 2).

(a)   (b)

Figure 2. View of the fabric EC pad: (**a**) upper water distributor, lower deposit, and water pump; and (**b**) connection to air inlet and outlet plenums.

2.2. Experimental Setup

The system is connected through flexible ducts to an Air Handling Unit (AHU) that provides the desired conditions of air Dry Bulb Temperature (DBT), Relative Humidity (RH) and volume flow (Figure 3). DBT and RH sensors at the outlet enable control of the AHU. The Airflow is measured with an orifice plate, by differential pressure measurements in the nozzle and six diameters far upstream. Pressure drop is also measured in the fabric EC system.

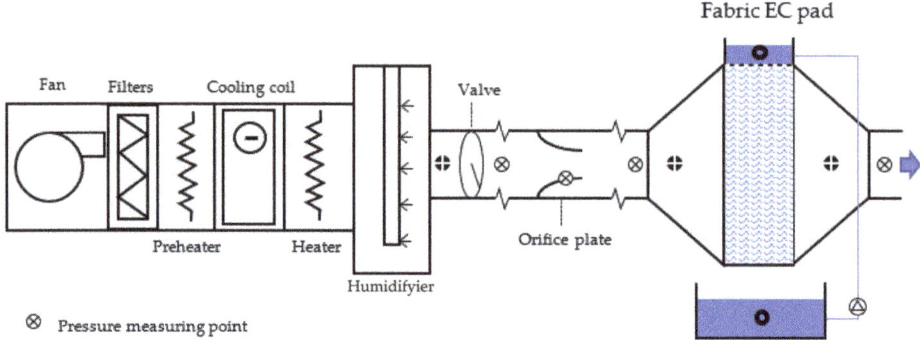

Figure 3. Scheme of the system connected to the Air Handling Unit (AHU) and position of measuring sensors.

DBT and RH are measured in triplicate at the fabric EC system inlet and outlet with 4-wire Pt100 and capacity sensors. The system characterization is based on the average measurements of these three-sensor measuring points. This triple measurement at each section also enables the checking of any possible stratification. At the air inlet, DBT standard deviation ranges from 0.52 to 0.57 °C, while RH standard deviation ranges from 0 to 2.4%. At the air outlet, DBT standard deviation ranges from 0.12 to 0.54 °C, and RH standard deviation ranges from 2.5 to 5.3%. Comparing these values to the uncertainty of the measurement given later (Section 3), it demonstrates that no representative stratification occurs.

Water temperature in both the lower tank and the upper distributor is measured in duplicate with ceramic Pt100 sensors. Temperature Pt100 sensors' accuracy is ± 0.1 °C, and they are calibrated with a Fluke 9103 dry well, while RH capacity sensors' accuracy is ± 2%, and they are calibrated with a Vaisala Humidity Calibrator HMK15. Pressure drop was measured with a Honeywell 163PC01D75 (160 PC series). The orifice plate was previously calibrated with a nozzle, using a Testo 435-4 for pressure drop measuring. Temperature, RH and differential pressure sensors are connected to an Agilent 34972A data acquisition system.

Additionally, airborne particles at the system outlet are characterized with a Lasair II Air particle counter. Measurements have been performed in duplicate at the three airflow levels and for the media wetted and dry (total of 12 tests).

2.3. Design of Experiments

To characterize the behaviour of the system, this work proposes a full factorial design of experiments. Target factors are three-level air volume flow (V), six-level DBT (T) and three-level specific humidity (w), the latter in kilograms of vapor to kilograms of dry air (kg_v / kg_{da}). Air DBT ranges from 25 to 50 °C to cover the most common summer conditions worldwide. Airflow and specific humidity are limited by the experimental setup: the air volume flow rate studied ranges from the lowest limit of the orifice plate accuracy and the maximum airflow provided by the AHU, while the minimum and maximum specific humidity considered are the air specific humidity at the laboratory and the maximum humidity achievable with the humidifier at 25 °C.

Tests are performed in Valladolid, Spain (approximately 700 m.a.s.l.). The experimental runs are performed once, randomized. Tests that did not reach a steady state were disregarded and repeated. Table 1 gathers the levels of the factors studied, tests performed, and their order.

Table 1. Factors, levels, and test code. Order of the test run is in parenthesis.

Airflow [m³/h]	Specific Humidity [kg_v/kg_{da}]	DBT [°C]					
		T1 **25 °C**	**T2** **30 °C**	**T3** **35 °C**	**T4** **40 °C**	**T5** **45 °C**	**T6** **50 °C**
	W1 0.0115	V1T1W1 (1)	V1T2W1 (8)	V1T3W1 (16)	V1T4W1 (10)	V1T5W1 (6)	V1T6W1 (13)
V1 210	W2 0.0150	V1T1W2 (9)	V1T2W2 (3)	V1T3W2 (5)	V1T4W2 (12)	V1T5W2 (14)	V1T6W2 (17)
	W3 0.0195	V1T1W3 (11)	V1T2W3 (7)	V1T3W3 (2)	V1T4W3 (4)	V1T5W3 (15)	V1T6W3 (18)
	W1 0.0115	V2T1W1 (43)	V2T2W1 (49)	V2T3W1 (37)	V2T4W1 (47)	V2T5W1 (52)	V2T6W1 (40)
V2 300	W2 0.0150	V2T1W2 (42)	V2T2W2 (39)	V2T3W2 (46)	V2T4W2 (50)	V2T5W2 (45)	V2T6W2 (53)
	W3 0.0195	V2T1W3 (54)	V2T2W3 (48)	V2T3W3 (51)	V2T4W3 (38)	V2T5W3 (44)	V2T6W3 (41)
	W1 0.0115	V3T1W1 (32)	V3T2W1 (22)	V3T3W1 (19)	V3T4W1 (27)	V3T5W1 (24)	V3T6W1 (30)
V3 450	W2 0.0150	V3T1W2 (36)	V3T2W2 (34)	V3T3W2 (21)	V3T4W2 (31)	V3T5W2 (26)	V3T6W2 (28)
	W3 0.0195	V3T1W3 (23)	V3T2W3 (33)	V3T3W3 (35)	V3T4W3 (20)	V3T5W3 (29)	V3T6W3 (25)

3. Results

Actual DBT and RH conditions of inlet air achieved in the AHU for the Design of Experiments shown in Table 1 are gathered in Table 2.

Table 2. Measured inlet conditions for each test.

Airflow [m^3/h]	Specific Humidity [kg$_v$/kg$_{da}$]	DBT [°C]					
		T1 25 °C	T2 30 °C	T3 35 °C	T4 40 °C	T5 45 °C	T6 50 °C
V1 210	W1 0.0115	25.2 °C 55.0%	29.6 °C 45.3%	36.4 °C 30.6%	39.7 °C 30.3%	46.5 °C 17.7%	48.5 °C 15.4%
	W2 0.0150	24.6 °C 84.8%	30.5 °C 55.9%	35.9 °C 43.0%	40.0 °C 32.3%	46.0 °C 21.9%	48.5 °C 17.1%
	W3 0.0195	25.2 °C 100 %	29.2 °C 71.4%	36.3 °C 55.0%	39.6 °C 48.8%	46.6 °C 29.2%	50.2 °C 22.3%
V2 300	W1 0.0115	25.8 °C 55.8%	30.0 °C 44.8%	34.6 °C 33.7%	39.9 °C 27.4%	45.3 °C 22.0%	51.5 °C 14.6%
	W2 0.0150	25.2 °C 71.7%	30.3 °C 60.2%	34.9 °C 45.5%	40.0 °C 39.5%	45.2 °C 29.8%	50.9 °C 20.2%
	W3 0.0195	25.3 °C 81.4%	30.0 °C 72.5%	34.9 °C 63.1%	40.8 °C 48.6%	45.9 °C 36.1%	51.6 °C 28.5%
V3 450	W1 0.0115	25.2 °C 58.0%	30.4 °C 43.9%	35.9 °C 31.5%	41.5 °C 30.6%	45.9 °C 16.9%	52.0 °C 18.6%
	W2 0.0150	25.0 °C 69.7%	30.0 °C 61.6%	35.7 °C 45.3%	41.5 °C 31.6%	46.8 °C 28.8%	48.0 °C 20.8%
	W3 0.0195	25.2 °C 93.0%	30.0 °C 69.2%	35.0 °C 58.7%	41.4 °C 47.4%	43.2 °C 39.3%	51.1 °C 23.3%

Outlet conditions and water temperatures are given in Tables 3 and 4, respectively.

Table 3. Measured air outlet conditions for each test.

Airflow [m^3/h]	Specific Humidity [kg$_v$/kg$_{da}$]	DBT [°C]					
		T1 25 °C	T2 30 °C	T3 35 °C	T4 40 °C	T5 45 °C	T6 50 °C
V1 210	W1 0.0115	20.1 °C 86.0%	20.5 °C 87.9%	23.7 °C 77.0%	25.2 °C 83.8%	27.0 °C 87.7%	29.3 °C 83.1%
	W2 0.0150	20.1 °C 100%	23.4 °C 97.8%	26.0 °C 95.7%	26.0 °C 91.1%	28.2 °C 88.7%	29.8 °C 74.4%
	W3 0.0195	22.7 °C 100%	23.6 °C 100%	26.6 °C 94.0%	27.7 °C 94.1%	30.2 °C 94.5%	32.0 °C 93.4%
V2 300	W1 0.0115	18.8 °C 91.9%	21.4 °C 87.3%	23.5 °C 85.8%	25.9 °C 80.1%	28.2 °C 76.6%	30.3 °C 74.0%
	W2 0.0150	20.9 °C 100%	22.9 °C 100%	25.3 °C 92.8%	27.8 °C 88.2%	29.4 °C 81.8%	31.3 °C 74.0%
	W3 0.0195	21.6 °C 100%	24.3 °C 100%	27.1 °C 100%	29.0 °C 100%	31.2 °C 90.8%	32.8 °C 84.7%
V3 450	W1 0.0115	19.6 °C 90.0%	22.2 °C 87.6%	25.9 °C 70.5 %	29.1 °C 72.8 %	29.5 °C 71.7 %	34.2 °C 69.5%
	W2 0.0150	21.0 °C 100%	23.8 °C 95.8 %	26.3 °C 89.7 %	29.5 °C 74.4 %	32.8 °C 73.5 %	32.1 °C 74.4%
	W3 0.0195	23.1 °C 99%	25.7 °C 100%	27.0 °C 98.1 %	30.7 °C 92.1 %	32.0 °C 89.0 %	37.4 °C 68.9%

Values provided in Tables 2–4 are the average values measured during the steady state periods for each test. The uncertainty of these values can be determined as the sum of the sensor accuracy (Section 2.2) and the standard deviation of the measured value along the period considered. The standard deviation of the DBT and RH measurements during steady state conditions is below ± 0.2 °C and ± 3% at the system inlet, respectively. These are larger than the DBT and RH standard deviations at the system outlet (± 0.1 °C and ± 1%, respectively), which is due to the effect of the AHU control. Consequently, uncertainty would be ± 0.3 °C for the DBT and ± 5% for the RH.

Next, the results of the obtained target parameters are described and discussed within existing research.

Table 4. Measured water temperatures for each test.

Airflow [m³/h]	Specific Humidity [kg$_v$/kg$_{da}$]	DBT [°C]					
		T1 25 °C	T2 30 °C	T3 35 °C	T4 40 °C	T5 45 °C	T6 50 °C
V1 210	W1 0.0115	17.1 °C	17.7 °C	20.0 °C	20.9 °C	21.6 °C	22.1 °C
	W2 0.0150	19.9 °C	21.3 °C	22.0 °C	22.6 °C	23.5 °C	23.7 °C
	W3 0.0195	22.2 °C	22.7 °C	23.0 °C	24.4 °C	26.0 °C	26.7 °C
V2 300	W1 0.0115	16.7 °C	18.0 °C	19.2 °C	20.4 °C	21.8 °C	22.4 °C
	W2 0.0150	20.0 °C	20.4 °C	21.6 °C	22.9 °C	22.7 °C	24.3 °C
	W3 0.0195	21.1 °C	21.8 °C	23.8 °C	24.6 °C	24.5 °C	25.6 °C
V3 450	W1 0.0115	16.9 °C	19.3 °C	19.6 °C	20.7 °C	21.4 °C	23.5 °C
	W2 0.0150	19.5 °C	20.7 °C	22.1 °C	22.3 °C	23.7 °C	23.7 °C
	W3 0.0195	21.1 °C	22.7 °C	22.9 °C	25.9 °C	24.9 °C	25.4 °C

3.1. Saturation Efficiency

Saturation efficiency, ε, relates the temperature drop achieved in the air between the inlet (T_{in}) and outlet (T_{out}) to the maximum temperature drop achievable, the latter being considered in terms of the adiabatic saturation temperature of inlet air ($T_{as\ in}$), as expressed in Equation (1):

$$\varepsilon = \frac{T_{in} - T_{out}}{T_{in} - T_{as\ in}} \tag{1}$$

Results obtained for the saturation efficiency are shown in Figure 4.

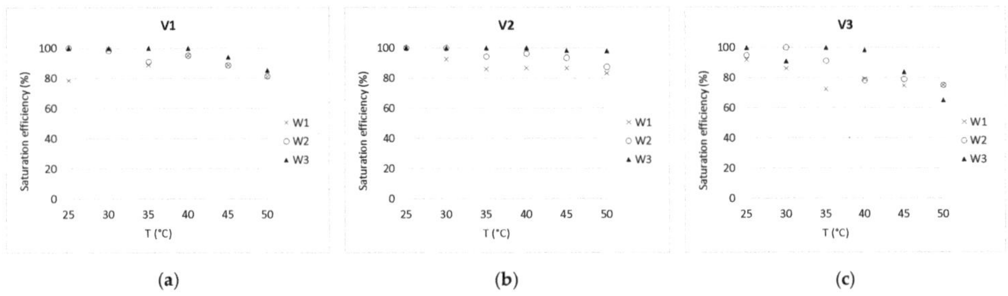

Figure 4. Values of the saturation efficiency in terms of inlet air Dry Bulb Temperature (DBT) (Tin) and specific humidity (W) for airflow levels: (**a**) V1, (**b**) V2, and (**c**) V3.

Saturation efficiency is almost constant with the inlet air DBT, though for extreme temperatures (over 40 °C) it decreases. Inlet air humidity level has also a minor effect. Saturation efficiency can be expected to be almost constant for the common range of inlet air hygrometric conditions. Nonetheless, the effect of the airflow level is more noticeable, as larger airflows incur into lower saturation efficiencies.

A more detailed analysis of this dependency is approached in the discussion section in comparison to existing research.

3.2. Air Humidification

Air humidification is studied through the increment of the specific humidity, w, as demonstrated in Equation (2):

$$\Delta w = w_{out} - w_{in} \tag{2}$$

Results shown in Figure 5 illustrate how the increment of the specific humidity strongly depends on the inlet air DBT. Higher inlet specific humidity slightly conditions a lower increment. Both results are due to the larger Wet Bulb Depression (WBD), this being the difference between the DBT and the Wet Bulb Temperature (WBT). Since saturation efficiency is fairly maintained for a particular design operating at a particular airflow, larger WBD also involves larger humidity increments.

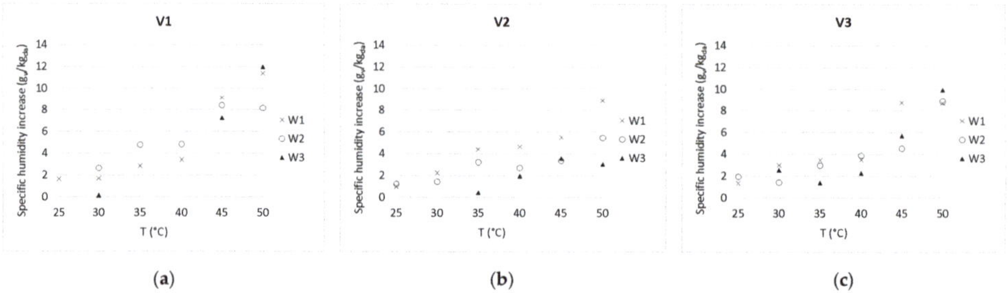

Figure 5. Air specific humidity increase in terms of inlet air DBT (Tin) and specific humidity (W) for airflow levels: (**a**) V1, (**b**) V2, and (**c**) V3.

Some tests (V1T1W2, V1T1W3, V2T1W3, V2T2W3 and V3T1W3) showed air slight dehumidification. This is because water temperatures (Table 2) did not reach the air saturation temperature of inlet air. Consequently, under inlet air conditions corresponding to high relative humidity, when adiabatic saturation and dew point (DPT) temperatures were similar, water temperature was slightly below air DPT. Tests V1T3W3 and V1T4W4 show the same behaviour, though inlet air relative humidity is low. In this case, water temperature remained far from the target adiabatic saturation temperature due to the thermal inertia of the 50 l water tank, thus not being representative.

3.3. Pressure Drop

Pressure drop in the fabric EC pad has been calculated through the experimental procedure shown in Figure 3, varying the potentiometer of the fan in four positions. Measurement of the pressure drop in the calibrated orifice plate enables determination of the air volume flow through the system (Table 5). Pressure drops obtained are low, as expected due to the design of the system and the low Reynolds numbers achievable under the air velocities tested for the airpath between wetted surfaces. Comparison to values obtained in the literature for conventional and other alternative pads is discussed in the next section.

Table 5. Airflow and pressure drop registered in the system.

Pressure Drop in the Orifice Plate (Pa)	Air volume Flow (m³/h)	Pressure Drop in the Fabric EC System (Pa)
766	543	9
531	451	6
179	261	3
95	190	2

3.4. Airborne Particles

DEC systems are known to perform some cleaning of treated air [2]. The study of particles airborne through the system provide insight into this issue and also into the water entrainment that can involve risk of dispersion of Legionella bacteria. To analyse the latter issue, special attention must be paid to 5 μm aerosols. This sizing is determinant because they are breathable aerosols of enough size to contain rod shaped 1 x 3 μm size bacteria [33]. Table 6 gathers the number of particles registered for all tests performed at each airflow level, with the dry and then wetted media.

Table 6. Airborne particles counted at the fabric EC pad outlet.

Media	Airflow (m³/h)	Test No.	Particle Size						
			0.3 μm	0.5 μm	1 μm	5 μm	10 μm	25 μm	Total
Dry	V1	1	300493	137149	58035	1704	290	18	497689
		6	304787	128504	53955	1694	229	7	489176
	V2	2	297348	129643	50795	1227	197	6	479216
		4	294970	122589	45750	1054	140	3	464506
	V3	3	298592	133134	52997	1010	103	9	485845
		5	294268	121133	44317	783	88	7	460596
Wet	V1	7	330227	112532	42305	1111	124	6	486305
		8	332127	111734	42688	1209	134	7	487899
	V2	9	327564	101634	33258	658	70	4	463188
		12	346768	106485	34215	598	72	2	488140
	V3	10	333924	110254	40466	798	86	4	485532
		11	334664	105435	35631	553	57	2	476342

The smallest registered particles of 0.3 μm account for 60 to 70% of the total number of particles counted. As well as this, almost all particles registered (99.6 to 99.9%) are equal to or smaller than 1 μm and therefore cannot contain Legionella bacteria.

Concerning air cleaning, the number of particles decreases through the wetted media in about 15–17% for 0.5 μm, 22–30% for 1 μm, 25–45% for 5 μm, and above 60% for the largest particles. However, the number of the smallest 0.3 μm particles increases in 9 to 14%. This demonstrates that wetted media actually performs air cleaning of particles equal to or above 0.5 μm. Water entrainment may also be restricted to small aerosols that cannot contain Legionella, though this needs further study using other measuring methods that distinguish between particles and aerosols. Nonetheless, proper maintenance of the system is strictly necessary to avoid proliferation of Legionella.

4. Discussion

Results for the air specific humidity increase correspond to those in existing research. Sheng & Nnanna [4] registered increments of the air specific humidity from 3. 4 g_v/kg_{da} at 25.5 °C to 7.4 g_v/kg_{da} at 45 °C, thus being of the same order of magnitude than the ones obtained in the present work.

Figure 6 shows the saturation efficiencies in terms of air velocity in comparison to previous results in existing literature on alternative pads [21,28,29]. For clarity and correspondence to the air hygrothermal conditions studied in the referred works, it only presents results obtained for the first three levels of temperature (T1, T2 and T3) at the intermediate humidity level (W2). Corresponding relative humidity is 70%, 50% and 40%, approximately. Only results for the largest thicknesses studied in the referred works are considered (15 cm). No information was given in the referred works as to the errors expected.

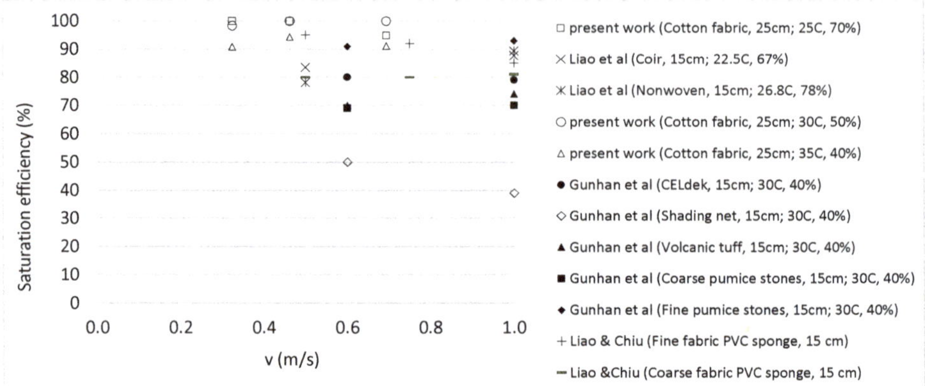

Figure 6. Saturation efficiencies obtained in the present work compared to existing studies [21,28,29].

Comparison to results obtained by Jain and Hindoliya [14] is given in Figure 7. To compare results obtained under the closest air conditions to the referred work, Figure 7 shows results from the present work obtained only at temperature levels T4 and T5, at the lowest humidity level (W1). No information was given in the referred work as to the errors expected. It can be observed that the proposed system using woven fabric can achieve larger saturation efficiencies than using other alternative materials made from vegetable fibres.

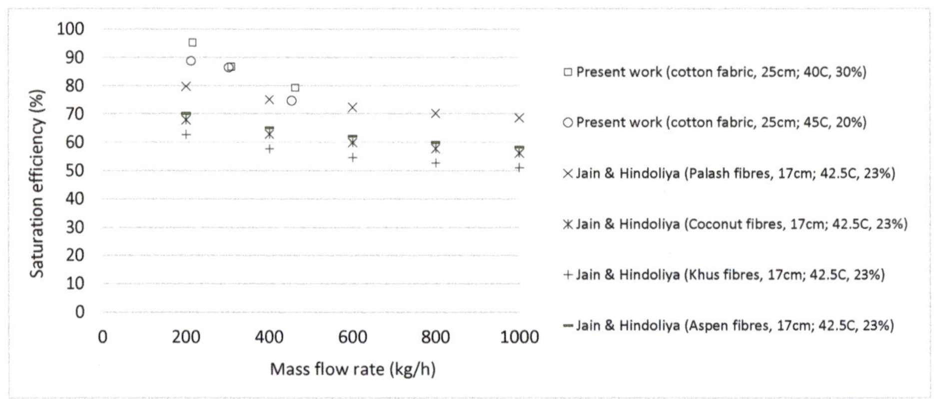

Figure 7. Saturation efficiencies obtained in the present work compared to existing studies [14].

Regarding the results compared in Figures 6 and 7, the new fabric-based EC cooling pad can achieve larger saturation efficiencies. This could be due to the larger thickness (25 cm) compared with the commonly largest thickness studied (15 cm) but also because of the wicking and capillary diffusion properties of the fabric used.

Taking a look at the results compared in Figure 8, larger thicknesses, in this case, does not incur excessive pressure drops. Pressure drops obtained by the new fabric 25 cm thick EC pad is of the same order of magnitude as that expected for conventional 15 cm cellulose pads [21] and other fabric-type pads in the literature [28]. This fact would be due to clearer air paths through the proposed pad compared with the compacity of the commercial CELdek®and configurations with coir and nonwoven fabric 15 cm pads.

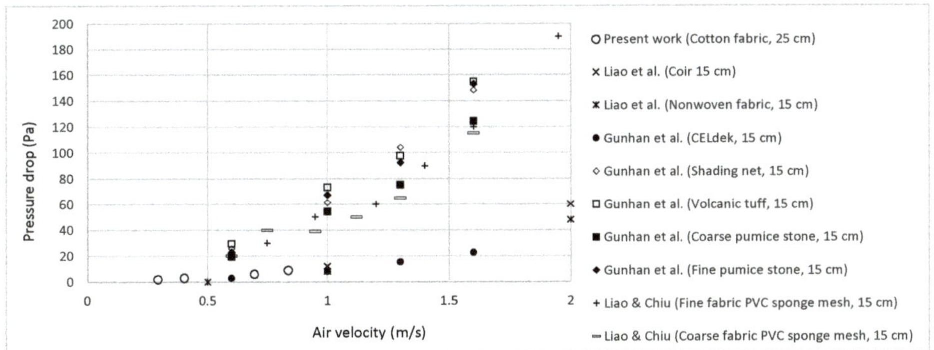

Figure 8. Pressure drop for the new fabric EC pad compared to existing studies [21,28,29].

Indeed, pressure drop is much lower than that obtained for alternative pads made of different stones [21] or fabric PVC sponge mesh [29]. Much larger pressure drops can be expected in pads made of different available fibres, such as khus, palash, coconut or aspen involve [13], and thus are not represented in Figure 8.

Concerning results for airborne particles, and to the knowledge of the authors, no previous research has approached this issue. Water entrainment is analysed in terms of the air velocity of cellulose and PVC pads by He et al. [10] by means of water-sensitive papers. They concluded that cellulose pads and low air velocities avoided risk of water entrainment.

5. Conclusions

Existing literature on the use of wetted fabrics for Evaporative Cooling purposes is scarce, despite demonstrated interest in these materials in terms of capillary diffusion and wicking ability. Related previous research has focused on its combination with passive architecture, and fabric-based EC pads are anecdotic among the large number of new pads built with alternative materials that can be found in the literature.

The present work designed and built an alternative EC pad using cotton fabric. The detailed description given of the system construction permits its reproduction.

The saturation efficiency registered is barely affected by air inlet hygrometric conditions, though larger airflows involve lower efficiencies. The values obtained are above those for other alternative pads studied in the literature. This could be due to both the pad design and the cotton fabric wicking ability and capillary diffusion.

Air humidification increases notably with inlet air DBT, slightly affecting specific humidity. This is due to larger Wet Bulb Depressions and stable saturation efficiencies in the same airflow.

The pressure drop generated for the 25 cm pad is of the same order of magnitude as results in existing research found for 15 cm conventional cellulose pads and 15 cm fabric-based pads, and much lower than other alternative pads made of vegetable fibres and different stones. Further research on the aerodynamic resistance of the pads could illustrate the optimum configuration.

Particle counting demonstrates that the wetted media performs air cleaning for particles above 0.3 μm. As well as this, more than 99.6% of the airborne particles counted cannot contain the Legionella bacteria, both for dry and wet media. However, proper maintenance of the system is strictly necessary to avoid microbial growth, as well as any risk of Legionella dispersion. In this sense, fabrics used as wetting media in simple configurations can be easily dismantled and washed for maintenance. Replacement of the media would be cost-effective. Further research on the durability of organic-based EC media would be necessary to enhance the use of these alternative systems.

The better performance obtained for the cotton fabric in this particular design, compared with other materials, can enhance the use, or reuse, of common fabrics for EC purposes.

Author Contributions: Conceptualization E.V.-G. and F.J.R.-M.; methodology, E.V.-G.; formal analysis, E.V.-G. and J.J.-R.; investigation, E.V.G., J.J.-R. and A.T.-G.; validation, visualization and writing, A.T.-G.; supervision, project administration and funding acquisition, F.J.R.-M. All authors have read and agreed to the published version of the manuscript.

Funding: This research was funded by the Education Department of the Regional Government of Castile and Leon and the European Regional Development Fund (ERDF) through the research project: "Análisis de tecnologías energéticamente eficientes para la sostenibilidad de los edificios" (Ref.: VA272P18).

Conflicts of Interest: The authors declare no conflict of interest. The funders had no role in the design of the study; in the collection, analyses, or interpretation of data; in the writing of the manuscript, or in the decision to publish the results.

References

1. Watt, J.R. *Evaporative Air Conditioning Handbook*; 2.; Chapman and Hall: New York, NY, USA, 1986.
2. Chapter 53 Evaporative Cooling. In *ASHRAE Handbook. HVAC Applications*; 2019. Available online: https://www.ashrae.org/technical-resources/ashrae-handbook/ashrae-handbook-online (accessed on 29 August 2020).
3. Chapter 41 Evaporative Air-Cooling Equipment. In *ASHRAE Handbook-Systems and Equipment*; 2020. Available online: https://www.ashrae.org/technical-resources/ashrae-handbook/description-2020-ashrae-handbook-hvac-systems-and-equipment (accessed on 29 August 2020).
4. Sheng, C.; Agwu Nnanna, A.G. Empirical correlation of cooling efficiency and transport phenomena of direct evaporative cooler. *Appl. Therm. Eng.* **2012**, *40*, 48–55. [CrossRef]
5. Malli, A.; Seyf, H.R.; Layeghi, M.; Sharifian, S.; Behravesh, H. Investigating the performance of cellulosic evaporative cooling pads. *Energy Convers. Manag.* **2011**, *52*, 2598–2603. [CrossRef]
6. Franco, A.; Valera, D.L.; Madueño, A.; Peña, A. Influence of Water and Air Flow on the Performance of Cellulose Evaporative Cooling Pads Used in Mediterranean Greenhouses. *Trans. ASABE* **2010**, *53*, 565–576. [CrossRef]
7. Warke, D.A.; Deshmukh, S.J. Experimental Analysis of Cellulose Cooling Pads Used in Evaporative Coolers. *Int. J. Energy Sci. Eng.* **2017**, *3*, 37–43.
8. Barzegar, M.; Layeghi, M.; Ebrahimi, G.; Hamzeh, Y.; Khorasani, M. Experimental evaluation of the performances of cellulosic pads made out of Kraft and NSSC corrugated papers as evaporative media. *Energy Convers. Manag.* **2012**, *54*, 24–29. [CrossRef]
9. Al-Badri, A.R.; Al-Waaly, A.A.Y. The influence of chilled water on the performance of direct evaporative cooling. *Energy Build.* **2017**, *155*, 143–150. [CrossRef]
10. He, S.; Guan, Z.; Gurgenci, H.; Hooman, K.; Lu, Y.; Alkhedhair, A.M. Experimental study of film media used for evaporative pre-cooling of air. *Energy Convers. Manag.* **2014**, *87*, 874–884. [CrossRef]
11. Franco, A.; Valera, D.L.; Peña, A. Energy efficiency in greenhouse evaporative cooling techniques: Cooling boxes versus cellulose pads. *Energies* **2014**, *7*, 1427–1447. [CrossRef]
12. Alam, M.F.; Sazidy, A.S.; Kabir, A.; Mridha, G.; Litu, N.A.; Rahman, M.A. An experimental study on the design, performance and suitability of evaporative cooling system using different indigenous materials. *AIP Conf. Proc.* **2017**, *1851*, 020075. [CrossRef]
13. Jain, J.K.; Hindoliya, D.A. Experimental performance of new evaporative cooling pad materials. *Sustain. Cities Soc.* **2011**, *1*, 252–256. [CrossRef]
14. Jain, J.K.; Hindoliya, D.A. Correlations for Saturation Efficiency of Evaporative Cooling Pads. *J. Inst. Eng. Ser. C* **2014**, *95*, 5–10. [CrossRef]
15. Anyanwu, E.E. Design and measured performance of a porous evaporative cooler for preservation of fruits and vegetables. *Energy Convers. Manag.* **2004**, *45*, 2187–2195. [CrossRef]
16. Rawangkul, R.; Khedari, J.; Hirunlabh, J.; Zeghmati, B. Performance analysis of a new sustainable evaporative cooling pad made from coconut coir. *Int. J. Sustain. Eng.* **2008**, *1*, 117–131. [CrossRef]
17. Al-Sulaiman, F. Evaluation of the performance of local fibers in evaporative cooling. *Energy Convers. Manag.* **2002**, *43*, 2267–2273. [CrossRef]

18. Ndukwu, M.C.; Manuwa, S.I. A techno-economic assessment for viability of some waste as cooling pads in evaporative cooling system. *Int. J. Agric. Biol. Eng.* **2015**, *8*, 151–158.

19. De Melo, J.C.F.; Bamberg, J.V.M.; MacHado, N.S.; Caldas, E.N.G.; Rodrigues, M.S. Evaporative cooling efficiency of pads consisting of vegetable loofah. *Comun. Sci.* **2019**, *10*, 38–44. [CrossRef]

20. Doğramacı, P.A.; Aydın, D. Comparative experimental investigation of novel organic materials for direct evaporative cooling applications in hot-dry climate. *J. Build. Eng.* **2020**, *30*. [CrossRef]

21. Gunhan, T.; Demir, V.; Yagcioglu, A.K. Evaluation of the Suitability of Some Local Materials as Cooling Pads. *Biosyst. Eng.* **2007**, *96*, 369–377. [CrossRef]

22. Laknizi, A.; Ben Abdellah, A.; Faqir, M.; Essadiqi, E.; Dhimdi, S. Performance characterization of a direct evaporative cooling pad based on pottery material. *Int. J. Sustain. Eng.* **2019**, *00*, 1–11. [CrossRef]

23. Korese, J.K.; Hensel, O. Experimental evaluation of bulk charcoal pad configuration on evaporative cooling effectiveness. *Agric. Eng. Int. CIGR J.* **2016**, *18*, 11–21.

24. Ahmed, E.M.; Abaas, O.; Ahmed, M.; Ismail, M.R. Performance evaluation of three different types of local evaporative cooling pads in greenhouses in Sudan. *Saudi J. Biol. Sci.* **2011**, *18*, 45–51. [CrossRef] [PubMed]

25. Inamdar, S.J.; Junghare, A.; Kale, P. Performance enhancement of evaporative cooling by using bamboo. *Int. J. Eng. Adv. Technol.* **2019**, *8*, 856–860.

26. Sudprasert, S.; Sankaewthong, S. Utilization of rice husks in a water-permeable material for passive evaporative cooling. *Case Stud. Constr. Mater.* **2018**, *8*, 51–60. [CrossRef]

27. Xu, P.; Ma, X.; Zhao, X.; Fancey, K.S. Experimental investigation on performance of fabrics for indirect evaporative cooling applications. *Build. Environ.* **2016**, *110*, 104–114. [CrossRef]

28. Liao, C.M.; Singh, S.; Wang, T. Sen Characterizing the performance of alternative evaporative cooling pad media in thermal environmental control applications. *J. Environ. Sci. Health-Part A Toxic/Hazardous Subst. Environ. Eng.* **1998**, *33*, 1391–1417.

29. Liao, C.M.; Chiu, K.H. Wind tunnel modeling the system performance of alternative evaporative cooling pads in Taiwan region. *Build. Environ.* **2002**, *37*, 177–187. [CrossRef]

30. Ghosal, M.K.; Tiwari, G.N.; Srivastava, N.S.L. Modeling and experimental validation of a greenhouse with evaporative cooling by moving water film over external shade cloth. *Energy Build.* **2003**, *35*, 843–850. [CrossRef]

31. Esparza L., C.J.; Escobar del Pozo, C.; Gómez A., A.; Gómez A., G.; Gonzalez C., E. Potential of a wet fabric device as a roof evaporative cooling solution: Mathematical and experimental analysis. *J. Build. Eng.* **2018**, *19*. [CrossRef]

32. Brown, W.K. Operation and Maintenance of Evaporative Coolers. *ASHRAE J.* **2000**, *42*, 27–31. Available online: https://technologyportal.ashrae.org/Journal/ArticleDetail/545 (accessed on 29 August 2020).

33. Puckorius, P.R.; Thomas, P.T.; Augspurger, R.L. Why Evaporative Coolers have not caused Legionnaires' disease. *ASHRAE J.* **1995**, *1*, 29–33.

 *sustainability*

Article

Sustainable Development of Human Society in Terms of Natural Depleting Resources Preservation Using Natural Renewable Raw Materials in a Novel Ecological Material Production

Cătălina Mihaela Grădinaru [1,*], **Radu Muntean** [2,*], **Adrian Alexandru Șerbănoiu** [1], **Vasilică Ciocan** [1] and **Andrei Burlacu** [1]

[1] Faculty of Civil Engineering and Building Services, "Gheorghe Asachi" Technical University of Iași, 700050 Iași, Romania; serbanoiu.adrian@tuiasi.ro (A.A.Ș.); vasilica.ciocan@tuiasi.ro (V.C.); andrei.burlacu@tuiasi.ro (A.B.)

[2] Faculty of Civil Engineering, Transilvania University of Brașov, 500152 Brașov, Romania

[*] Correspondence: radu.m@unitbv.ro (R.M.); catalina.gradinaru@tuiasi.ro (C.M.G.); Tel.: +40745183892 (R.M.)

Received: 23 February 2020; Accepted: 25 March 2020; Published: 27 March 2020

Abstract: In the last few years, the building industry experienced a significant development as a response to the demographic growth of human society and to the increasing demand for housing. Their construction involves the traditional use of concrete as a material that provides added strength to the finished building. This is manufactured respecting standard recipes depending on the way of its use. Anyway, all concrete recipes involve the use of mineral aggregates extracted from the riverbed, as is happening in Romania, or rock blocks crushing, as reported in other countries. Under these conditions, the rationalization of the use of natural mineral resources and the identification of new possibilities to reduce their consumption through their replacement with vegetal waste has become an important research issue. In this study, two types of vegetal waste—namely, shredded corn cobs and sunflower stalks—were used to manufacture novel ecological concretes. The vegetal wastes, both in untreated and treated forms (with 20% and 40% of sodium silicate solution), were used to replace 50% of the river (mineral) aggregate volume. The obtained concretes were tested, and the values of some important parameters in the concrete characterization (such as bulk density, water adsorption capacity, compressive strength and splitting tensile strength) were compared with the concrete contains cement CEM II/A-LL 42.5R. The obtained results show that these vegetal wastes have the potential to be used in the manufacturing of new ecological concrete. In addition, this alternative material meets the requirements for the sustainable and healthy development of the environment, offering low-polluting solutions in the context of an increasing demand for constructions.

Keywords: vegetal waste; shredded corn cobs; sunflower stalks; green concrete

---

1. Introduction

Besides various benefits, the traditional materials used in the building industry are considered costly and polluting through the process of obtaining them. Concrete, for example, is a widely used material in the constructions industry, with increasing demand due to the global demographic growth. This is because, a concrete building system is considered energy efficient due to its special ability to combine a high thermal mass, very low air infiltration, and a high R-value (thermal resistance) in the exterior walls [1]. In addition, concrete buildings have fewer joints than those made of wood or steel frame, which means a tighter system in which the material itself absorbs and stores heat energy, with a delayed impact of the exterior temperature on the inside environment [2].

The major component of the concrete is represented by the mineral aggregates, extracted mainly from the riverbed (as in Romania) [3], or obtained from rock blocks crushing like in other countries [4]. These mineral aggregates are quite widespread natural resources, but this does not mean they are always available for use, and there is no need to be cautious in their extraction.

For example, marine resources (sand and gravel deposited by rivers and glaciers during the ice period) are an important source of mineral aggregates, and their use reduces the need for mine extraction and avoids the use of agricultural areas, which are important benefits for environment [5]. In addition, the delivery distances of the dredged mineral aggregates are generally shorter due to the fact they are landed in ports and their transport is made mainly by sea and not by road [6]. However, the removal of marine sediments will change the height, direction, and distance between waves and in tidal currents, with important consequences for marine wildlife, ecosystems, and navigation activities [7].

In the regions that do not have sand and gravel, and where mineral aggregates are obtained from crushed stone extracted from very large depths, which involves very high costs [5]. In this case, the obtaining of mineral aggregates is a fairly polluting process for the environment, mainly due to the greenhouse emissions. The environmental impacts associated with mineral aggregate mining are represented by acidic mine drainage, conversion of land use accompanied by loss of habitat for humans and animals, noise, dust, blasting effects, erosion, sedimentation, and relief changes. These are related to the fact that mineral aggregates such as sand, gravel, and bedrock are extracted from pits and quarries (open-pit mining and quarrying are commonly used and, in special cases, underground mining). Their presence means the loss of agricultural land and the shift of land use from terrestrial to lake habitats, by-products resulting from aggregate processing, and an impact on physical infrastructure (i.e., buildings, roads, and dams) [4,8–10]. In some cases, land intended to be developed for mineral aggregates extraction can have rare geomorphic characteristics or can be the habitat for rare and endangered species. Mining in such areas can lead to serious and long-lasting environmental consequences in the site vicinity or even very far away locations form it. Mineral aggregate sites can disturb surrounding areas, mainly via truck traffic and air pollution, with a negative impact for residents in the form of noise and visual disamenities. Mineral aggregate mining can lead to the release of deleterious sediments, salt, and chemicals into the watercourses, soil, and air [6,9,11]. To estimate the real impact of this activity, it is necessary to consider that, in Europe, there are over 30,000 extraction sites, which produce around seven tons of aggregate per EU citizen [10]—the global request for mineral aggregates is about 22 billion tons per year [12].

Therefore, even if the dredging of marine mineral aggregates is more advantageous than digging or mining, it must adhere to the principles of sustainable management of long-term availability of resources, minimizing the negative impact on the environment and the socio-economic impact of the extraction process. This is because both methods deplete a category of natural resources, alter the landscape, and affect the natural habitat of different plants and animals.

However, the environmental ethics principles in the building industry have to consider a close relationship between nature and buildings, a harmony of human' society development in the context of all that nature means. This means that a sustainable building industry should take into account environmental problems such as the natural resources consumption, pollutants, and excessive energy usage, which means that the industry needs to work simultaneously in three interdependent directions—social, environmental, and economic [13].

This is the main reason why, in recent years, various alternatives such as waste fly ash; blast furnace slag; silica fume; plants ashes such as corn cob ash, wheat ash, and rice husk ash; recycled aggregates from demolitions; expanded perlite; expanded clay or expanded polystyrene; PET (polyethylene terephthalate) fibers; steel fibers; glass; plastics; lignocellulosic materials; or agricultural waste have been used as alternatives to reduce the amount of mineral aggregates needed to manufacture the concrete [14–24]. Among all these materials, the use of agricultural waste has become increasingly popular due to the improved characteristics of the obtained concrete, which are mainly related to thermal and acoustic insulation, lower costs, and environmental protection by reducing the use of the conventional raw materials [1,25–30].

Among many agricultural wastes studied for their suitability in building sector, corn cobs and sunflower have the advantage of worldwide available [31,32]. In addition, these materials are easy to process using local resources without generating extra costs because harvesting and storage is done using common agricultural equipment. Moreover, it has been shown that the use of such agricultural waste can result in high-quality aggregates for building materials using simple processing and low sintering temperature, further contributing to the limitation of environment pollution [33]. For example, the addition of corn cobs and sunflowers to concrete manufacturing improves thermal and acoustic insulation [1,26–30], with good resistance to sodium sulphate [34]. The authors of reference [35] used corn cob as pore-forming agent in lightweight clay bricks, obtaining apparent and bulk density, flexural strength, decreased linear shrinkage, and increased water absorption and porosity. Shredded corn cobs were used in the production of lightweight concrete as an alternative sustainable solution compared to the light aggregates used today, such as those of expanded clay, cork, or expanded polystyrene. The authors of reference [36] made a lightweight concrete from corn cobs in a 6/1/1 ratio (crushed corn chips/Portland cement/water) and an expanded clay concrete in the same ratio (expanded clay/Portland cement/water), obtaining lower mechanical characteristics than the classical concrete but also very low densities, which makes it usable in non-structural applications. The authors of reference [28] conducted a research on shredded corn cobs used as aggregates to produce lightweight concrete masonry blocks. The density of the obtained products was 1680 kg/m^3, a value that places the concrete obtained into lightweight concrete category, according to the regulations in force. To reduce the water absorption and to improve the adhesion between concrete and aggregates, the corn cobs were coated with a cement paste. The tests performed showed a good resistance to repeated cycles of freeze-thaw. The authors of reference [27] studied the properties of concrete with sunflower aggregates, compared to those of concrete with hemp aggregates, in a binder/aggregate ratio of 2. The tests performed on concrete with sunflower aggregates showed a thermal conductivity of 0.096 W/mK and an average compression strength of 0.5 N/mm^2 determined at 60 days. Also, [8] showed that aggregates from sunflower and hemp stalks have a similar honeycomb structure and similar chemical composition. The authors developed a concrete with sunflower aggregates and binder based on limestone and pumice, with the ratio binder/aggregates equal to 18. The compressive strength at 28 days of this type of concrete was 2.52 N/mm^2. The same concrete recipe, but which included hemp aggregates, had a compressive strength of 2.77 N/mm^2. [37] developed a composite material of chitosan and particles of sunflower stalks, with a thermal conductivity of 0.056 W/mK and a compressive strength of 2 N/mm^2, in the case of a recipe with 4.3% chitosan and sunflower particles bigger than 3 mm. All these research findings suggest that the use of such agricultural wastes in concrete production can be a viable alternative to help in reducing the environmental impact associated with the mineral aggregate extraction industry in the context of the sustainable development of human society.

In this study, shredded corn cobs and sunflower stalks were used to manufacture novel ecological concretes. These vegetal wastes were used to replace 50% of the river (mineral) aggregate volume. The obtained concretes were tested, and the values of some important parameters in the concrete characterization (such as bulk density, water adsorption capacity, compressive strength, and splitting tensile strength) were compared with the concrete contains cement CEM II/A-LL 42.5R. The aim of this research was to make a material with alternative applicability in building elements where ordinary concrete would be less efficient. This ecological alternative complies with the criteria of sustainable development, offering non-polluting variants in the context of a highly industrialized society.

2. Materials and Methods

This research implied the development of concrete recipes with corn cobs and sunflower stalks as partial replacements for the mineral aggregates. As a starting point, we established a reference concrete (RC) recipe with conventional components of C30/37 strength class and 8 mm the maximum size of the aggregates, calculated according to reference [38].

The reference concrete contains CEM II/A-LL 42.5R cement, sand (0–4 mm diameter), and river gravel (sort 4–8 mm). A superplasticizer based on policarboxilateter (Sika Plast 140, produced by Sika Romania) was used to reduce the water requirements while maintaining workability, and an accelerator based on rhodanid (Sika BE 5, produced by Sika Romania) was used for hardening the concrete and accelerating the cement hydration process. The water/cement ratio was 0.43.

The formulas for vegetal concrete implied the use of shredded corn cobs and sunflower stalks as 50% by volume replacement of the river (mineral) aggregates. The procedure for obtaining the vegetal aggregates from corn cobs and sunflower stalks, respectively, was as follows:

From maize plants, only corn cobs (no husks or grains) (Figure 1a) were harvested, and only stalks without leaves and hats were harvested from sunflower plants (Figure 1b).

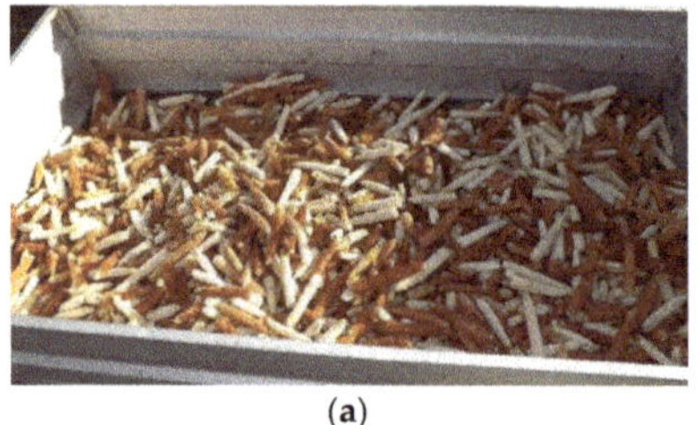

(**a**)   (**b**)

Figure 1. Vegetal material after harvesting: (**a**) corn cobs; (**b**) sunflower stalks.

After harvesting, the sunflower stalks were left to dry in stacks in the environment (Figure 2).

Figure 2. Sunflower stalks drying.

Corn cobs were shredded in granules smaller than 6 mm using a mill for grinding animal feed (Figure 3a). Sunflower stalks were shredded with the same mill, resulting in granules with a diameter of less than 5–6 mm and fibers less than 25 mm long (Figure 3b).

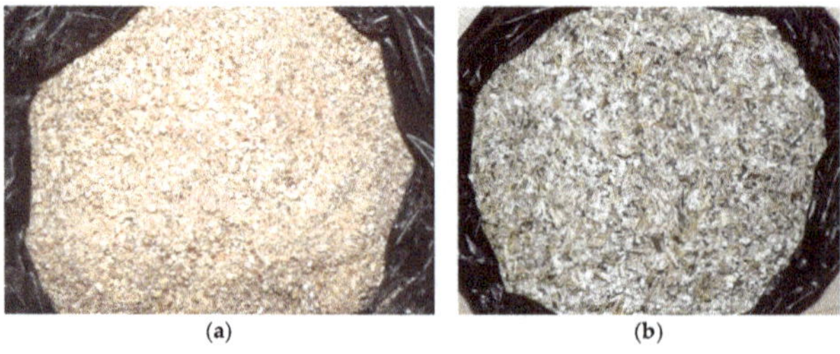

(**a**)   (**b**)

Figure 3. The aspect of vegetal material after shredding: (**a**) corn cobs; (**b**) sunflower stalks.

After shredding, the vegetal material was treated with sodium silicate solution by immersion (Figure 4).

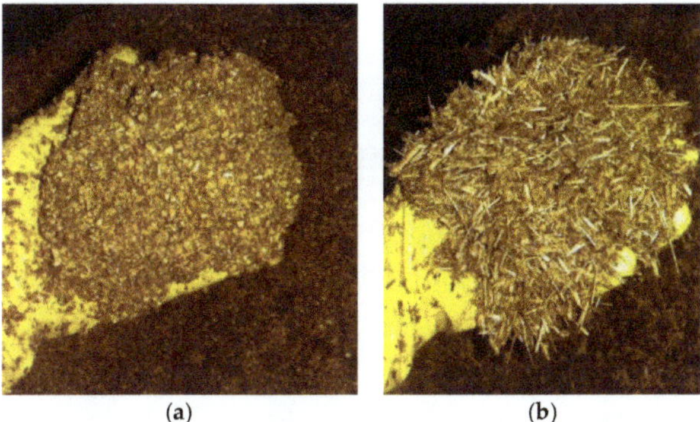

Figure 4. The aspect of vegetal material after shredding and treatment with sodium silicate solution, in a wet state: (**a**) corn cobs; (**b**) sunflower stalks

After immersion in the sodium silicate solution, the vegetal material was left to dry at a temperature of about 25–27 °C (Figure 5).

Figure 5. The drying of the vegetal material at a temperature of about 25–27 °C after the treatment with a sodium silicate solution.

After drying to a constant mass, the plant material was used as a plant aggregate in the concrete composition (Figure 6).

Figure 6. The aspect of vegetal material after shredding and treatment with sodium silicate solution, in a dry state: (**a**) corn cobs; (**b**) sunflower stalks.

The vegetal aggregates were used in three forms—untreated, treated with 20% sodium silicate solution, and treated with 40% sodium silicate solution. Their visual aspect was the same no matter the solution concentration (Figure 7c,f).

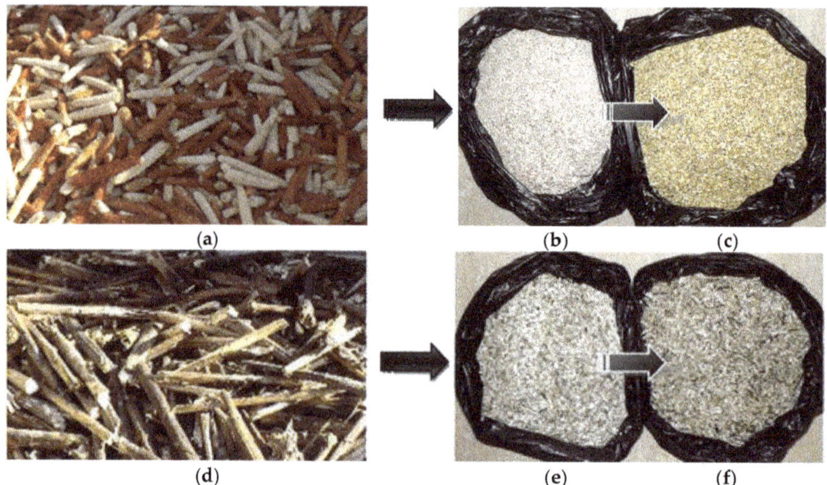

Figure 7. Corn cobs and sunflower stalks aspect—from stalk to treated granules: (**a**) corn cobs; (**b**) untreated shredded corn cobs; (**c**) treated shredded corn cobs; (**d**) sunflower stalks; (**e**) untreated shredded sunflower stalks; (**f**) treated shredded sunflower stalks.

Sodium metasilicate compound (Na_2SiO_3), also known as glass water or liquid glass, is available in liquid and solid forms and is used as passive fire protection for concrete and to protect against termites, rot, and degradation in refractories, textiles, woodworking, and automobiles.

To reduce the water absorption capacity of the vegetal aggregates, the shredded granules were treated first with 20% sodium silicate solution. Then, they were dried, and their bulk density and water absorption capacity were measured.

Because the objective was to reduce the water absorption capacity for the plant aggregates as much as possible, other vegetal aggregates were immersed in a 40% concentrated solution of sodium silicate, dried, and then their bulk density and water absorption capacity measured, as in the first case.

An additional amount of water, varying according to the specific absorption capacity of the plant aggregates, was added to the concrete recipes with vegetal aggregates at a rate calculated to provide a water/cement ratio of 0.43. We developed the following concrete recipes:

- RC—reference concrete;
- CUCC—concrete with untreated granules of corn cob;
- CTCC20—concrete with treated granules of corn cob with 20% sodium silicate solution;
- CTCC40—concrete with treated granules of corn cob with 40% sodium silicate solution;
- CUSF—concrete with untreated sunflower granules;
- CTSF20—concrete with treated sunflower granules with 20% sodium silicate solution;
- CTSF40—concrete with treated sunflower granules with 40% sodium silicate solution.

For testing their compressive strength, $150 \times 150 \times 150$ mm cubic molds were used, while for testing their flexural and splitting tensile strength, $100 \times 100 \times 500$ mm prism molds were used. There were three replicates for each test. All the tests were performed at the age of 28 days, according to the current technical norms [39–42].

3. Results and Discussions

3.1. Evaluation of Vegetal Aggregate Bulk Density and Water Absorption Capacity

Table 1 shows the determined values for bulk density and water absorption capacity for the vegetal aggregates.

Table 1. The bulk density and water absorption capacity of the vegetal aggregates.

Vegetal Aggregates	Treatment Applied	Bulk Density [kg/m^3]	Water Absorption Capacity [%]
	no treatment	281.25	294
Corn cobs	20% sodium silicate solution	242.19	181
	40% sodium silicate solution	398.43	127
	no treatment	207.03	402
Sunflower stalks	20% sodium silicate solution	164.06	292
	40% sodium silicate solution	328.12	100

In the case of corn cob granules, the treatment with 20% sodium silicate solution resulted in a decrease of water absorption by 38.44%; in the case of corn cob granules treated with 40% sodium silicate solution, it was 127%, which means a 56.80% reduction compared to the untreated granules. In the case of sunflower stalk granules, the treatment with 20% sodium silicate solution resulted in a reduction of water absorption of 52.24%; in the case of sunflower stalk granules treated with 40% sodium silicate solution, it was 100%, meaning a 75.12% reduction comparative to the untreated granules.

3.2. Evaluation of the Vegetal Concrete Density

The density evolution curves for the obtained concrete are showed in Figure 8. The highest slope of the curve was registered by CUCC. This concrete variant involved the greatest quantity of added water besides the quantity calculated for cement hydration and, implicitly, the greatest quantity of water lost during concrete curing. This loss of water led to the lowest mechanical strengths resulting from the weaker connection at the interfacial transition zone between the vegetal aggregates and the cement paste.

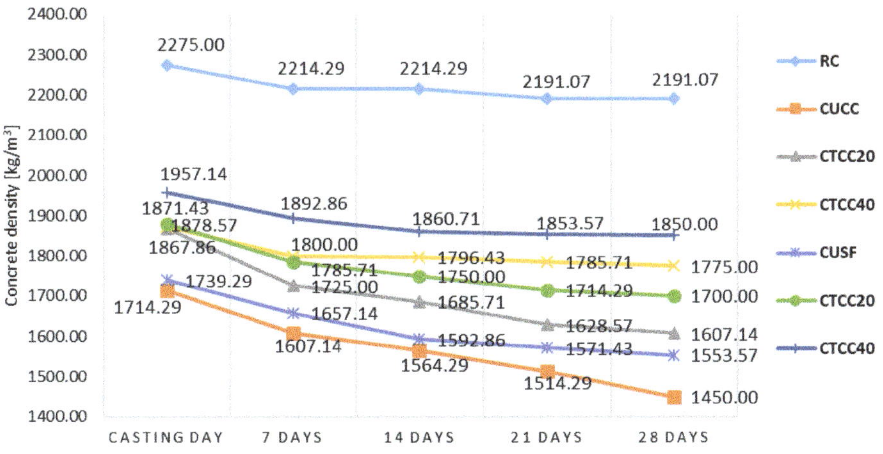

Figure 8. The evolution of concrete density, from casting to the age of 28 days (kg/m^3).

The increase in the density of the concrete with treated corn cobs in both versions was about the same intensity as to the sunflower concrete: CTCC20 and CTSF20 had an increase in density of approximately 10%, and CTCC40 and CTSF40 had an increase of approximately 20% (values reported to CUCC and CUSF, respectively).

3.3. Evaluation of the Compressive Strength of Concretes with Vegetal Aggregates

The results obtained after testing the compressive strength for the concrete described above are presented in Figure 9.

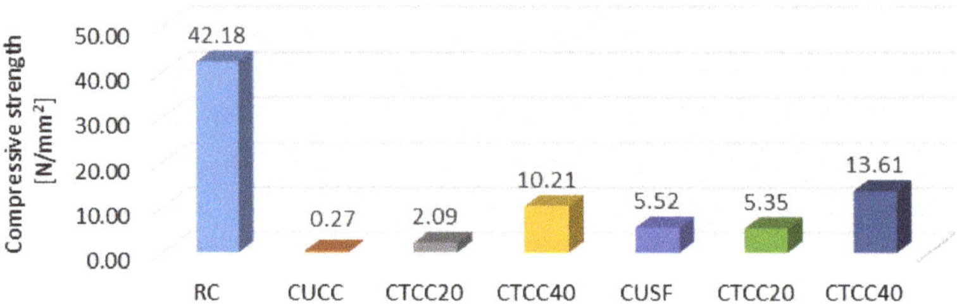

Figure 9. Compressive strength for the developed concrete recipes (N/mm^2).

A replacement of 50% of mineral with vegetal aggregates led to a significant decrease compared to the reference concrete. On the other hand, treating the vegetal aggregates with a 40% sodium silicate solution led to an important improvement in terms of compressive strength compared to the concrete with untreated vegetal aggregates.

The concrete with untreated corn cobs was shown to have a very low compressive strength, recording a value of only 0.27 N/mm^2 compared to the one of reference concrete of 42.18 N/mm^2. This particularly low value is due to the presence of a significant quantity of water) the inadequate moisture of the specimens was visible to the naked eye). The water absorbed by the vegetal part of the concrete was highest compared to the other concrete recipes, and, as shown in Figure 9, it was the recipe with the highest decrease in density. This drop means that it had the largest quantity of evaporated water (leaving the concrete structure), which left behind an increased volume of air and very poor interface links between aggregates and the cement paste.

The corn cobs treatment with 40% sodium silicate solution increased the compressive strength of concrete 30 times comparative to the concrete with untreated plant aggregates, and about 5 times comparative with the first treatment variant, resulting in a value of 10.21 N/mm^2. This significant improvement was due to the dramatic decrease in corn cobs absorption capacity (from 294% in the untreated variant up to 127% in the variant treated with 40% solution).

Regarding the concrete with sunflower plant aggregates, it recorded a compressive strength much higher than that with corn cobs, especially in the untreated aggregate variants (5.52 N/mm^2) and that treated with 20% sodium silicate solution (5.35 N/mm^2), despite having an absorption capacity higher than that of corn cobs. This leads to the conclusion that a high compressive strength of CUSF compared to CUCC (by approximately 15 times) contributed to the strength of the plant aggregates themselves. In the composition of the sunflower stem, the bark has a more rigid, woody, and fibrous structure with mechanical characteristics superior to corn cobs. Contrary to expectations, treatment in the first variation of sunflower aggregates led to a small decrease in compression strength, a result that can be explained by the additional amount of water.

Treatment of sunflower aggregates with 40% sodium silicate solution resulted in a 146.5% increase in compressive strength of concrete with these aggregates compared to the concrete with untreated sunflower aggregates, and it was 33.3% higher than of the concretes with corn cob aggregates treated with the same concentration of sodium silicate solution. In conclusion, sunflower concrete registered improved performance in compressive strength compared to corn cob concrete.

By analyzing the concrete compressive strength in comparison with the capacity of water absorption for the aggregates involved (Figure 10), the reduction of the last parameter determined the increase of the compressive strength, both for the corn cob concrete and for the sunflower concrete.

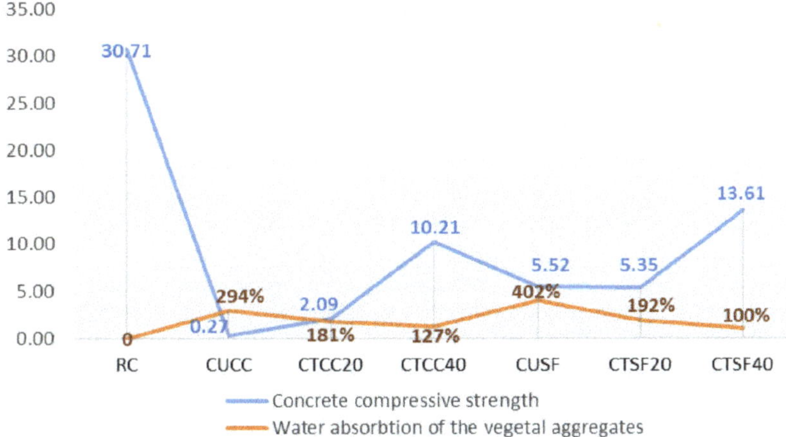

Figure 10. Concrete compressive strength after 28 days of curing (N/mm^2), comparatively presented with water absorption capacity of the vegetal aggregates (%).

In its natural form, the sunflower stalk is made up of a fibrous, rigid coat (the bark) and a marrow with very porous structure and a much smaller density than the bark. During the shredding process, the marrow underwent some compaction of its volume, and the ratio between it and the fibrous part of the stem was altered in the opposite direction to that of the normal, natural state of the sunflower stalk. This led to a predominant share of the fibrous part in the composition of the shredded sunflower granules, which gives superior mechanical strength to these plant aggregates.

In the corn cob structure, there is no great difference between its component layers, and the mechanical strengths of the cobs are noticeably lower than those of the sunflower stalks. Thus, in the variants involving untreated vegetal granules, although the water absorption of the sunflower aggregates was higher (402%), the concrete with these aggregates was characterized by much higher compressive strength values than those recorded for the corn cob concretes due to the more rigid structure of the sunflower stalk. This characteristic of sunflower aggregates compensated for and exceeded its increased rate of water absorption due to the more porous structure of the marrow of the sunflower stalk.

3.4. Evaluation of the Flexural Tensile Strength of Concretes with Vegetal Aggregates

The concrete specimens with untreated corn cob aggregates suffered damage during the curing process due to the concrete shrinkage as a consequence of the water loss from its structure and the decrease in the volume of the plant aggregates, thus developing aggregate–cement interface forces that were opposite and parallel to the prism length and high enough to cause a continuous crack across the entire cross section of the prisms.

Treating the corn cobs with 20% sodium silicate solution prevented the occurrence of that shrinkage concrete crack but showed a very low bending strength (0.68 N/mm^2) compared to that determined for RC (3.44 N/mm^2). A higher concentration for the sodium silicate solution applied to the corn cob aggregates led to a significant increase for tensile strength of the afferent concrete, CTCC40, by 188.2% (1.96 N/mm^2) (Figure 11).

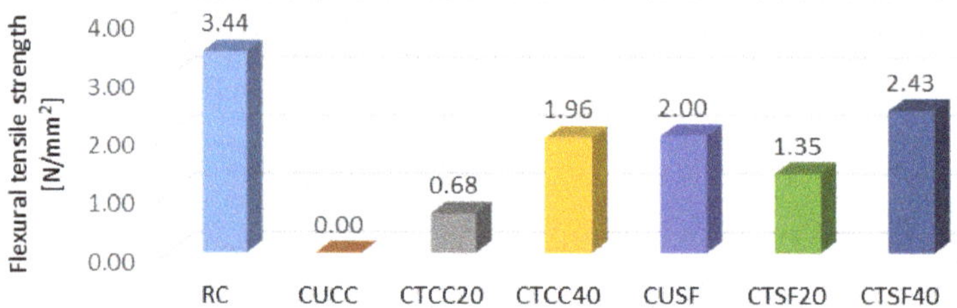

Figure 11. Concrete flexural tensile strength (N/mm^2).

The sunflower concrete in its variant with untreated aggregate showed a flexural tensile strength slightly superior to the one of the concrete with corn cobs treated with the more concentrated sodium silicate solution (2.0 N/mm^2), due to the fiber form of the sunflower aggregates, which somehow reinforced the structure of the concrete. Treating sunflower aggregates with 20% glass water resulted in an unexpected drop in the flexural tensile strength by 32.5% (1.35 N/mm^2), which was probably due to the wood fibers stiffening, which led to decreased flexibility and the development of a casual character. Applying 40% glass water solution to the sunflower aggregates further increased the bending strength up to 2.43 N/mm^2 (Figure 11). This increase is determined by higher wood fiber stiffening (increasing the concentration of the glass water solution leads to a weight gain for the aggregates that can better oppose the bending forces).

In conclusion, in terms of flexural tensile strength, sunflower concrete had the best values in all variants of aggregate treatment, with the highest strength obtained by CTSF40 of 2.43 N/mm^2 being below the value recorded by RC (3.44 N/mm^2).

3.5. Evaluation of the Splitting Tensile Strength of Concretes with Vegetal Aggregates

Concrete recipes with plant aggregates recorded values well below those of the RC in the case of splitting strength. The most resistant variant (with sunflower treated with 40% sodium silicate solution) recorded values that were approximately 58% lower than the reference (1.27 N/mm^2) (Figure 12).

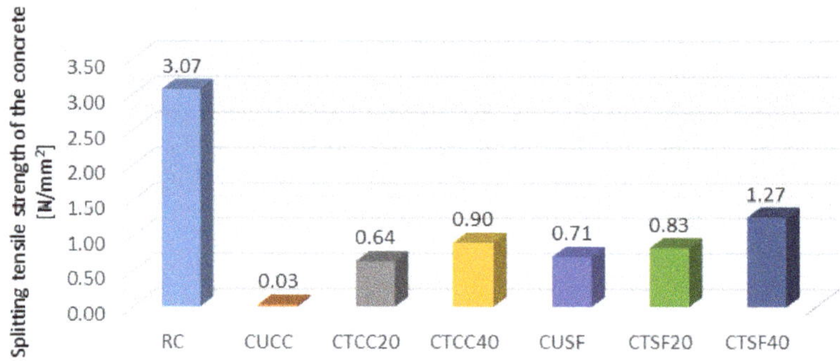

Figure 12. Concrete splitting tensile strength (N/mm^2).

CUCC showed a very low splitting strength, which was almost impossible to test (0.03 N/mm^2) (Figure 12). The explanation is similar with that regarding the compressive strength: the specimens showed inadequate moisture, which was visible to the naked eye. The water absorbed by the vegetal part of the concrete was higher for this concrete recipe than for the other vegetal aggregate concrete recipes. Thus, the amount of water absorbed led to the greatest decrease in density during the curing period of 28 days, causing a considerable evaporation of water from the concrete structure, which left behind an increased volume of air and a very weak interface between the aggregates and the cement paste.

The treatment of corn cobs led to an increased splitting tensile strength of the concrete compared to the one with untreated aggregates, but it was not more than 30% of the value obtained by RC (0.64 N/mm^2 for CTCC20 and 0.90 N/mm^2 for CTCC40) (Figure 12).

Sunflower concretes recorded superior splitting tensile strengths to corn cob concretes, in addition to compressive and flexural tensile strength tests. In the case of concrete with untreated sunflower granules, a strength of 0.71 N/mm^2 was obtained, which was higher than the concrete with corn cobs treated with the 20% sodium silicate solution. The first treatment of sunflower granules led to a slight increase in concrete splitting tensile strength, while the second treatment variant resulted in a much higher value of about 79% compared with that of concrete with untreated aggregates.

4. Application Areas of the Novel Concrete with Plant Aggregates

According to our presented test results, plant aggregates may be successfully used for concrete, leading to a reduction of mineral aggregates using of around 900 kg per m^3 of concrete when compared to conventional concrete. If we consider a replacement of 50% by volume ratio of mineral aggregates in the developed concrete recipes, this means a 50% reduction of the mineral aggregates used for concrete production. Indeed, the plant aggregates can be used only in specific concrete types due to their mechanical strength limitations; in general, concrete with plant aggregates cannot be used in structural purposes. The concrete with corn cob or sunflower aggregates can be used for lightweight concrete screeds, lightweight substrates/subfloors, or acoustic mats. This is in line with the results presented in a previous made with two concrete recipes with the best mechanical strengths, CTCC40 and CTSF40, developed by the authors of reference [25]. The aforementioned study presented some acoustic characteristics, such as the noise reduction coefficient (NRC) and real sound absorption coefficient, of the CTCC40 and CTSF40, which were noted in that paper as mc-corn and mc-sun, respectively. The authors obtained significantly better results for the plant concrete comparing to conventional concrete. CTCC40 had a NRC of 0.193 in the case of the 40 mm thickness sample, comparing to 0.157 registered by conventional concrete used as a reference. Its real sound absorption coefficient measured on a frequency of 2000 Hz was 0.427, compared to 0.279 for the reference. These values were obtained on an 80 mm thick sample. The CTSF40 (mc-sun) registered a NRC of 0.177 and a real sound absorption coefficient of 0.418 in the same conditions of analysis.

It is well known that lightweight concretes of any kind provide better thermal insulation than normal weight concrete. By using lightweight concrete with corn cob or sunflower aggregates as screeds in a building, a reduction of the thermal bridging at the slabs level can be obtained. This means a reduction of heat transfer and a cold sensation at floor level.

Vegetal concrete can be used in modular elements for buildings closures, such as bricks or panelized walls manufactured off site and then assembled with minimal site disturbance and no waste. As blocks, the lightweight vegetal concrete can be used to develop sound-insulating blocks or lightweight blocks for partition walls. Also, it can be used to make panels for partition walls or sound-absorbent panels due to the material porosity ensured by the chopped corn cobs or sunflower stalks. The production of building elements in the form of cast blocks made of lightweight concrete with vegetal aggregates has lower energy consumption compared to the production of bricks (ceramic blocks), which require thermal treatments/burning in furnaces.

The use of lightweight concrete with vegetal aggregates can increase work productivity by replacing partition walls made of bricks with walls cast in formworks or prefabricated walls while keeping the main features of sound and thermal insulation as well as mechanical characteristics at a satisfactory level.

If the vegetal lightweight concrete is used in mortar for masonry and/or plastering, it will help to correct small-scale thermal bridging. It can ensure a continuity of the thermal insulation when it is used concomitant as screed and plaster in a building.

Lightweight concrete with corn cob or sunflower aggregates can also be used on flat roofs to level and regularize the support under waterproofing membranes and provide thermal insulation against heat and cold.

By using lightweight concrete, obtained by replacing the usual aggregates with vegetal ones (like corn cob or sunflower aggregates), the realization of non-structural elements of a building (concrete screeds, closing walls or partitions, finishes, etc.) leads to the reduction of the total weight of the building, resulting in favorable effects on its behavior during earthquake.

It is also possible to increase the efficiency of the machines and equipment used in the production, transport, and casting of concrete (concrete plants, concrete mixers, cranes, pumps, etc.) and a decrease in their energy consumption due to the smaller weight per unit volume of the concrete used in non-structural purposes.

The use of concretes with vegetal aggregates can help to increase the sustainability of buildings and reduce their total carbon footprint by replacing some raw materials that are increasingly difficult to obtain with natural, renewable materials that obtained with much smaller energy consumption.

Taking all of this into account, it can be concluded that CTCC40 and CTSF40 can widely be used in construction industry.

5. Conclusions

The present study aimed to present the benefits of mineral aggregates replacement in concrete by plant aggregates in the context of sustainable development and environmental protection. The developed vegetal concrete represents a variant of innovative and sustainable building material that could result in mineral aggregate savings and a positive effect on global health and environment.

In our research, we used corn cobs and sunflower stalks as plant aggregates, which were previously treated with sodium silicate solution. Some physical properties of vegetal concrete, especially thermal and acoustic properties, were reported to be greatly altered by the presence and the amount of absorbed water. The increased porosity and internal structure of the porous plant aggregates are responsible for the high absorption of water and their retention capacity. Due to high moisture absorption capacity, the durability of vegetal concrete represents a challenge. Using the shredded corn cobs and sunflower stalks as partial replacement for aggregates in concrete production requires some treatments to be applied in order to reduce their water absorption capacity and increase the interface compatibility with the concrete cement paste. To accomplish this objective, we adopted a treatment of vegetal granules with sodium silicate solution. As a result of the performed treatment with sodium silicate solution, we obtained a significant reduction in water absorption capacity.

The case study results showed an increase in concrete density when treated vegetal aggregates were used of approximately 10% in the case of using a solution of sodium silicate of 20% concentration and of approximately 20% in the case of 40% sodium silicate solution compared to the concrete with untreated vegetal aggregates. The treatment of vegetal aggregates with 40% solution of sodium silicate solution led to a significant improvement in the compressive strength of the concrete compared to the that with untreated plant aggregates. Sunflower concrete's mechanical strengths is superior to corn cob concrete.

By analyzing the compressive strength in comparison to the water absorption capacity of the aggregates involved, it was observed that reducing the water absorption capacity of the vegetal aggregates resulted in an increase of the compressive strength for both the corn cob concrete and the sunflower concrete.

From a mechanical properties point of view, the investigated concrete in our study can be used for concrete screeds or for modular elements for buildings closures, such as bricks or prefabricated walls that are manufactured off site and then assembled with minimal site disturbance and no waste. The obtained results also showed that vegetal aggregates may be successfully used in the manufacturing of concrete, leading to a reduction of mineral aggregate use of around 900 kg per m^3 of concrete, as comparing to the conventional concrete. The use of such plant aggregates is an essential component of sustainable development by reducing the usage of mineral aggregates. This is an ecological alternative that meets the sustainable requirements for a healthy environment, offering low-pollutant alternatives in the context of an increasing demand for construction materials.

Author Contributions: Conceptualization, A.A.Ş.; Funding acquisition, R.M.; Investigation, C.M.G.; Methodology, C.M.G., V.C. and A.B.; Writing—original draft, C.M.G.; Writing—review & editing, C.M.G. and R.M. All authors have read and agreed to the published version of the manuscript.

Funding: This research did not receive any specific grant from funding agencies in the public, commercial, or not-for-profit sectors.

Acknowledgments: Special thanks to the Sika Romania representatives for their technical support and for providing the necessary additives in order to accomplish this research.

Conflicts of Interest: The authors declare no conflict of interest.

References

1. Pinto, J.; Paiva, A.; Varum, H.; Costa, A.; Cruz, D.; Pereira, S.; Fernandes, L.; Tavares, P.; Agarwal, J. Corn's cob as a potential ecological thermal insulation material. *Energy Build.* **2011**, *43*, 1985–1990. [CrossRef]
2. USGBC (United States Green Building Council). LEED v4 Reference Guide for Homes Design and Construction. 2013. Available online: https://www.usgbc.org/sites/all/assets/section/files/v4-guide-excerpts/Excerpt_v4_HOMES.pdf (accessed on 15 October 2019).
3. Gradinaru, C.M. The environmental impact of concrete production and its greening necessity. In *Resilient Society Environment and Human Action Series A*; Ozunu, A., Nistor, I.A., Petrescu, D.C., Burny, P., Petrescu-Mag, R.M., Eds.; Les Presses Agronomiques: Gembloux, Belgium; Bioflux: Cluj-Napoca, Romania, 2017; pp. 81–94.
4. Hicks, L. *Aggregates Supply in England: Issues for Planning*; Open Report no. 059; British Geological Survey: Nottingham, UK, 2008; Available online: http://nora.nerc.ac.uk/id/eprint/5215/1/OR08059.pdf (accessed on 19 November 2019).
5. Langer, W.H.; Drew, L.J.; Sachs, J.S. Aggregate and the environment. In *American Geological Institute Environmental Awareness Series*; American Geological Institute: Silver Spring, MD, USA, 2004; Volume 8, p. 64.
6. Grant, A. Estimating the Marginal Effect of Pits and Quarries on Rural Residential Property Values in Wellington County, Ontario: A Hedonic Approach. Master's Thesis, University of Guelph, Guelph, ON, Canada, 2017.
7. BMAPA (British Marine Aggregate Producers Association); The (England) Crown Estate; Defra, Marine Management Organisation; Natural England; JNCC; Historic England; Cefas. Good Practice Guidance: Extraction by Dredging of Aggregates from England's Seabed. 2017. Available online: https://bmapa.org/documents/BMAPA_TCE_Good_Practice_Guidance_04.2017.pdf (accessed on 19 November 2019).
8. Nozahic, V.; Amziane, S.; Torrent, G.; Saïdi, K.; De Baynast, H. Design of green concrete made of plant-derived aggregates and a pumice–lime binder. *Cem. Concr. Compos.* **2012**, *34*, 231–241. [CrossRef]
9. Campbell, B.A. Aggregate Resource Extraction: Examining Environmental Impacts on Optimal Extraction and Reclamation Strategies. Master's Thesis, University of Alberta, Edmonton, AB, Canada, 2014. Available online: https://era.library.ualberta.ca/items/a562f94b-dbd1-4c77-b4e8-2c5b53f5ff17/view/f2d8eabe-5f45-468b-9f3c-2791b66db455/Campbell_Brett_A_201409_MSc.pdf (accessed on 19 November 2019).
10. European Environment Agency. *Effectiveness of Environmental Taxes and Charges for Managing Sand, Gravel and Rock Extraction in Selected EU Countries*; EEA Report no 2; European Environment Agency: Copenhagen, Denmark, 2008; p. 64. Available online: https://www.eea.europa.eu/publications/eea_report_2008_2/file (accessed on 19 November 2019).
11. Langer, W.H.; Arbogast, B.F. Environmental impacts of mining natural aggregate. In *Deposit and Geoenvironmental Models for Resource Exploitation and Environmental Security*; Nato Science Partnership Subseries 2; Fabbri, A.G., Gaál, G., McCammon, R.B., Eds.; Springer: Dordrecht, Switzerland, 2002; Volume 80.
12. Danielsen, S.W.; Kuznetsova, E. Resource management and a best available concept for aggregate sustainability. In *Sustainable Use of Traditional Geomaterials Construction Practice*; Geological Society: London, UK, 2016; Volume 416, pp. 59–70.

13. Na, H. Environmental Ethics and Sustainable Design: A Case Study on the Traditional Korean Residential Building Type: Han Oak. Master's Thesis, School of Engineering Technology, Eastern Michigan University, Ypsilanti, Michigan, 2013.

14. Gradinaru, C.M.; Babor, D.; Gradinaru, A.C. Poly-ethylene terephthalate waste recycling for environment protection and sustainable building activities. In Proceedings of the 17th Edition of National Technical-Scientific Conference, Modern Technologies for the 3rd Millenium, Oradea, Romania, 22–23 March 2018; pp. 285–290.

15. Gradinaru, C.M.; Barbuta, M.; Ciocan, V.; Antohie, E.; Babor, D. The effects of sodium silicate on corn cob aggregates and on the concrete obtained with these agricultural waste. *IOP Conf. Ser. Mater. Sci. Eng.* **2018**, *399*, 012021. [CrossRef]

16. Gradinaru, C.M.; Barbuta, M.; Ciocan, V.; Serbanoiu, A.A. A study on the effects of the cement and mineral aggregates replacement with waste materials. *IOP Conf. Ser. Mater. Sci. Eng.* **2018**, *399*, 012020. [CrossRef]

17. Gradinaru, C.M.; Barbuta, M.; Ciocan, V.; Serbanoiu, A.A.; Gradinaru, A.C. Health and environmental effects of heavy metals resulted from fly ash and cement obtaining and trials to reduce their pollutant concentration by a process of combining-exclusion. In Proceedings of the 17th International Multidisciplinary Scientific Geoconference SGEM 2017, Albena, Bulgaria, 29 June–5 July 2017; Volume 17, pp. 441–447.

18. Gradinaru, C.M.; Serbanoiu, A.A.; Serbanoiu, B.V. Concrete with thermal insulating properties—A double benefit in terms of money and environmental protection. *Adv. Environ. Sci.* **2018**, *10*, 56–61.

19. Gradinaru, C.M.; Serbanoiu, A.A.; Serbanoiu, B.V. Fibre reinforced concrete—A sustainable material in the context of building industry and environmental challenges. *Adv. Environ. Sci.* **2018**, *10*, 1–6.

20. Mercader Moyano, P.; Requena García de la Cruz, M.V.; Yajnes, M.E. Development of new eco-efficient cement-based construction materials and recycled fine aggregates and Eps from Cdw. *Open Constr. Build. Technol. J.* **2017**, *11*, 381–394. [CrossRef]

21. Mercader Moyano, P.; Yajnes, M.; Caruso, S. Experimental characterisation of a cement-based compound with recycled aggregates and Eps from rehabilitation work. *Revista de la Construcción* **2016**, *15*, 97–106. [CrossRef]

22. Muntean, R.; Cazacu, C.; Chitonu, G.C.; Galatanu, T. Lost shuttering for columns as possible applications of dispersely reinforced concrete with polypropylene fibers. *IOP Conf. Ser. Mater. Sci. Eng.* **2018**, *399*, 12040. [CrossRef]

23. Muntean, R.; Cazacu, C. Using PET (polyethylene terephthalate) waste for buildings. *J. Appl. Eng. Sci.* **2011**, *1*, 73–80.

24. Serbanoiu, A.A.; Barbuta, M.; Burlacu, A.; Gradinaru, C.M. Fly ash cement concrete with steel fibers—Comparative study. *Environ. Eng. Manag. J.* **2017**, *16*, 1123–1128.

25. Oancea, I.; Bujoreanu, C.; Budescu, M.; Benchea, M.; Gradinaru, C.M. Considerations on sound absorption coefficient of sustainable concrete with different waste replacements. *J. Clean. Prod.* **2018**, *203*, 301–312. [CrossRef]

26. Chabriac, P.A.; Gourdon, E.; Gle, P.; Fabbri, A.; Lenorman, H. Agricultural by-products for building insulation: Acoustical characterization and modelling to predict micro-structural parameters. *Constr. Build. Mater.* **2016**, *112*, 158–167. [CrossRef]

27. Chabannes, M.; Nozahic, V.; Amziane, S. Design and multi physical properties of a new insulating concrete using sunflower stem aggregates and eco-friendly binders. *Mater. Struct.* **2015**, *48*, 1815–1829. [CrossRef]

28. Faustino, J.; Silva, E.; Pinto, J.; Soares, E.; Cunha, V.M.C.F.; Soares, S. Lightweight concrete masonry units based on processed granulate of corn cob as aggregate. *Materiales de Construcción* **2015**, *65*, e055.

29. Binici, H.; Eken, M.; Dolaz, M.; Aksogan, O.; Kara, M. An environmentally friendly thermal insulation material from sunflower stalk, textile waste and stubble fibres. *Constr. Build. Mater.* **2014**, *51*, 24–33. [CrossRef]

30. Pinto, J.; Cruz, D.; Paiva, A.; Pereira, S.; Tavares, P.; Fernandes, L.; Varum, H. Characterization of corn cob as a possible raw building material. *Constr. Build. Mater.* **2012**, *34*, 28–33. [CrossRef]

31. Amziane, S.; Sonebi, M. Overview on Biobased building material made with plant aggregate. *RILEM Techn. Lett.* **2016**, *1*, 31–38. [CrossRef]

32. Leff, B.; Ramankutty, N.; Foley, J.A. Geographic distribution of major crops across the world. *Glob. Biogeochem. Cycles* **2004**, *18*. [CrossRef]

33. Farias, R.D.; García, C.M.; Cotes Palomino, T.; Andreola, F.; Lancellotti, I.; Barbieri, L. Valorization of agro-industrial wastes in lightweight aggregates for agronomic use: Preliminary study. *Environ. Eng. Manag. J.* **2017**, *16*, 1691–1699. [CrossRef]

34. Binici, H.; Zengin, H.; Zengin, G.; Kaplan, H.; Yucegok, F. Resistance to sodium sulphate attack of plain and blended cement containing corncob ash and ground granulated blast furnace slag. *Sci. Res. Essays* **2009**, *4*, 98–106.
35. Nkayem Njeumen, D.E.; Mbey, J.A.; Kenne Diffo, B.B.; Njopwouo, D. Preliminary study on the use of corn cob as pore forming agent in lightweight clay bricks: Physical and mechanical features. *J. Build. Eng.* **2016**, *5*, 254–259. [CrossRef]
36. Pinto, J.; Vieira, B.; Pereira, H.; Jacinto, C.; Vilela, P.; Paiva, A.; Pereira, S.; Cunha, V.M.C.F.; Varum, H. Corn cob lightweight concrete for non-structural applications. *Constr. Build. Mater.* **2012**, *34*, 346–351. [CrossRef]
37. Mati-Baouche, N.; De Baynast, H.; Lebert, A.; Sun, S.; Lopez-Mingo, C.J.S.; Leclaire, P.; Michaud, P. Mechanical, thermal and acoustical characterization of an insulating bio-based composite made from sunflower stalks particles and chitosan. *Ind. Crop. Prod.* **2014**, *58*, 244–250. [CrossRef]
38. *Practice Code for Execution of Concrete Works, Reinforced Concrete and Prestressed Concrete. Part 1: Concrete Production*; NE 012/1; National Institute of Research-Development in Constructions and the Economy of Constructions: Bucharest, Romania, 2007.
39. *Testing Hardened Concrete, Part 5: Flexural Strength of Test Specimens*; SR EN 12390-5; Standards Association of Romania: Bucharest, Romania, 2019.
40. *Testing Fresh Concrete, Part 6: Density*; SR EN 12350-6; Standards Association of Romania: Bucharest, Romania, 2019.
41. *Testing Hardened Concrete, Part 6: Split Tensile Strength of Test Specimens*; SR EN 12390-6; Standards Association of Romania: Bucharest, Romania, 2010.
42. *Testing Hardened Concrete, Part 3: Compressive Strength of Test Specimens*; SR EN 12390-3; Standards Association of Romania: Bucharest, Romania, 2019.

 *sustainability*

Article

Ecofibers for the Reinforcement of Cement Mortars for Coating Promoting the Circular Economy

Carolina Piña Ramírez [1], Alejandra Vidales Barriguete [2], Julián García Muñoz [1], Mercedes del Río Merino [1,*] and Patricia del Solar Serrano [1]

1   Department of Architectural Constructions and their Control, Superior Technical School of Building, Polytechnic University of Madrid, 28040 Madrid, Spain; carolina.pina@upm.es (C.P.R.); julian.garciam@upm.es (J.G.M.); patricia.delsolar@upm.es (P.d.S.S.)
2   Department of Building Technology, Superior Technical School of Building, Polytechnic University of Madrid, 28040 Madrid, Spain; alejandra.vidales@upm.es
*   Correspondence: mercedes.delrio@upm.es

Received: 17 February 2020; Accepted: 28 March 2020; Published: 2 April 2020

Abstract: Nowadays, nobody can deny that climate change is a reality and that the life cycle of buildings contributes greatly to that reality. Therefore, proposals such as the circular economy must be integrated into the construction sector. This article shows part of the results of a research project whose objective is to introduce circular economy criteria in building materials, seeking new uses for construction and demolition waste from buildings. In particular, this article analyses the possibility of replacing fibres currently used to reinforce cement mortars with recycled fibres. After consulting the bibliography, we can conclude that some studies analyse the behaviour of cement mortars reinforced with different types of fibres, but none has been found that analyses the behaviour of these mortars for the application of continuous coatings. For this purpose, a two-stage experimental plan is designed to test cement mortar samples with different types of fibres, recycled fibres and commercial fibres, taking into consideration the characteristics that these mortars have to comply to be applied as continuous coatings. Moreover, a detailed study about the porosity of these mortars and its influence on how the mortars behave with regard to compression, water vapour permeability and impermeability has been conducted. From the results obtained, it can be concluded that the mortars containing recycled fibres have very similar resistance, absorption and permeability values to those containing commercial fibres, so that they might be suitable for application as external coatings.

Keywords: mineral wool; cement mortar; recycling; circular economy; construction and demolition waste

1. Introduction

It is true that the climate has always been changing, but the problem is that currently the pace is faster than humans have ever seen. If this pace is maintained over time, if we do not stop it, climate change threatens to make parts of the planet uninhabitable for life as we know it.

It has been scientifically proven that the industry, and in particular the construction industry, is partially responsible for this problem since it is one of the sectors that consumes more resources and generates more emissions and waste [1]. Therefore, working to reduce the impact it generates is essential [2].

According to the UK Green Building Council (UKGBC), the construction sector uses more than 400 million tons of material each year. In particular, the products that are used during the construction process can damage the environment due to intensive extraction of raw materials, transporting to manufacturing plant and site, energy consumption in manufacture and in use, waste generation and so on [3].

In Spain, in 2007, there was an expenditure of 20.1 tons of resources and materials per person and per year, compared to the 16.5 tons of the EU average, although it is true that during 2007 there was a high level of building [4]. In addition, the manufacture of construction materials consumes a large amount of energy, especially in the case of materials such as glass or plastics [5].

On the other hand, the production of waste from construction and demolition waste (CDW), occurs in each and every phase of the building life cycle [6]. These CDW cause damage, not only in their immediate surroundings [7], but also because of the contamination of the soil where they are deposited, the contamination of the runoff water or nearby rivers, atmospheric pollution due to volatile elements and odours, degradation of the surroundings and consumption of transport energy and maintenance of landfills [8].

Therefore, moving from a linear economy to a circular economy is one of the strategies that will reduce both the consumption of resources and the waste generated [9]. In addition, this strategy must be viewed from a global vision of the entire life cycle of buildings [10].

In 2015, the European Commission adopted an ambitious Circular Economy Action Plan, which includes measures that will help stimulate Europe's transition towards a circular economy, boost global competitiveness, foster sustainable economic growth and generate new jobs.

The EU Action Plan for the Circular Economy establishes a concrete and ambitious program of action, with measures covering the whole cycle: from production and consumption to waste management and the market for secondary raw materials and a revised legislative proposal on waste [11].

Therefore, in the last decade there have been many studies whose objectives have been to incorporate circular economy criteria in the manufacture of construction materials, specifically on cement mortars, for example the studies by Aciu et al. [12], which conclude that it is possible to manufacture mortars with PVC waste, or the studies by Piña et al., [13], that confirm the feasibility of replacing, in mortars, glass or polypropylene reinforcing fibres by fibres from mineral wool residues, or the studies by Morales Conde et al. [14], that analyse the Physical and mechanical properties of wood-gypsum composites from demolition material in rehabilitation works or the works done by Gadea et al. [15], about lightweight mortar made with recycled polyurethane foam or the research projects done by Muñoz-Ruiperez et al. [16], about mortars made with expanded clay and recycled aggregates.

On the other hand, coatings are finishes of building façades that perform a dual function: providing a specific appearance (colour and texture) and improving some of the façade's characteristics.

These coatings are usually cement-based mortar (plaster and render) or mixed cement and limestone and are used to be mixed on site with approximate dosages, which means that throughout the work, the composition of the coating might not be homogeneous as it depends on the experience and skill of the workers. Therefore, ready-to-use coatings have been on the market for a number of years, and only the water recommended by the manufacturer has to be added on site, eliminating the drawbacks of on-site preparation and hence the lack of homogeneity of traditional coatings. This type of mortar is known as single layer coating [17]. These mortars are normally quite fluid, as they are usually projected, and a significant yield is sought when they are administered.

The composition of coating mortars depends on the rendering properties required but, in general, they are based on cement (white or grey), rarely lime, and contain carbonate/silica aggregates [18]. They can also incorporate lightweight aggregates such as perlite, vermiculite, etc., and other additives such as pigments, water retainers, water repellents, aerators, accelerators, retardants, synthetic resins, fibres, etc. These aggregates are more typical in single layer coatings, as they improve ease of application, adhesion, mechanical strength or surface hardness, water impermeability, or water vapour permeability. The typical composition of cement mortars used in coatings is around 1:3:0.6 (cement: sand: water), [19–21], with the percentages of additions varying widely.

Despite the advantages of using single layer coatings, it is still very common to apply cement mortar coatings on walls in the traditional way, since the mortars used provide significant durability and high adhesion, which has made them historically suitable for outdoor use.

Coating mortars are defined in Standards UNE-EN 998-1:2003, [22], executed according to UNE-EN 13914-1:2019 [23] and UNE-EN 13914-2:2019 [24] and are categorised according to two basic characteristics: Compressive strength (CS), calculated according to UNE-EN 1015-11:2000/A1:2007 [25] and Water Absorption (W) calculated according to UNE-EN 1015-18:2003 [26].

The properties of mortars depend not only on their composition but also on certain external conditions such as the preparation of the substrate, the application of the product and the protection of the coating.

Cement-based coatings require compressive strength and water absorption that are achieved with relatively low cement dosages. In fact, it can be affirmed [27] that the proportion of cement added to the mortar is directly related to their impermeability and rigidity. Mortars with larger proportions of cement are more rigid, which means a greater probability of cracks appearing, which are their main setback. Mortars used in these study trials seek a balance between rigidity and impermeability without sacrificing good results in both aspects, which are fundamental in a good coating mortar.

Cracking can occur for different reasons, which have been studied in detail by several authors [28]. The primary causes range from microcracks caused by shrinkage during the setting or drying of the binder to the low plastic deformability of the mortar [29] and are usually combined with others of a secondary nature, such as sensitivity to thermal or mechanical actions [30].

A key consequence of the cracking of coating mortars is the increase in their permeability to water [31]. The appearance of cracks in the mortar causes major permeability problems both in the short and long term. One of the immediate problems is the appearance of dampness inside enclosures; one of the long-term problems is the logical increase (by attrition) in the size of the cracks and the consequent degradation of the mortar.

Different strategies are used to prevent this cracking. These include the use of additions to improve mortar plasticity [32], the incorporation of different types of fibres in their preparation [33], or the insertion of meshes or reinforcements between coating layers. These techniques can be used together to reduce the possibility of cracks but do not always eliminate them completely.

However, there is agreement in the academic community that the appearance of a crack implies a change in the permeability of the material. This leads to a reduction in its durability, since once the crack has occurred the properties of the material change drastically, and it cannot be assumed that the estimated durability for an uncracked surface can be maintained [28].

2. Purpose

This article shows part of the results of a research project that analyses the viability of replacing, in mortars for coatings, glass and polypropylene fibres (fibres commonly used today) by fibres from the recycling of mineral wool insulation panels [13] and analyses their cracking and water behaviour.

3. Experimental Plan

A two-phase experimental plan was designed to analyse the behaviour of these mortars with respect to the requirements of current regulations: resistance to compression and water absorption. The relationship between shrinkage cracks and mortar porosity with the different types of fibres used was also included in the experimental plan.

3.1. Materials

Cem II/B-L 32.5 N cement and natural river sand sieved through a 4 mm mesh screen from a gravel pit or quarry were used.

The water used comes from the Canal de Isabel II of the Community of Madrid. Its technical characteristics are established in standard UNE EN 13279-2 [34].

Three types of mineral wool waste were also used (Figure 1). These were collected manually from the Centre for the Integral Treatment of Construction and Demolition Waste on the N-1 motorway to Irún, Km 40, in the town of El Molar, Community of Madrid and managed by Gestión y Desarrollo del Medio Ambiente de Madrid (GEDESMA S.A.) They were stored outdoors and mixed with other waste materials. After selecting this waste, it was separated into piles of hand rock wool (RLR) and fibreglass (RFV). A sample of unselected mixed mineral wool waste, as it was found in the landfill, was also collected (RMIX). The three types of mineral fibre residues were subjected to grinding in a 550 W machine at a nominal voltage of 50 Hz, for 3 intervals of 3 min each, to obtain a format that was suitable to be included in the mortars with a length of 500–1000 μm.

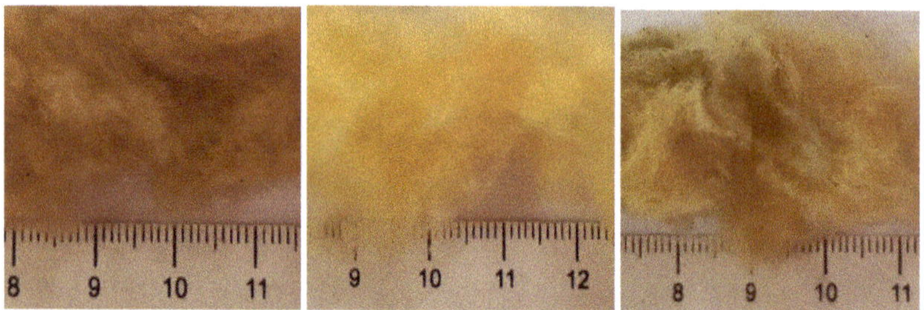

Figure 1. Mineral fibre waste: rock wool (left); fibreglass (centre); mixed (right).

The commercial polypropylene fibres used were 6-metre long SikaCim® Fibres-6 for concrete and mortars from Sika, S.A.U. and 12 mm long AR Tecnicret® fibreglass (Figure 2).

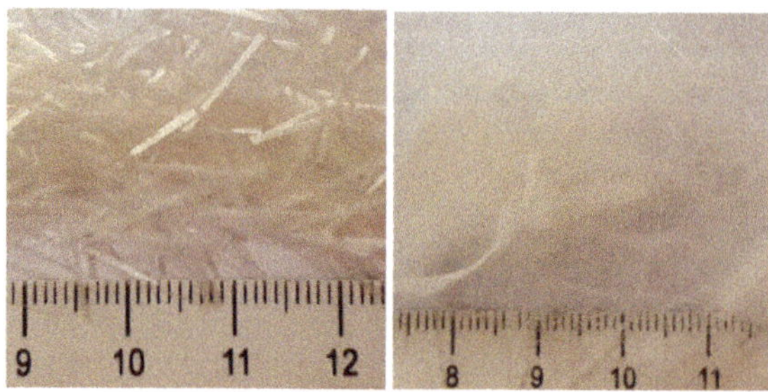

Figure 2. Commercial fibreglass fibres (left); polypropylene (right).

3.2. Composition of the Mixtures

Six different cement mortar mixtures were prepared in the laboratory: a reference mortar without fibres (MREF), three mortars with mineral wool waste from insulation: mortar with rock wool waste (MRLR), mortar with fibreglass waste (MRFV) and mortar with a combination of mineral wool waste (MMIX), and two mortars with commercial fibres to compare their behaviour with the recycled samples: mortar with fibreglass (MFV), mortar with polypropylene fibre (MPP).

The dosage maintained in all mixtures is 1:3:0.6 (cement: sand: water), (Table 1); this dosage was established based on the bibliography consulted [19–21] and the results can be compared with other previous works [13].

Table 1. Sample dosage.

Name	Cement (g)	Sand (g)	Water (g)	Fibres (g)
MREF	1000	3000	600	—-
MRLR	1000	3000	600	1.70
MRFV	1000	3000	600	1.70
MRMIX	1000	3000	600	1.70
MPP	1000	3000	600	1.70
MFV	1000	3000	600	1.70

3.3. Tests

The experimental plan was conducted in two phases:
Phase 1
In the first phase, 4 × 4 × 16 cm specimens are made of the reference mortar and the mortars reinforced with commercial fibres and fibres from the recycling of RCD (Figure 3).

Figure 3. Cement mortar mixtures specimens prepared in the laboratory.

The physical, mechanical and behavioural tests against water were carried out on these specimens, as listed below:

Bulk density of fresh mortar (UNE-EN 1015-6) [35]: it has been calculated as the mass of each of the mixtures divided by the volume occupied by the mortar introduced and compacted in a container. The mass considered has been that necessary to fill the 1-L container. The filling of the container has been carried out in two stages. In the first one, half of the container is filled and tapped 10 times for compaction; in the second stage, the container is filled to its limit and compacted again by striking it 10 times.

Compressive strength: (UNE-EN 1015-11) [25]: The Ibertest Autotest 200 cement strength testing machine was used in 4 × 4 × 16 cm^3 specimens. With constant speed, a load was gradually applied at a rate between 50 and 500 N/s until failure occurs and the value obtained was recorded.

Consistence (UNE-EN 1015-3) [36]: The flow table method was used, according to the procedure described in part 3 of the standard. The fresh mortar is poured in two layers on a frustoconical mould placed on the shaking table; each layer is compacted with ten tapping strokes. After 15 s, the mould is removed, and 15 vertical shakes are carried out with a constant frequency of approximately 1 s. using the calliper, the diameter of the mortar is measured in two perpendicular directions and the average value is calculated.

Water absorption by capillarity (UNE-EN 1015-18): [26] It was calculated by obtaining the amount of water absorbed by the 160 × 40 × 40 mm prismatic samples cured after 28 days and dried in an oven. The test specimens are impregnated with paraffin and then broken into two halves. On the break side they are submerged in water with a depth of between 5 and 10 mm.

Total absorption: The specimens are dried in a stove for 24 h, cooled in a desiccator to ambient laboratory temperature, and subsequently immersed in water, weight increases by water absorption after 24 and 48 h were recorded.

Water vapour permeability (UNE-EN 1015-19): [37]: The amount of water that has passed per unit area of the specimen in a unit of time when there is a pressure difference of one unit between its walls.

For this purpose, circular specimens are made with fresh mortar. After curing, these specimens are placed on perimeter sealed containers. These vessels contain a saline solution of potassium nitrate that maintains the relative humidity at 93.2%. The difference in relative humidity between the two faces of the test piece causes the water to gradually evaporate.

Phase 2

In the second phase, a detailed study of the shrinkage cracking and porosity was conducted on the previous mortars, as described below:

Contraction test (UNE 80112) [38]: (shrinkage): Changes in length of each sample were measured compared to the original length. The specimens used, measuring 25x25x285 mm3, had stainless steel cylinders at their ends, to be placed in a vertical position, on the length comparator, at the time of measurement. This measurement has been carried out at mould release, at 7 days and at 14 days.

Air content (UNE-EN 1015-7) [39]: Obtained by means of the pressure method in which pressurized air was applied to the chamber containing the mortar and, when it decreased, the air content present in the sample was indicated on the measuring device. This test was carried out by filling, in four layers, the container with fresh mortar. Each layer was compacted with 10 short strokes of the piston. Once the container—to which a manometer had been connected—was levelled, air was injected into the air chamber until it reached a stable pressure, which was determined in a first calibration test. The valves were closed and, once equilibrium was reached, the air content value was read from the manometer.

Porosimetry by injection of Hg: The AutoPore IV 9500 porosimeter was used to characterize the porosity of the material by applying several pressure levels to the sample submerged in mercury, with the pressure required to produce the penetration of the mercury being inversely proportional to the size of the pores. The specimens made for this test were cylindrical, with a diameter of 10 mm and a height of 10 mm. The test was carried out at room temperature, bringing the sample to a vacuum of 50 μmHg that was kept for five minutes. Subsequently, each sample was subjected to increasing pressures that force the mercury percolation into the pores up to 206MPa.

Scanning electron microscopy: The JEOL JSM-840 high-resolution environmental scanning microscope was used to study of the crystalline texture and morphology of the 6 different mortar mixtures. This microscope allows the observation of samples of sizes up to 50-100-150 mm in the motorized stage. The samples tested, small fragments of approximately 10x10 mm2, were taken from rests of the specimens that were subjected to flexion and compression tests. Once placed under the microscope, "the beam passes through the condenser and objective lenses, and is swept across the sample by the scanning coils, while a detector records the number of low-energy secondary electrons emitted for each point on the surface ", so that an SEM image is formed [40].

4. Results and Discussion

4.1. Results Obtained in Phase 1

The results obtained from the physical, mechanical and water behaviour tests are shown.

4.1.1. Apparent Density of Fresh Mortar and Consistency

Figure 4, shows the average of the results obtained in three test specimens of each mixture of the bulk density test on the fresh mortar and the consistency of the mixtures. It can be seen that adding fibres, whether they be from waste or commercial sources, decreases their density with regard to the reference mixture (MREF) but very slightly due to the small amount of fibres added.

Figure 4. Results of the test of the apparent density of the fresh mortar (kg/m^3) and consistency of the mixtures (cm).

The mortar containing commercial fibres MPP has the lowest density (1201 kg/m^3) very similar to that containing MRFV fibre waste (1204 kg/m^3), representing only a decrease of 2.50% in comparison with the reference density (1232 kg/m^3). MFV and MRLR mixtures have the highest density value (1216 kg/m^3), although this is 1.30% lower than the reference mortar. It should be noted that all of them could be considered as lightweight mortars as they have a density of less than 1300 kg/m^3 according to UNE EN 1015-6 [35].

Regarding consistency, all mixtures are within the range established for plastic mortars in the standard mentioned in the previous paragraph. The values of the mortars containing fibres are between 10.2% and 3.2% lower than those of MREF. Plasticity values showed that the most fluid mixture was MRLR (166 mm) similar to that of MFV (167.5 mm) and the less fluid one was MRFV (179 mm) similar to MMIX (178 mm). The MPP mixture showed intermediate values (170 mm).

It should be pointed out that MRFV and MFV mortars behave differently, while the former has a lower density value and a drier consistency (within the plasticity range), the latter has a higher density value with a higher, more fluid consistency.

4.1.2. Compressive Strength Test

Compressive strength of the test pieces after the test is shown in Figure 5. These results are the average of those obtained in 3 test specimens of each mixture considered.

Compressive strength for mortars MRLR-MRFV decreased by 0.04%–10.53% in relation to the reference and, on the contrary, mortars MMIX-MPP-MFV increased by 9.34%–1.06%–7.56%. Similar to what occurs in the compressive strength test in the previous section, the mortar containing the fibre waste with better compression resistance was MMIX (25.75 N/mm^2), and the worst was MRFV (21.07 N/mm^2). In the case of mortars with commercial fibres, MFV (25.33 N/mm^2) showed an improved performance of 3.39% compared to MPP (23.80 N/mm^2), as stated by other researchers [41,42].

In all cases, the minimum values established in standard UNE-EN 1015-11:2000/A1:2007 [25] were exceeded for the case of mortars R1 and/or R3, whose compressive strength should be 3.5–7.5 N/mm^2 and/or $\geq$ 6 N/mm^2.

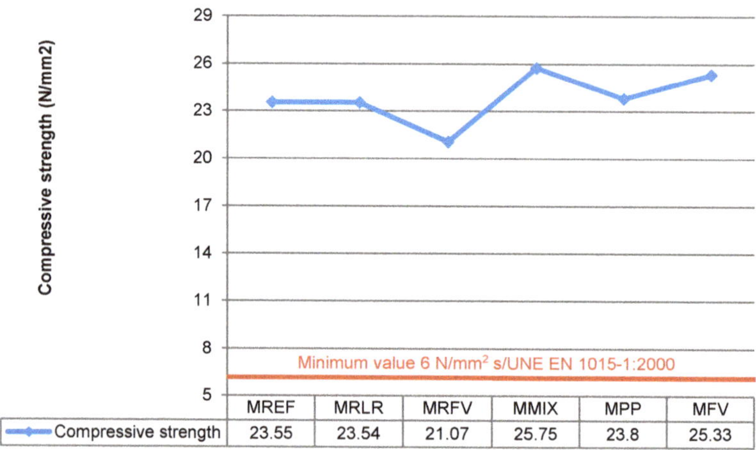

	MREF	MRLR	MRFV	MMIX	MPP	MFV
Compressive strength	23.55	23.54	21.07	25.75	23.8	25.33

Figure 5. Result of the compressive strength test (N/mm^2).

4.1.3. Absorption of Water by Capillarity

The average of the results of the tests carried out on 3 test pieces of each mixture for the absorption of water by capillarity test is shown in Figure 6. These mortars must absorb as little as possible due to their application. This particularly affects mortars exposed to rainwater and those located at the base of the foundations. The greater the water absorption, the greater the possibilities of the appearance of moisture by filtration. In any case, the standard UNE-EN 1015-18 [26]: requires for mortars R1 c $\leq$ 0.4 kg/m^2 min$^{0.5}$ and mortars R3 c $\leq$ 0.2 kg/m^2 min$^{0.5}$.

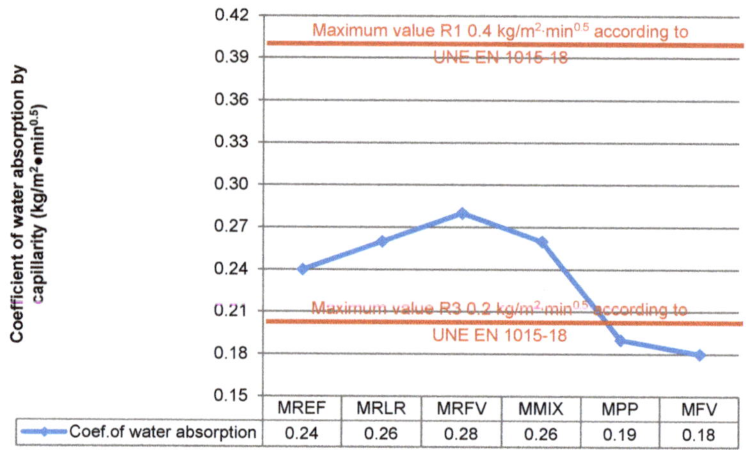

	MREF	MRLR	MRFV	MMIX	MPP	MFV
Coef.of water absorption	0.24	0.26	0.28	0.26	0.19	0.18

Figure 6. Result of the water absorption test by capillary action (kg/m^2·min$^{0.5}$).

While mortars containing fibre waste had a higher absorption coefficient in relation to the reference mortar, MRFV 16.7%, MRLR and MMIX 8.3%, this was considerably lower for mortars containing commercial fibres, MPP by 20.8% and MFV by 25%. The mortar with the greatest water absorption capacity by capillarity was MRFV (0.28 kg/m^2·min$^{0.5}$ and the lowest capacity MFV (0.18 kg/m^2·min$^{0.5}$). Therefore, all mortars in this study would comply with the specifications of standard mortars (R1) with medium resistance to filtration, and only mortars containing commercial fibres would comply with the provisions of mortars (R3), that is, with very high resistance to filtration. The reference mortar would not meet the values established for mortars (R3) either, as occurs in other studies [43].

4.1.4. Total Water Absorption

Figure 7 shows total water absorption values of the mixtures under study. These results are the average of the data obtained in the 3 test specimens of each mixture. It can be seen that, after 24 h, the samples are practically saturated since the increase is between 0% and 0.02%.

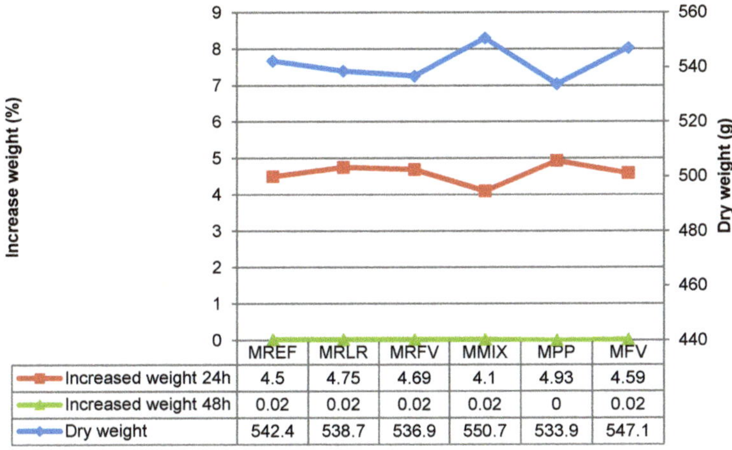

Figure 7. Result of the total water absorption test (g).

In this test, all the mixtures exceeded the 4.5% increase in weight obtained by MREF, except for MMIX, which decreased its total absorption by approximately 8.90% with respect to said reference. It should be noted that, at 24 h, the mortar containing fibre waste with the greatest total absorption is MRLR (4.75%), compared to that containing commercial fibre, MPP (4.93%). MMIX (4.1%) and MFV (4.59%), with fibre waste and commercial fibres, respectively, have a lower total absorption capacity.

4.1.5. Permeability to Water Vapour

This section contains an analysis of the degree of resistance to the passage of water vapour through ready-made mortars. The previous section showed that mortars for outdoor use should be as impermeable as possible, and it should be noted that they must maintain their water vapour permeability to allow for transpiration of walls and prevent the appearance of condensation on the inside.

Table 2 shows the results of the weights obtained during the 6-week duration of the water vapour permeability test in one test specimen of each mixture; Figure 8 shows the water vapour permeability values, which are very similar to those of the reference mortar.

Table 2. Weights of the samples for 6 weeks (g).

Name.	Initial weight (g)	Weight week 1 (g)	Weight week 2 (g)	Weight week 3 (g)	Weight week 4 (g)	Weight week 5 (g)	Weight week 6 (g)
MREF	2095.60	2074.20	2067.00	2060.20	2053.90	2048.70	2043.40
MRLR	2035.30	2014.60	2007.70	2001.10	1995.00	1989.90	1984.60
MRFV	2079.50	2058.60	2050.90	2044.00	2037.80	2032.70	2027.50
MMIX	2061.50	2039.70	2032.60	2025.90	2020.00	2015.00	2099.90
MPP	2133.60	2113.60	2106.40	2099.60	2093.70	2088.90	2084.20
MFV	2149.50	2127.40	2119.60	2112.20	2105.90	2100.80	2095.30

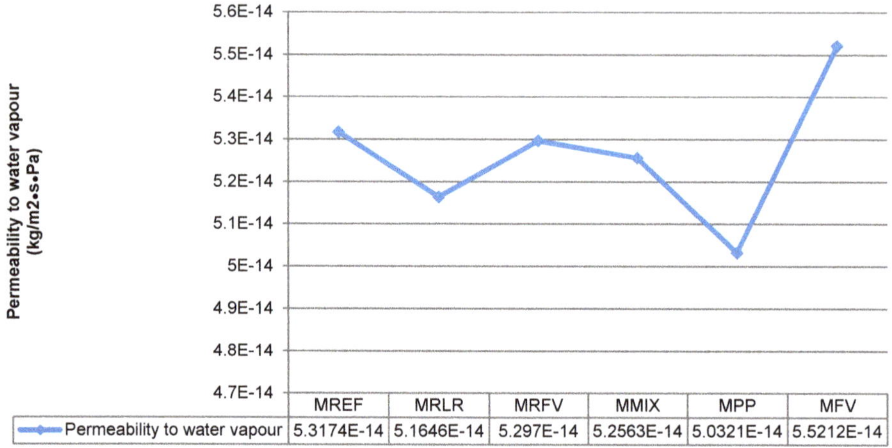

Figure 8. Result of the water vapour permeability test (Kg/m^2·s·Pa).

Mortars with fibre waste, MRLR-MRFV-MMIX, have the lowest values of water vapour permeability, i.e., 2.87%-0.38%-1.15% less than the reference mortar. In the case of commercial fibres, MPP also showed a reduction of 5.37% in relation to the reference mortar but MFV increased its value by 3.83% (Figure 8). The mortar with the lowest water vapour permeability was MPP with a value of 5.0321*10^{-14}. This is due to the fact that, contrary to water absorption values, higher values of water vapour permeability are obtained in mortars with a greater amount of aggregate [44], since CO2 diffuses better in the aggregate than in the matrix, favouring permeability.

4.2. Results Obtained in Phase 2

4.2.1. Shrinkage

Figure 9 shows the values obtained in the contraction test in 24 h, 7 days and 14 days. They represent the average of 4 test specimens of each mixture. After the first day, there were significant differences between the reference mortar and some of the prepared samples.

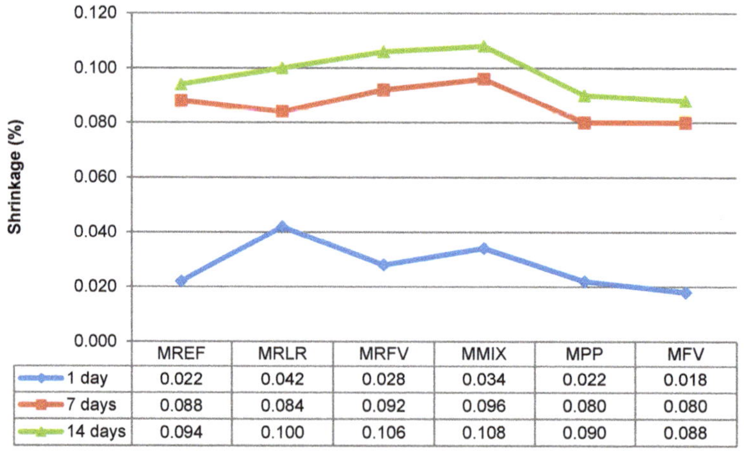

Figure 9. Result of retraction test (%).

After 24 h, mortars containing fibre waste, MRLR-MRFV-MMIX, reached 90.91%–27.27%–54.55% respectively, in relation to the reference mortar. The commercial fibre mortar MPP had the same shrinkage as MREF, but MFV had 18.18% less shrinkage.

After 7 days, the mortar containing fibre waste, MRLR, and commercial fibre mortars, MPP-MFV, retracted 4.55%–9.09%–9.09% less, respectively, than the reference mortar. MRFV and MMIX showed greater shrinkage than MREF, more specifically 4.55% and 9.09%.

Finally, after 14 days, only the mortars containing commercial fibres, MPP-MFV, showed the least shrinkage [45], 4.26%–6.38% less, in relation to the reference mortar, respectively. The mortars with fibre waste residues, MRLR-MRFV-MMIX, showed greater retraction than MREF, i.e., 6.38%–12.77%–14.89%, respectively.

4.2.2. Air Content and Porosimetry

The amount of air content from fresh mortars directly influences the porosity of the material, a greater amount of air causes greater porosity, which improves workability but reduces mechanical resistance.

The results of both tests are shown in Figure 10. This test was carried out on one test specimen of each mixture. When the fibre waste or commercial fibres were incorporated into the mortars, the air content decreased in relation to the reference sample [46].

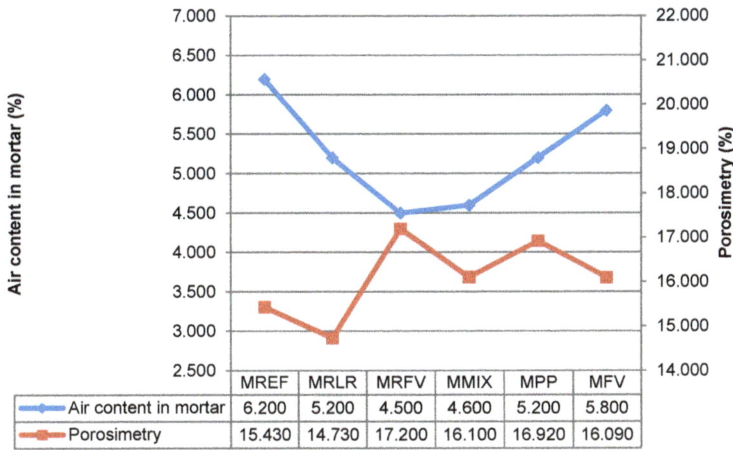

	MREF	MRLR	MRFV	MMIX	MPP	MFV
Air content in mortar	6.200	5.200	4.500	4.600	5.200	5.800
Porosimetry	15.430	14.730	17.200	16.100	16.920	16.090

Figure 10. Result of the test on air content in fresh mortar (%) and porosimetry (%).

Mortar containing fibre waste, MRLR, showed a greater amount of air content (5.2%) and still represented 16.13% less than MREF; MRFV showed the least amount of air content (4.5%), 27.42% less than MREF. Regarding commercial fibres, MFV showed the highest air content (5.8%), 6.45% less than the reference sample.

The results of the air content tests in the fresh mortar did not coincide exactly with those obtained in the porosimetry test on dry mortars. The mortar containing fibre waste that showed the highest percentage of pores is MRFV (17.2%), 11.47% more than MREF; the lowest percentage was shown by MRLR (14.73%), with 4.54% less pores than the reference sample. Amongst mortars containing commercial fibres, MPP showed the least amount of pores (16.92%), 9.66% more than MREF, and MFV showed the lowest amount (16.09%), 4.28% more than the reference sample.

4.2.3. Scanning Electron Microscopy

Photomicrographs of selected compounds are shown in Figures 11–13. It can be seen in the images that there is a good dispersion of the fibres in the mortars, although it is also obvious that the distribution may not be homogeneous, preventing control of surface properties.

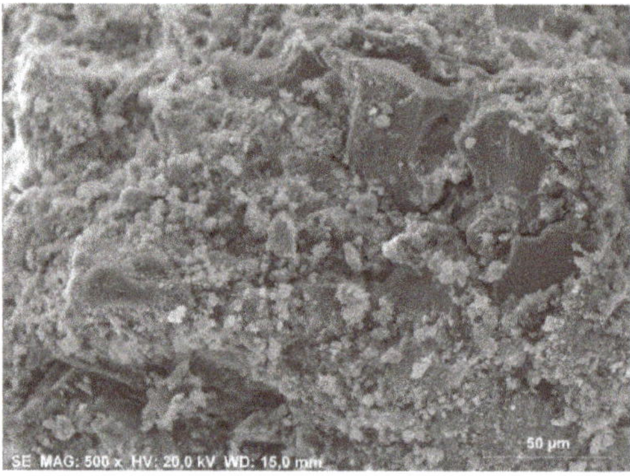

Figure 11. Microscopic image of the reference mixture (MREF).

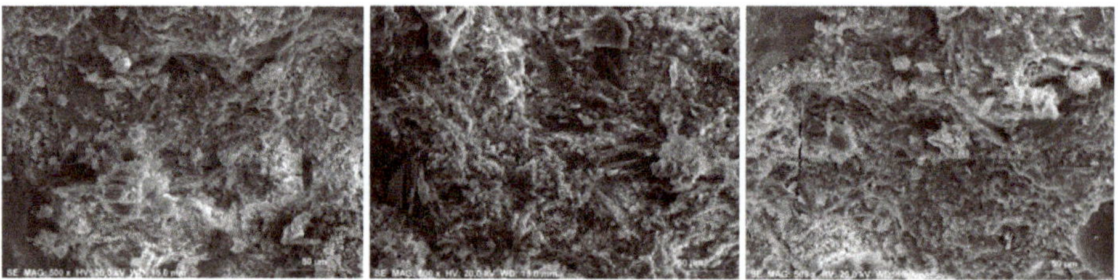

Figure 12. Microscopic image of mixtures containing fibre waste: mortar with rock wool waste (MRLR) (left), mortar with fibreglass waste (MRFV) (centre), mortar with a combination of mineral wool waste (MMIX) (right).

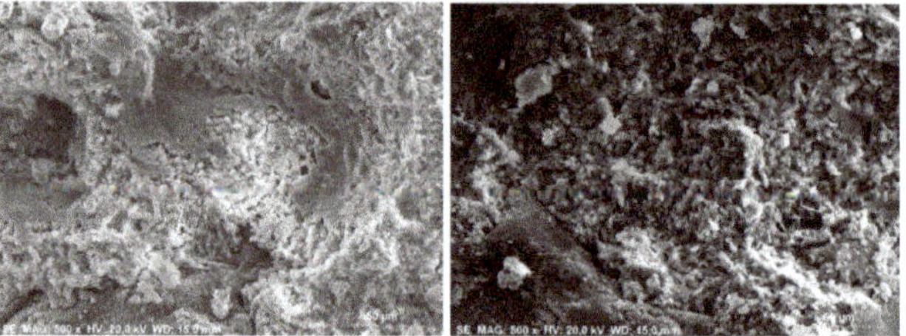

Figure 13. Microscopic image of mixtures containing commercial fibres: mortar with polypropylene fibre (MPP) (left), mortar with fibreglass (MFV) (right).

It should be noted that ready-made mortars do not have a high percentage of fibres and that these quantities are not enough to produce significant changes in the microscopic structure of the composite material [20].

Therefore, these results indicate that the presence of fibres does not affect the setting process of the mortar, where the mortar mass sets around the fibres without any additional or significant porosity occurring in the pore size range covered by these experimental techniques. In other words, the fibres are "surrounded" by the set mass, indicating good adhesion between the mortar and the fibres, which is also proven by good mechanical behaviour.

5. Conclusions

As a general conclusion, it can be confirmed that it is possible to replace in coatings the fibres usually used by fibres from recycling.

As specific conclusions:

All the mortars studied show that fibres from recycling do not affect the setting of the mortar, nor do they produce significant changes in the microscopic structure of the compound, as their behaviour is very similar to that of the mortars reinforced with commercial fibres.

In addition, and because the amount of fibres added is not very large, the density of the mortars reinforced with fibres from recycling is similar to the density without fibres or reinforced with commercial fibres. These can be considered lightweight mortars, since their density is lower than 1300 Kg/m^3 according to UNE EN 1015-6 [35].

Regarding consistency, it should be noted that mortars reinforced with recycled fibres have consistencies similar to those reinforced with commercial fibres, such as plastic mortars, and are within the range established in the standard mentioned in the previous paragraph.

These mortars are found to be viable for exterior coatings as, similar to all of those with recycled fibres, they exceed the minimum values of compressive strength established in standard UNE-EN 1015-11:2000/A1:2007 [25] for mortars R1 and/or R3, 3.5–7.5 N/mm2 and/or $\geq$ 6 N/mm2 and with water absorption values that are typical of mortars designated as R1, medium resistance to filtration. They can be suitable for application as exterior coatings even on walls exposed to rain and moisture. At this point, it should be noted that mortar reinforced with commercial fibres complies with the provisions of the standard for mortars (R3), i.e., mortars with very high resistance to filtration would be obtained.

Finally, it has been found that the type of fibre does not affect the water vapour permeability of mortars, since the permeability values obtained in all reinforced types are very similar to those of the reference mortar. However, shrinkage of mortars is lower in those reinforced with commercial fibres, which is probably due to the greater length of these, so it would be interesting to work with longer recycled fibres to confirm that their length plays a decisive role in the reduction of shrinkage.

Author Contributions: M.d.R.M. conceived the original idea and supervised the project. C.P.R. and A.V.B. planned the experiments and carried out the experimental Plan. J.G.M. and P.d.S.S. contributed to the interpretation of the results and wrote the manuscript. All authors have read and agreed to the published version of the manuscript.

Funding: This research received no external funding.

Conflicts of Interest: The authors declare no conflict of interest.

References

1. Liu, T.; Song, W.; Zou, D.; Li, L. Dynamic mechanical analysis of cement mortar prepared with recycled cathode ray tube (CRT) glass as fine aggregate. *J. Clean. Prod.* **2018**, *174*, 1436–1443. [CrossRef]
2. Salah, M. The dilemma of developing architectural minds aiming for sustainability. In Proceedings of the International Conference for Academic Disciplines Al Ain University of Science and Technology (AAU), and the American-Based International Journal of Arts and Sciences (IJAS), Al Ain, United Arab Emirates, 31 January 2016.
3. Resource Use. UKGBC's Vision for a Sustainable Built Environment Is One That Eliminates Waste and Maximises Resource Efficiency. Available online: https://www.ukgbc.org/resource-use/ (accessed on 14 August 2019).
4. Cerrillo, A. España Modera el Uso de Recursos Naturales. *La Vanguardia*. Available online: https://www.lavanguardia.com/medio-ambiente/20120219/54256331741/-espana-modera-uso-recursos-naturales.html (accessed on 19 February 2012).
5. Ling, T.C.; Poon, C.S. A comparative study on the feasible use of recycled beverage and CRT funnel glass as fine aggregate in cement mortar. *J. Clean. Prod.* **2012**, *29*, 46–52. [CrossRef]
6. Martínez, P.S.; Cortina, M.G.; Martínez, F.F.; Sánchez, A.R. Comparative study of three types of fine recycled aggregates from construction and demolition waste (CDW), and their use in masonry mortar fabrication. *J. Clean. Prod.* **2016**, *118*, 162–169. [CrossRef]

7.	Xuan, W.; Chen, X.; Yang, G.; Dai, F.; Chen, Y. Impact behavior and microstructure of cement mortar incorporating waste carpet fibers after exposure to high temperatures. *J. Clean. Prod.* **2018**, *187*, 222–236. [CrossRef]

8.	Butera, S.; Christensen, T.H.; Astrup, T.F. Life Cycle Assessment of construction and demolition waste management. *Waste Manag.* **2015**, *44*, 196–205. [CrossRef]

9.	Sanjuán-Barbudo, M.A. Cemento y hormigón en la economía circular. *Cem. Hormig.* **2016**, *976*, 6–16.

10.	Barrios, Á. Impactos Ambientales en Construcción Sostenible. Construcción Sostenible. 2012. Available online: https://www.eoi.es/wiki/index.php/Construcción_sostenible (accessed on 30 May 2019).

11.	Mercader Moyano, P.; Ramírez de Arellano Agudo, A. Selective classification and quantification model of C&D waste from material resources consumed in residential building construction. *Waste Manag. Res.* **2013**, *31*, 458–474.

12.	Aciu, C.; Ilutiu-Varvara, D.A.; Manea, D.L.; Orban, Y.A.; Babota, F. Recycling of plastic waste materials in the composition of ecological mortars. *Procedia Manuf.* **2018**, *22*, 274–279. [CrossRef]

13.	Ramírez, C.P.; Sánchez, E.A.; del Río Merino, M.; Arrebola, C.V.; Barriguete, A.V. Feasibility of the Use of Mineral Wool Fibres Recovered from CDW for the Reinforcement of Conglomerates by Study of Their Porosity. *Constr. Build. Mater.* **2018**, *191*, 460–468. [CrossRef]

14.	Morales Conde, M.J.; Rodriguez Liñan, C.; Pedreño Rojas, M.A. Physical and mechanical properties of wood-gypsum composites from demolition material in rehabilitation works. *Constr. Build. Mater.* **2016**, *114*, 6–14. [CrossRef]

15.	Gadea, J.; Rodríguez Saiz, A.; Campos, P.L.; Garabito, J.; Calderón, V. Lightweight mortar made with recycled polyurethane foam. *Cem. Concr. Compos.* **2010**, *32*, 672–677. [CrossRef]

16.	Muñoz-Ruiperez, C.; Rodríguez Saiz, A.; Gutiérrez-González, S.; Calderón, V. Lightweight masonry mortars made with expanded clay and recycled aggregates. *Constr. Build. Mater.* **2016**, *118*, 139–145. [CrossRef]

17.	Sandin, K. Mortars for Masonry and Rendering, Choice and Application. *Build. Issues* **1995**, *7*.

18.	Junco, C.; Gadea, J.; Rodríguez, A.; Gutiérrez-González, S.; Calderón, V. Durability of lightweight masonry mortars made with white recycled polyurethane foam. *Cem. Concr. Compos.* **2012**, *34*, 1174–1179. [CrossRef]

19.	Cheng, A.; Lin, W.T.; Huang, R. Application of Rock Wool Waste in Cement-Based Composites. *Mater. Des.* **2011**, *32*, 636–642. [CrossRef]

20.	Lin, W.T. Improved Microstructure of Cement-Based Composites through the Addition of Rock Wool Particles. *Mater. Charact.* **2013**, *84*, 1–9. [CrossRef]

21.	Wang, R.; Meyer, C. Performance of Cement Mortar Made with Recycled High Impact Polystyrene. *Cem. Concr. Compos.* **2012**, *34*, 975–981. [CrossRef]

22.	*UNE-EN 998-1:2010, Specification for Mortar for Masonry—Part 1: Rendering and Plastering Mortar*; Spanish Association for Standardization: Madrid, Spain, 2010.

23.	*UNE-EN 13914-1:2019, Design, Preparation and Application of External Rendering and Internal Plastering—Part 1: External Rendering*; Spanish Association for Standardization: Madrid, Spain, 2019.

24.	*UNE-EN 13914-2:2019, Design, Preparation and Application of External Rendering and Internal Plastering—Part 2: Design Considerations and Essential Principles for Internal Plastering*; Spanish Association for Standardization: Madrid, Spain, 2019.

25.	*UNE-EN 1015-11:2000/A1:2007, Methods of Test for Mortar for Masonry—Part 11: Determination of Flexural and Compressive Strength of Hardened Mortar*; Spanish Association for Standardization: Madrid, Spain, 2000.

26.	*UNE-EN 1015-18:2003, Methods of Test for Mortar for Masonry—Part 18: Determination of Water Absorption Coefficient due to Capillary Action of Hardened Mortar*; Spanish Association for Standardization: Madrid, Spain, 2003.

27.	Veiga, M.R.; Velosa, A.L.; Magalhães, A.C. Evaluation of mechanical compatibility of renders to apply on old walls based on a restrained shrinkage test. *Mater. Struct.* **2007**, *40*, 1115–1126. [CrossRef]

28.	Lepech, M.; Li, V.C. Water permeability of cracked cementitious composi-tes. In Proceedings of the 11th International Conference on Fracture 2005 (ICF11), Turin, Italy, 20–25 March 2005; pp. 1464–1469.

29.	Lange, D.A.; Jennings, H.M.; Shah, S.P. Relationship between Fracture Surface Roughness and Fracture Behavior of Cement Paste and Mortar. *J. Am. Ceram. Soc.* **1993**, *76*, 589–597. [CrossRef]

30.	del Olmo Rodríguez, C. Los morteros, Control de calidad. *Inf. Constr.* **1994**, *46*, 57–73. [CrossRef]

31.	Glória Gomes, M.; Flores-Colen, I.; Manga, L.M.; Soares, A.; de Brito, J. The influence of moisture content on the thermal conductivity of external thermal mortars. *Constr. Build. Mater.* **2017**, *135*, 279–286. [CrossRef]

32.	Menéndez, I.; Triviño, F.; Hernández, F. Correlaciones entre plasticidad, resistencias mecánicas, relaciones agua/cemento y proporciones de filler calizo en morteros de cemento. *Mater. Constr.* **1993**, *43*, 39–52. [CrossRef]

33. Araya-Letelier, G.; Antico, F.C.; Carrasco, M.; Rojas, P.; García-Herrera, C.M. Effectiveness of new natural fibers on damage-mechanical performance of mortar. *Constr. Build. Mater.* **2017**, *152*, 672–682. [CrossRef]

34. *UNE-EN 13279-2:2014, Gypsum Binders and Gypsum Plasters—Part 2: Test Methods*; Spanish Association for Standardization: Madrid, Spain, 2014.

35. *UNE EN 1015-6:1999, Methods of Test for Mortar for Masonry—Part 6: Determination of Bulk Density of Fresh Mortar*; Spanish Association for Standardization: Madrid, Spain, 1999.

36. *UNE-EN 1015-3:2000, Methods of Test for Mortar for Masonry—Part 3: Determination of Consistence of Fresh Mortar (by Flow Table)*; Spanish Association for Standardization: Madrid, Spain, 2000.

37. *UNE-EN 1015-19:1999, Methods of Test for Mortar for Masonry—Part 19: Determination of Water Vapour Permeability of Hardened Rendering and Plastering Mortars*; Spanish Association for Standardization: Madrid, Spain, 1999.

38. *UNE 80112:2016, Test Methods of Cements. Physical Analysis. Determination of Contraction in Air and Expansion in Water*; Spanish Association for Standardization: Madrid, Spain, 2016.

39. *UNE-EN 1015-7: 1999, Methods of test for mortar for masonry—part 7: Determination of air content of fresh mortar*; Spanish Association for Standardization: Madrid, Spain, 1999.

40. Vidales Barriguete, A. Caracterización Fisicoquímica y Aplicaciones de yeso con Adición de Residuo Plástico de Cables Mediante Criterios de Economía Circular. Ph.D. Thesis, Universidad Politécnica de Madrid, Madrid, Spain, 2019.

41. Puertas, F.; Gil-Maroto, A.; Palacios, M.; Amat, T. Morteros de Escoria Activada Alcalinamente Reforzados Con Fibra de Vidrio AR, Comportamiento y Propiedades. *Mater. Constr.* **2006**, *56*, 79–90. [CrossRef]

42. Qin, X.C.; Li, X.M.; Cai, X.P. The Applicability of Alkaline-Resistant Glass Fiber in Cement Mortar of Road Pavement: Corrosion Mechanism and Performance Analysis. *Int. J. Pavement Res. Technol.* **2017**, *10*, 536–544.

43. Henkensiefken, R.; Castro, J.; Bentz, D.; Nantung, T.; Weiss, J. Water Absorption in Internally Cured Mortar Made with Water-Fi Lled Lightweight Aggregate. *Cem. Concr. Res.* **2009**, *39*, 883–892. [CrossRef]

44. Arizzi, A.; Cultrone, G.L. Influencia de La Interfase Árido-Matriz (ITZ) En Las Propiedades de Morteros de Cal. *Rev. Soc. Esp. Minerol.* **2012**, *16*, 60–61.

45. Puertas, F.; Amat, T.; Vázquez, T. Comportamiento de Morteros de Cementos Alcalinos Reforzados Con Fibras Acrílicas y de Polipropileno. *Mater. Constr.* **2000**, *50*, 69–84. [CrossRef]

46. Jennings, H.M.; Bullard, J.W. From Electrons to Infrastructure: Engineering Concrete from the Bottom Up. *Cem. Concr. Res.* **2011**, *41*, 727–735. [CrossRef]

Article

Evaluating Environmental Impact in Foundations and Structures through Disaggregated Models: Towards the Decarbonisation of the Construction Sector

Pilar Mercader-Moyano [1,*] **and Jesús Roldán-Porras** [2]

1   Department of Building Construction I, University of Seville, 41012 Seville, Spain
2   Department of Industrial and Civil Engineering, University of Cadiz, 11519 Cádiz, Spain; jesus.roldan@uca.es
*   Correspondence: pmm@us.es

Received: 26 May 2020; Accepted: 22 June 2020; Published: 24 June 2020

Abstract: Having a tool in Spanish regulations to evaluate the sustainability of the construction process in a simple and efficient way (Annex 13 of the Structural Concrete Instruction EHE-08) means an advance with respect to regulations in other countries. However, the complexity of homogenising the conditions that affect the execution of each structure, which are of a very heterogeneous and variable nature, in order to be able to evaluate their contribution to sustainability within the same reference framework, is the greatest obstacle and can have a great influence on the representativeness of the obtained results. However, there are variables that, given their specificity and nature, are not contemplated in this methodology (dust, noise and vibration emission, transportation). This paper proposes a complementary disaggregated model to evaluate the sustainability of variables that are not considered, namely the transportation of materials to the worksite, the commute of workers, the construction process, the emissions of dust, noise and vibrations, as well as the necessary load tests. The results of the application of this model to the real case of the foundations of two singular buildings, show the importance that these previously unexamined variables can have when choosing the most sustainable technical solution in terms of CO_2 emissions.

Keywords: CO_2 emissions; disaggregated model; dust; noise and vibration emission; environmental impact studies; foundations and structures; sustainability; transportation

---

1. Introduction

Understanding the concept of sustainability is the first step in the study and analysis of the current tools used to evaluate environmental impact at a very reduced scale, which includes structures and foundations in general, and in particular those developed in urban areas since construction sites constitute major sources of pollutants provoking negative impacts on the environment [1]. In comparative terms, sustainability is a broader concept than that of sustainable development, since it can be applied effectively to scales below the regional level, as well as to products, processes and services. Therefore, sustainability is applicable to different levels or objects, such as a certain sector, a specific activity, a product, a process, a group or an entity. In this context, Badi et al. conduct a thorough review on the sustainability agenda in the construction sector [2] while Ahmad researches empirical interactions among the construction sector, energy consumption, urbanisation and carbon emissions [3]. In fact, the building sector in general has shown increased interest in environmental initiatives [4].

A first approach to the quantifiability of sustainability can be obtained from the use of indicators. Note that environmental indicators are not strictly indicators of sustainability. They indicate the state and changes of the environment; but they must also indicate the state and changes of the human system in relation to the natural system [5]. Environmental indicators can be applied in a simple way to specific dimensions of sustainable development, such as economic sustainability, or in a complex manner, bringing together ecological, social and economic elements into the concept of comprehensive sustainability [6]. However, sustainability indicators have been designed to condense complex information into signals and are able to selectively support polarised sides of the same debate [7].

Among all the studied indicators, those that are considered to be applicable are worth highlighting, such as the Life Cycle Assessment (LCA) and the Ecological Footprint (EF) [8]. In general terms and classified by levels, depending on the country of origin, there currently are commercial tools based on the LCA that are specific to building (Athena Estimator, Catalog Construction CH, Metabase, OFEN, etc.), tools for the evaluation of materials and building solutions (Athena Estimator, LCAid, Legep 1.2, Lisa, Metabase, TCQ2000, etc.) and tools to certify the global sustainability of a building during its occupation stage (BREEAM, CASBEE, Enlace, GBTool, Green Globes, LEED, Verde, etc.) [9].

The LCA encompasses the full cycle of the product, process or activity, taking into account the stages of extraction and processing of raw materials, production, transportation and distribution, use, reuse and maintenance, recycling and its final disposal [10–12]. It is a design tool that researches and evaluates the environmental impact of a product or service during all of the stages of its existence: extraction, production, distribution, use and end of life (reuse, recycling, recovery and disposal of waste), making it particularly difficult to apply at such a small scale [13].

On the other hand, geotechnical engineering explores new ways of sustainable work for the coming years, based on the increasing importance of energy saving and carbon dioxide reduction, based on a better study of the behaviour of materials, analysis of the site and the development of the design process [14], such as energy foundations and other thermoactive structures, where the concrete elements in contact with the ground, which are necessary for structural reasons and work simultaneously as heat exchangers, contributing to the protection of the environment and providing significant long-term savings as a result of reduced maintenance costs [15]. Similarly, other emerging fields of research such as the consideration of the soil as a living ecosystem, show enormous potential for innovative and sustainable solutions to the different geotechnical problems that arise in the field of foundations and structures [16].

Regarding the problem of scale, some research studies carried out by the University of Seville establish methodologies that allow the evaluation of the ecological footprint applicable to the residential building sector at the regional level [8] and at an international level [17], focusing on the urban scale, without extrapolating their results to the scale of building construction.

The particularities of the construction process complicate the definition of the units of measurement of the indicators. Time becomes a fundamental factor and it conditions the hypotheses that must be taken into account at this level of scale. The time factor is understood here as the time for the execution and implementation that conditions the potential sources of CO_2 emissions (dust, noise, vibrations, greater or lesser need for labour and therefore transportation, materials, etc.). At this scale, the partial execution time of any technical solution is one of the aspects to be taken into account in the decision making process, and the effects on the environmental impact of these are almost never taken into account.

In the field of foundations, in general, and in particular when it comes to the improvement of the terrain, project designers should incorporate the goals of sustainable development in their designs focused on soil improvement and construction methods, through the quantitative evaluation of the environmental impact.

At this scale, where most of the structures and foundations of buildings in urban environments are designed and executed without any type of sustainability requirement, it is necessary for them to incorporate previous evaluations such as the one included in Instruction EHE-08. The application of these evaluations is not currently mandatory. The application of these evaluations is not currently obligatory. In the future and in future revisions of this Instruction, it should incorporate the mandatory nature of the sustainable evaluation of the implemented building solution, incorporating the variables that are currently not taken into account as established in this research, and establish control tools in the case of changes in the solutions planned and evaluated in the project phase.

Simplified models for the evaluation of emissions and energy allows the estimation of the incorporated energy during the life cycle and the CO_2 emissions during the design process, which allows us to take into account the effects that soil improvements have in the environment [18,19].

On the other hand, regarding structures and foundations in Spain, Annex 13 of the Spanish Structural Concrete Instruction (EHE-08) incorporates a methodology that allows us to the evaluate the so-called Index of Contribution of the Structure to Sustainability (ICSS, ICES in Spanish). This methodology corresponds to the Integrated Value Method for Sustainable Assessments (IVMSA, MIVES in Spanish), developed by a multidisciplinary group led by the Polytechnic University of Catalonia, LABEIN-TECNALIA and the University of the Basque Country. This index is obtained from various parameters pertaining to the three basic levels of sustainability: environmental, social and economic, and it determines the structure's contribution to sustainability. From an environmental point of view, we must look to the Environmental Sensitivity Index (ISMA in Spanish), which measures the decrease in the usage of natural resources and the emission of pollution, and energy saving and recycling, among other aspects. Regarding social and economic sustainability, workers' training and safety are taken into account, as well as the application of new research findings or the extension of the life cycle of the structure, among others [20]. At the urban planning level, the urban resilience index provides a socio-ecological perspective to favour more sustainable, resilient and liveable cities in the 21st century [21].

The tools for the quantification of the usage of resources and CO_2 emissions produced in construction [9], along with the current development of tools based on this same principle, provide possible pathways for the quantitative evaluation of the level of emissions, a resource that can be used by anyone involved in the construction process [20]. In this context, other research shows a method to calculate a sustainability index that helps the decision-making process by quantifying the environmental impact of all the different alternatives that are commonly chosen in the installation of underground service infrastructure [22]. As a matter of fact, infrastructure works produce a high environmental and social impact mainly due to the pollutant emissions from heavy equipment [23,24]. Moreover, weight indicators of materials and resources are commonly used by transportation infrastructure rating systems to quantify sustainability [25]. In this regard, Paschalidou et al. assess air quality in metro-railway construction works in Greece [26].

Other studies done at a smaller scale within the structural building process focus on the placement of concrete [27]. They allow us an estimated quantification of the energy consumption and CO_2 emissions that result from the preparation of concrete in the plant, the transport of fresh concrete in concrete mixers, pumping on site and vibrating. Said estimate is based on the technical information given by the manufacturers of the equipment; this allows us to evaluate the impact of concrete structures. On the other hand, Luo et al. compares precast and in situ cast piles from GHG emissions life cycle [28] whilst Li et al. explores on-site industrialisation to improve sustainability towards a cleaner form of production [29].

Due to the particularity of their features and special circumstances, the application of the methodology mentioned in Annex 13 of the EHE-08 to singular foundations in general, and in particular to the lowered foundation of the Cilíndrico and Casa del Coronel buildings during the construction of the new Faculties of Law and Labour Studies of the University of Seville in 2007 [30], allows us to identify certain aspects or variables that where not examined and that may be of special importance when it comes to determining the contribution of their structures or foundations to their sustainability; these variables range from the transportation of materials, the workers' commute, the construction process, temporary traffic diversions, to joint emissions of dust, noise and vibrations; the variables also include carrying out the mandatory load test, as is required by law in most cases [31].

This is a completely pioneering building solution due to the difficulty stemming from the interference with the works execution of the subway line.

This action is very representative in two aspects which are not contemplated in the ICSS, such as the temporary traffic diversions (which were carried out in a sustained manner over time, and as shown in the research, may have had special impact for or against the sustainability of the implemented solution) or the load tests to be carried out given the special characteristics of the adopted solutions (bridge-type foundation). The University of Seville was the promoter of the works developed in this paper and the author of said paper participated actively in their execution, as the project manager.

The disaggregation of the construction process is essential to establish tools that allow us to evaluate the contribution of structures and foundations to sustainability. In the aggregated models, as is the case of the IVMSA methodology, a single value is set for all the parameters of the model. Therefore, the model predicts outputs for the provided inputs without reporting what takes place within the system (obtaining a certain level of sustainability according to the method). However, the disaggregation of the construction process at the required scale (structure and foundations) allows us to research the specific conditioning factors of the construction process, along with its relationship with the environment in which it is located. Sandanayake et al. develop a methodology based on an analytical hierarchy process to weight materials, transportation and equipment carbon emissions in regard to five impact categories: Global Warming, Photochemical Oxidant Formation, Acidification and Eutrophication Potentials, as well as Human Toxicity [32].

Despite the fact that the methodology proposed in Annex 13 of the EHE-08 signifies an important advance in the incorporation of sustainability criteria at the project level and regarding concrete structures compared to other countries' regulations, some deficiencies have been detected, which should be addressed, as there is scarce research in this area. One of the few examples can be found in a Martinez et al., who propose a multi-objective genetic algorithm optimisation method to assess road and infrastructure construction processes minimising carbon emissions, embodied energy and economic costs [33].

We can find references to singular interventions in the foundations of buildings in cases where emergency situations took place, such as those produced by earthquakes, which have put evaluation protocols into place [34] or processes for decision-making in later works [35]. More research has been done when foundations affected by settlements have been in need of repair [36–39], some of which have meant a leap forward in the study of techniques, processes, solutions and results in real cases [40]. Aspects such as intervention on site, derived costs, or the restoration of buildings in situations of social emergency, have also not been extensively researched or at least not disseminated, given the reluctance to publish technical data that may be liable [40]. In general terms, this means that technicians who face situations of this type do not have references on which to base their decisions and that, in these scenarios of uncertainty, conservative and oversized solutions are preferred, for reasons both economic and technical. A notable exception can be found in Pineda et al., who performed an environmental and structural assessment of strengthening works in a Spanish church [41].

In this paper, the University of Seville aims to disseminate to the scientific community the research done during the construction works carried out in the practical case, just as it did with the foundation repairs in 40 homes in the "RENFE" neighbourhood, whose foundations were affected by the adjacent excavation of the new Faculty of Medicine. A standard model has been created by this research for its use in similar situations from the perspective of sustainability [42].

The main goal of this paper is to formulate a disaggregated model for the evaluation of the environmental impact of building solutions for structures and foundations (CDM-SABSSF), which allows for a global quantification of their sustainability, considering and consequently complementing any variables not examined in the methodology proposed by EHE-08. Neither the methodology proposed in the model that we reference in this paper nor the complementary one proposed in our research are based on LCA.

The application of the methodology proposed in the model will in turn pave the way to international regulations and laws pertaining to the design and building of concrete-made elements; among them, Eurocode 2, applicable in European Union member states, which does not include a section on sustainability criteria during the design phase. The ACI-318-14 standard in the US, the CIRSOC-201-2005 regulation in Argentina or the SP63.13330.2012 standard in Russia.

The proposed methodology is tested by applying it to a real case study, in order to contrast the obtained results if it were applied, considering the model proposed in this research.

The case study shows that the sustainability of a building solution can change depending on the indicators or analytic tools used.

2. Materials and Methods

The diversity of possible building solutions or methods for structures and foundations according to their type, operation and procedures for their implementation is one of the main reasons to propose and define a new model to complement the information obtained after determining the index (ICSS). This model, which complements the one established in Annex 13 of the aforementioned regulation, constitutes an independent tool for the objective assessment of the sustainability of construction solutions or methods in the structures and foundations phase, in order to guarantee a unique and homogeneous comparative framework within the reach of all the agents involved in the building process. For the development of the methodology brought forward by this paper, we need to develop an overview of the index (ICSS) and its scope, as well as any variables not considered in accordance to the required scale. This index focuses on the three approaches to sustainability (environmental, social and economic), with different scopes and criteria depending on the case. In the attached figure (Figure 1) we can see the scope of the methodology of the Integrated Value Model for Sustainable Assessment (IVMSA, MIVES in Spanish), on which Annex 13 of -08 is based, as well as the unrelated aspects that hold relevance in the case of single foundations.

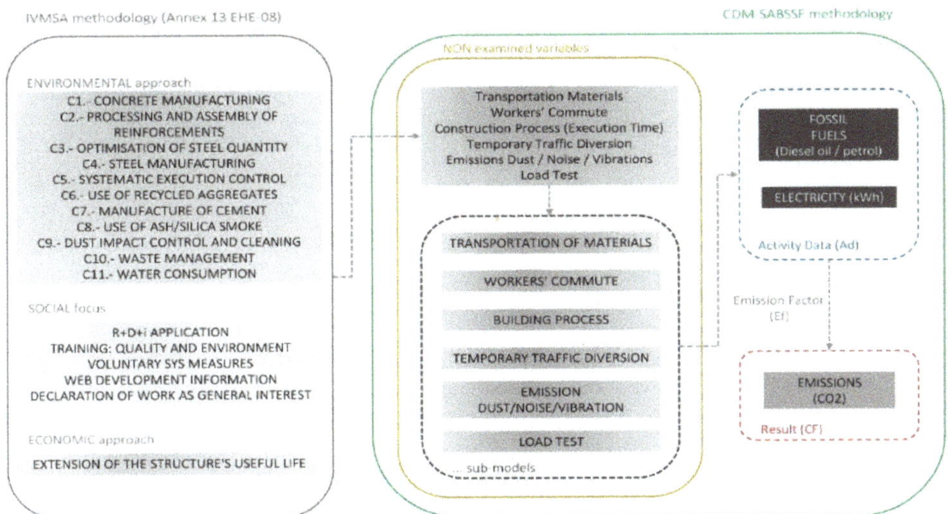

Figure 1. IVMSA methodology, Non Examined Variables. Methodological Basis CDM-SABSSF.

The aim of this paper is not to go into detail regarding the IVMSA methodology on which the ICSS (value added) is based. Also, the fact that IVMSA is included in the Spanish Instruction does not prevent it from being applied as a method for the evaluation of the sustainability of structures and foundations in other countries. In fact, the methodology proposed in this paper (CDM-SABSSF) complements the decision making process derived from obtaining the ICSS, and can be applied in any structural area (national and international).

It is worth noting that with regards to the unidentified aspects and the required scale, the building activity and the construction process at this level in general present homogeneous patterns regarding energy consumption (electric and fossil fuels) and the use of manpower. There is a clear relationship between unrelated aspects and the use of energy resources. However, in some of them, as is the case of noise and vibrations, this relationship is much more difficult to establish.

The proposed model uses an indicator that has been sufficiently contrasted by the Spanish Climate Change Office [43] easy to use, and applicable to all agents involved at this scale. The carbon footprint, which will allows us to know the energy consumption levels that are predominant in this evaluation phase, identifying the amount of GHG emissions that are released into the atmosphere (expressed in CO_2eq) as a consequence of the development of said activity, the work execution of the foundation or structure of a building, using Formula (1); understanding by "manufactured product" in situ the manufacture of a product from elements and materials through a system that allows it to be built through the use of necessary auxiliary and human means with the required technical capacity.

$$\text{Carbon Footprint (CF)} = \text{Activity Data (Ad)} \times \text{Emission Factor (Ef)} \tag{1}$$

in which: Activity data (Ad) stands for the parameter that defines the degree or level of activity that generates GHG emissions. The emission factor (Ef), refers to the amount of GHG emitted by each unit of the parameter (Ad).

At this scale, we consider the data generated from the transportation of the factory, main suppliers of concrete, steel and hot-rolled steel, to carrying out the load test, including the auxiliary and personal means necessary to implement said activity according to their technical attributes and design.

The proposed model and its scope, named as Disaggregated Model for the Sustainable Evaluation of Solutions in the Structures and Foundations phase (CDM-SABSSF in Spanish), requires the development of different sub-models according to the nature of each of the "non examined variables" (by annex 13 of the EHE-08) which constitute the focus of this research, based on the disaggregation of the building process. It allows us to establish and accurately discern the specific singularities of each of these variables that remained unexamined, in order to integrate the most appropriate variables and elements in each sub-model, which will constitute the methodological basis of the model that is developed below (Figure 1).

Based on this general approach, each sub-model must establish the mechanisms and tools to determine the representative activity data (Ad) according to the nature of the unexamined variables, in order to obtain the level of CO_2 emissions from the corresponding conversion factors (Ef) as appropriate. The activity data that should be obtained, in accordance with the goals of the model, are based on the quantity of fossil fuels and electricity consumed during the work execution process according to the technical solution adopted in each of the scenarios considered.

Independently from the specific sub-model that would be considered along with its particularities, we can set the emission factors associated with fossil fuel consumption included in the above-mentioned guide as a reference tool [43], due to its simplicity and easy application by the agents intervening in the construction process.

With respect to emissions associated with energy consumption, the emissions data provided by the electric company during the time period corresponding to the study, or the data provided by the Ministry of Agriculture, Fisheries and Food [44] should be applied.

Regarding the practical applications of this model, we have considered the period of works execution of the mentioned works (2007) and the data corresponding to the electric company (ENDESA ENERGÍA, S.A.) and to the joint venture of the construction companies (FACULTADES UTE) that led the works execution during the construction of the foundations that we are analysing. Shown below are the emission factors considered with respect to the usage of fossil fuel (Table 1) and electricity usage (Table 2), respectively.

Table 1. Fossil Fuel Emission Factors.

Fuel	Emission Factor (Ef)	
Petrol	2.196	kg CO_2/L
Diesel A	2.471	kg CO_2/L
Diesel C	2.786	kg CO_2/L
Generic LPG	1.656	kg CO_2/L
Gas	0.202	kg CO_2/kWh
Butane Gas	2.964	kg CO_2/kg
Propane	2.938	kg CO_2/kg
Fuel Oil	3.054	kg CO_2/kg
Coal (National)	2.300	kg CO_2/kg
Coal (Imported)	2.530	kg CO_2/kg
Petroleum Coke	3.195	kg CO_2/kg

Table 2. Energy Sources (ENDESA ENERGÍA, S.A) and Emission Factors for Electricity (2007).

Source	Endesa Energia, S.A
Renewable (Pure + Hybrid)	27.70%
High Efficiency Cogeneration	2.00%
Cogeneration	6.20%
Coal	22.50%
Fuel/Gas	3.20%
Nuclear	16.80%
Others	0.90%
Ef CO_2 (kg CO_2/kWh)	0.37

3. Development of the Disaggregated Model

Based on the proposal of the general model, we must establish the different sub-models for each of the "non examined variables" in order to obtain the activity data needed to determine the carbon footprint, after applying the corresponding emission factors.

3.1. Transportation of Materials

The materials that pertain to this specific phase (foundations and structures) are mainly fresh concrete, corrugated steel and formwork material. In order to determine the fuel consumption associated with the transportation of the necessary materials, we take into account the data provided by the Ministry of Development [45], which determines the average fuel consumption calculated for the most common types of vehicles used to transport goods (Table 3), which allows us to evaluate the CO_2 emissions produced during the transportation of materials from factories to the building site.

The sub-model proposed to determine the volume of emissions derived from the transportation of materials is based on the following expression (2):

$$E_{TM} = \sum_{i=1}^{n} \left[\left(\frac{W_i}{LC_i} \right) \cdot D_i \cdot C_i \right] \cdot Ef_i \tag{2}$$

where W_i represents the weight of the material to be transported, LC_i the load capacity of the type of vehicle considered, D_i the distance between the factory or distribution centre and the building site, C_i the energy consumption associated with the type of fuel used and Ef_i the emissions that corresponds to each of these.

Table 3. Average Fuel Consumption f or Vehicles Transporting Goods.

Specialty	Vehicle Type Analysed	Consumption [1] (L/100 km)
General Cargo	Articulated Vehicle	36.5
	3-axle vehicle	28.0
	2-axle vehicle	26.0
Chillers [2]	Articulated Vehicle	36.5
	3-axle vehicle	26.0
Tanks transporting Hazardous Goods: Gas/Chemicals/Food/Powder	Articulated Vehicle	36.0
Carrier	Road Train	42. 0
Industrial Carriers	Road Train	48.0
Road Train	2-axle vehicle + 3-axle Trailer	40.0
Container/Bulk	Articulated Vehicle	36.5
Works	Articulated Vehicle	40.0
Livestock Transportation	3-axle vehicle	28.0
	2-axle vehicle	26.0
Concrete mixers	3-axle vehicle	64.0
	4-axle vehicle	66.0
Intermediate Vehicle Van–	2-axle vehicle	20.0
2 Axles Van	2-axle vehicle	12.0

[1] Referring to fuel consumption of Type A diesel; [2] Not including fuel consumption for the cooling unit = 3–4 L/h (additional).

3.2. Workers' Commute

In order to objectively determine the CO_2 emissions derived from the displacement and commute of workers during the execution of the foundations and structures, due to the specificity of the work and the sectorial grouping of its workers from the geographical area, we establish the sub-model based on the knowledge of the subcontracting companies that develop them; as well as their execution time. The real data obtained will be determined by the type of vehicle, fuel used and the number of occupants, based on the following Formula (3):

$$E_{DT} = \sum_{i=1}^{n} [T_i \cdot D_i \cdot C_i] \cdot Ef_i \qquad (3)$$

where T_i represents the number of trips necessary for the commute of the workers, D_i the distance between the workers' place of residence, C_i the energy consumption depending on the type of fuel used and Ef_i the corresponding factor of emission.

The electricity consumption by workers during breaks for breakfast, lunch and activities carried out after the end of the working day (lights, use of microwaves, hot water heaters for showers, etc.) is not taken into account, due to the heterogeneous nature of these factors and their lack of repercussions.

3.3. Construction Process

Any building solution implies the use of human resources (workers), a series of additional means and specific techniques that determine the runtime of a building unit. The consumption of fossil fuels by the human resources (site personnel and workers) has been taken into account in the previous sub-model.

The differential element of these "non examined variables" is the use of additional machinery associated with a specific construction method, which will involve a level of consumption of fossil fuels and electricity energy depending on the hours of use of the machinery (tower cranes, mobile cranes, small, manual machinery, etc.). The data on actual consumption of this additional machinery is determined by the manufacturer or by the Association of Infrastructure Construction and Concession Companies (SEOPAN in Spanish) [46], which allows us to determine the main and secondary energy consumption depending on the type of fuel and the power allocated to the additional machinery, as shown in the table below (Table 4).

Table 4. Main and Secondary Energy Consumption: Construction Machinery (Source: SEOPAN, 2008 [46]).

Main Energy Consumption		Secondary Energy Consumption
Diesel oil (L/h per kW)	0.15–0.20	15% of the main energy consumption
Petrol (L/h per kW)	0.30–0.40	8% of the main energy consumption
Electricity (kWh per kW)	0.60–0.70	5% of the main energy consumption

The following formula represents the proposed sub-model to determine the volume of emissions associated with the construction process (4):

$$E_{PC} = \sum_{i=1}^{n} [(T_i \cdot D_i \cdot C_i) + (H_i \cdot C_i)] \cdot Ef_i + \sum_{j=1}^{m} (H_j \cdot C_j) \cdot Ef_j \tag{4}$$

where, in the case of fossil fuels, T_i represents the number of trips necessary for the return trip of the machinery in question to and from the construction site, D_i the distance to transport the machinery from the company's facilities, C_i the energy consumption associated with the type of fuel used, H_i the hours of usage of the machinery and Ef_j the emission factor corresponding to the consumption of fossil fuels, while H_j represents the hours of use of machinery running on electricity, C_j represents their electricity consumption and Ef_j represents the emission factor corresponding to electric energy consumption.

3.4. Temporary Traffic Diversions

Provisional traffic diversions in civil engineering projects related to road transportation infrastructure are in themselves one of the most important variables to be taken into consideration during the execution phase of the works. However, this is not the case for construction works in urban environments. Approaches aimed at assessing the environmental impact that these traffic diversions may have in urban environments are rare, and said measures do not necessarily result in a negative impact on CO_2 emissions.

A provisional traffic diversion that requires less commute time will result in a lower consumption of fossil fuels by regular users of the affected route, constituting a reduction in the usual levels of CO_2 emissions from an environmental point of view, as has been shown in the period of confinement caused by the COVID-19 pandemic, in which cities have reached the lowest level of CO_2 emissions in 50 years.

In addition to the length of the route, it is essential to assess the traffic levels in the routes affected by these diversions in order to determine average fuel consumption in accordance to the vehicles that are taken into account.

Average Daily Traffic (ADT), defined as the total number of vehicles that have passed through a section of a road during a given year divided by 365 days, is the most commonly used index to assess traffic volumes on urban and interurban roads. The data can usually be obtained from local traffic agencies. Once the affected traffic has been assessed and the time period had been estimated, the greater or lesser level of emissions derived from the consumption of fossil fuels can be established, during the period in which the traffic diversions exist, according to the following Formula (5):

$$E_{DPT} = \sum_{i=1}^{n} [L \cdot N_i \cdot T_i \cdot C_i] \cdot Ef_i \tag{5}$$

where L stands for the length travelled by the vehicles affected by the traffic diversion (and which may be greater or less than the originally intended length), N_i the number and type of vehicle affected (obtained from a previous assessment of the traffic) T_i the effective duration of the provisional diversion, C_i the energy consumption associated with the type of fuel used and Ef_i the corresponding emission factor.

3.5. Dust, Noise and Vibrations Emissions

Any specific activity of the scale mentioned in this research involves a joint emission of dust, noise and vibrations to the immediate environment. The need for tools that allow us to evaluate this impact in a global, simplified way and within the reach of all the agents intervening in the process, has prompted us to propose a model that is somewhat ambitious in order to be able to determine an order of magnitude that is sufficiently representative and that in turn allows us to establish a comparison in homogeneous terms with the results obtained in the evaluation of other non examined variables. In other words, it is very difficult to establish a single quantitative variable that allows us to determine, with a sufficient degree of representativeness, the effects produced by the joint emissions of dust, noise and vibration.

With regards to the joint effect that is produced by certain construction works (emission of dust, noise and vibrations) it should be noted that, among others, the typical reaction of people who carry out their activity in nearby buildings (such as homes, work or education centres, etc.) consists of "closing the windows" as a response to dusty and noisy environments (or even vibrations). Such a common action, which is the go-to response of any citizen in this type of situation, can lead to considerable variations in the patterns of use and operation of air conditioning systems at certain times of the year. These variations in the energy consumption patterns of air conditioning and heating systems is linked to a variation in the consumption of electricity and by extension, in the level of CO_2 emissions associated with it. In short, there is a more direct relationship between dust and noise emissions (and to a lesser extent, vibrations) and CO_2 emissions.

The proposed sub-model is based on the main variable involved in the effects generated by the emission of dust, noise and vibrations, which is related to the distance or proximity of the buildings near the area where the construction works are taking place. In general terms, the hypothesis that we establish explains that the levels of dust, noise and vibrations that are created affects the first proximity radius (first level of buildings surrounding the location of the construction works), since the following rings are protected by the first level, which acts as a protective barrier against said emissions.

Conversely, once the number and composition of the buildings that make up the proximity ring has been determined, it is necessary to assess them based on the available data (type of building, type of housing and commercial premises, percentage of occupancy and use, opening hours, nature of the indirect services, etc.) in order to be able to determine the percentage of air conditioning equipment to be considered in the calculation. The assessment of air conditioning demand can be extracted from data provided by the Institute for Energy Diversification and Saving (IDAE in Spanish) [47], which establishes the values of reference to be taken into account with regards to the demand for heating and air conditioning by surface area in different Spanish cities.

Another central aspect of the sub-model is undoubtedly the runtime of execution of the works and the time of year to be considered, which has a direct proportional effect on the summer season (increased need for air conditioning), and to a lesser extent in autumn and spring depending on the area under consideration. For this reason, it is necessary to consider a coefficient of effect that takes into account the time of execution of the works in the different seasons of the year. The sub-model proposed to determine the volume of emissions associated with these variables corresponds to the formula below (6):

$$E_{PRV} = \sum_{i=1}^{n} [DA_i \cdot IC_i \cdot T_i \cdot CL_i \cdot CEA_i] \cdot Ef_i \tag{6}$$

where DA_i stands for the affected demand or air conditioning and refrigeration needs corresponding to each household or property affected (primary ring), IC_i the percentage of increase in energy consumption considered over the affected demand associated with the action of "closing windows" and therefore to be considered as net needs for the overall calculation of emissions, T_i the works execution period, CL_i the length coefficient to calculate the greater or lesser proximity of the property to the centre of the emissions, CEA_i the specific activity coefficient that takes into account the specifics of the activity carried out in the property in question and Ef_i corresponds to the emission factor.

3.6. Load Test

The sub-model proposed to evaluate the volume of CO_2 emissions derived from the necessary activities that must be carried out during the mandatory load test depends to a great extent on the type of tests to be performed and may be very different if we consider load tests on foundations or on structures at slab level, where the needs and the means to be used are radically different depending on the case (buckets with water, extraction of cores in the case of structural rehabilitation and reforms, non-destructive tests, etc.).

In this regard, we have focused on the specific conditions for a unique foundation, such as the one in the case study mentioned in this paper, a shored foundation, where the load test consists of carrying out four static load assumptions by using 4-axle trucks loaded with a maximum total weight, including tare, of 40 T and carrying out a dynamic load test consisting of the transit of one of the trucks on a standardised board in order to induce a vibration in the foundation and thus determine the fundamental frequency of vibration of the foundation. The proposed sub-model, which would also be valid for the types of load tests carried out in the case of road bridges, allows us to determine the volume of emissions according to the following expression (7):

$$E_{PPC} = \sum_{i=1}^{n} [(T_i{\cdot}D_i{\cdot}C_i) + (H_i{\cdot}C_i)]{\cdot}Ef_i \tag{7}$$

where, in the case of fossil fuels, V_i represents the number of trips necessary for the return journey of the vehicles and machinery used, D_i the transport distance between the premises of the company that owns it and the building site, VT_i represents the number of trips needed for carrying out the test, DT_i the internal distance travelled during the execution of the test, C_i the energy consumption associated with the type of fuel used, H_i the hours of usage and Ef_i the emission factor corresponding to the consumption of fossil fuels.

4. Applications of the Case Study

In this section, we proceed to apply the proposed model to a real case study. A specific case that had a great social and economic impact: the works carried out on the foundations of the Edificio Cilíndrico and the Casa del Coronel, during the execution of the works for the new Faculties of Law and Labour Studies of the University of Seville, that took place between 2005 and 2008 in said city.

The case study was selected because it represents a unique opportunity to obtain research knowledge and to serve as an example for other similar cases that may arise in the future when faced with a similar real case that may take place.

This is a totally pioneering solution where two entire buildings needed to be completely shored due to the interference with the execution of the construction works in the subway line, instead of being demolished for the subway line to run through that space.

During the construction of both buildings, Line 1 of the Seville subway system was also under construction. Many problems arose as the alignment of the subway's layout caused a diversion of the original line, which meant that the space allocated for the foundations of the two university buildings was invaded. Ferrocarriles Andaluces, which belongs to the Department of Public Works and Transport of the regional government (Junta de Andalucía), acting as the promoter of the works on Line 1 of the

Seville Metro subway system, decided that the line should pass right through emblematic buildings during their execution of works and rehabilitation phase; one of the authors of the current research participated in this process as head of production of the joint venture in charge of the execution of the works (FACULTADES UTE, in which FERROVIAL-AGROMAN, S.A. has a majority shareholding).

The building solution that was applied was a clear case of knowledge transfer to the production sector, and was resolved by the execution of an innovative foundation supported by these buildings to make it compatible with the transport infrastructure, so Line 1 of the Seville METRO subway system could cross right below these buildings.

The proposed model aims to show the possible improvements to be considered in similar cases from an environmental point of view, in order to seek initiatives that push for the decarbonisation of the construction sector, as a plan of action against Climate Change.

The building solution that was carried out to lay the foundations of both buildings over the subway tunnel was a pioneering solution in Spain, given the size of the buildings affected. It was an engineering challenge of great magnitude and it was based on the use of 35 pre-stressed double "T" beams with edges of varying length depending on the areas (2.00 m in the area of the Edificio Cilíndrico and 1.70 m in the area of the Casa del Coronel), arranged at the top, with an upper 60 cm slab of reinforced concrete acting as the compressed head of the set. The goal of this solution was to completely pile up the foundations of both buildings and preserve the subway's tunnel's original route (as seen in Figure 2).

Figure 2. Execution of the works on the Cilíndrico and Casa del Coronel buildings over the Seville subway line. Source: The authors.

The first step is based on a detailed study of the solution, based on the use of prefabricated double "T" beams (VDT solution), with special attention to the load test due to its special features and unique nature. The complexity of completely overhanging two buildings to preserve the METRO tunnel created the need to carry out a load test that would guarantee the adequate response of the structure under a load of this level.

With this in mind, we developed a finite element model of the VDT solution in order to be able to compare it with the aforementioned applied model (the one included in the load test project) and with the actual response of the structure during the test (according to the results included in the final reports issued by the laboratory that carried out the project). Figure 3 shows the finite element model made for the foundation, using the ANSYS program.

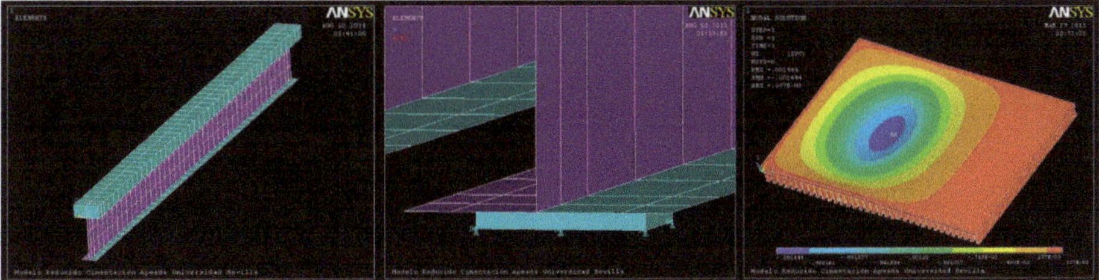

Figure 3. Foundation Model (ANSYS) – Simulation of a Static Load Assumption of the works on the Cilíndrico and Casa del Coronel buildings over the Sevilla subway line.

The figure, showed below represents the modelling of the double "T" beam, the upper slab and the support equipment (neoprene), as well as the simulation of one of the static load assumptions performed.

Regarding the load test, it is worth stressing that given the features of the loads in question (axle loads from the pillars), it was necessary to carry out four static load assumptions, combining the number and arrangement of fully loaded 4-axle trucks (total weight of 40 T) together with a dynamic assumption consisting of driving one of these trucks on a plank (standard RILEM-type), thus causing an excitation of the foundation board.

The instrumentation that was carried out consisted of the arrangement of different extensometers to measure the descents at the beam's span centres during the performance of the different static assumptions, as well as the arrangement of two accelerometers to record the vertical accelerations during the dynamic load assumption. In the case of the static load assumptions, represented in the following figures (Figure 4), the decreases measured at the span centre of the beams can be compared, together with the results of the model included in the load test project, and the model that was performed in this paper for the VDT solution, which is the one that was actually implemented.

The results clearly show that the developed model is much closer to the real response of the structure (greater convergence of results when comparing the decreases in the real case and those obtained in the ANSYS model), and where the response of the model included in the load test project is much less rigid than the real case. The divergence observed in static load assumption No. 1 (Figure 4a) is especially noteworthy.

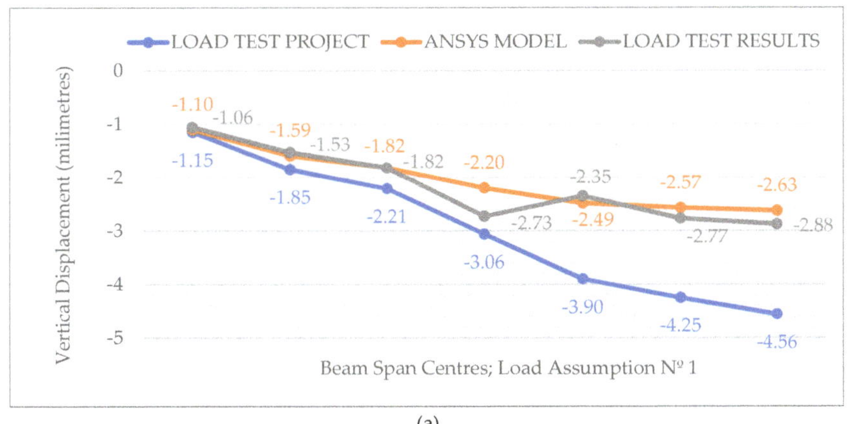

(a)

Figure 4. *Cont.*

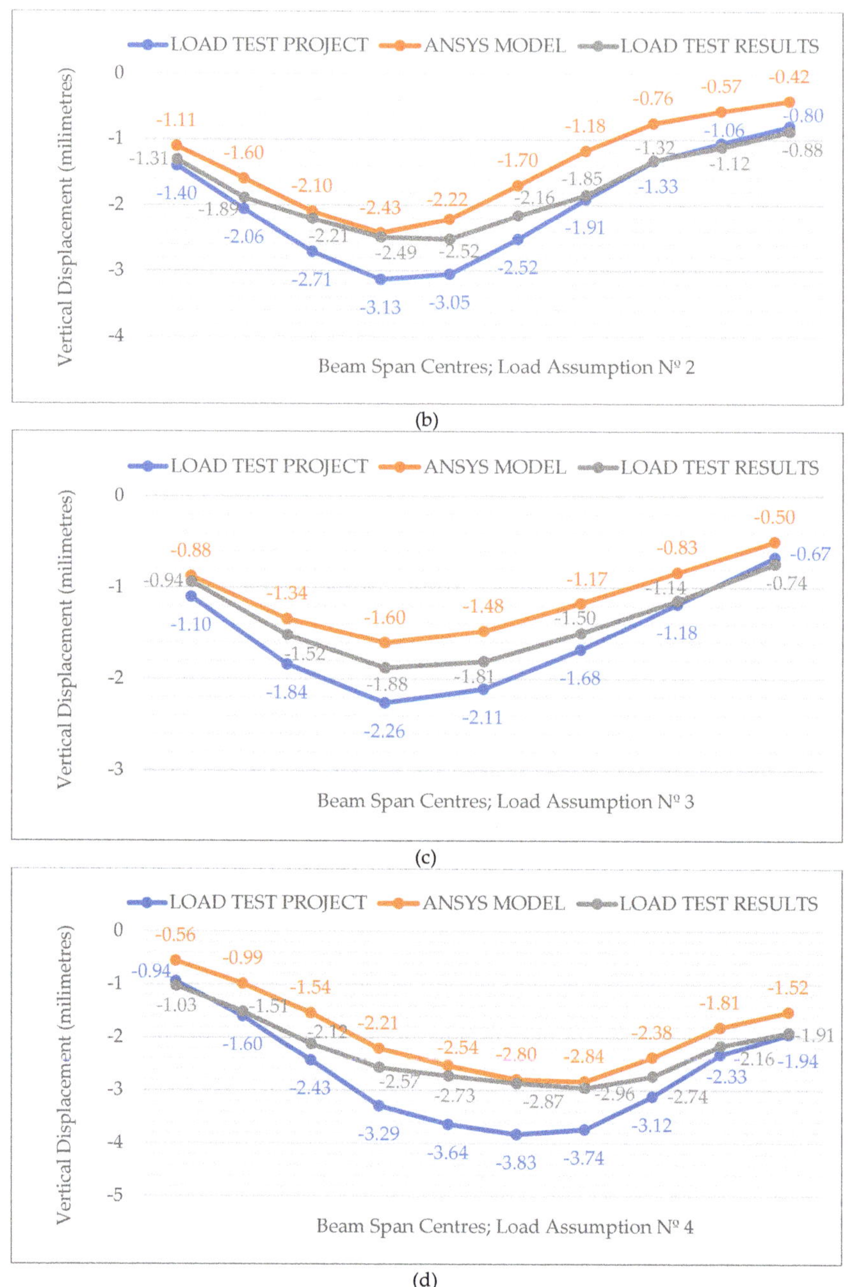

(b)

(c)

(d)

Figure 4. Vertical Displacements in Beam Span Centres; Load Assumptions Nº1, Nº2; Nº3 and Nº4. (**a**) static load test hypothesis Nº1; (**b**) static load test hypothesis Nº2; (**c**) static load test hypothesis Nº3; (**d**) static load test hypothesis Nº4.

Secondly, a study is carried out to explore other technically possible alternatives to the implemented building solution (VDT). Among the examined alternatives, three different solutions are selected and developed: firstly, two solutions based on the use of tamped beams; and a third one based on the use of post-tensioned concrete (HPO Solution). Regarding the solutions based on the use of tamped beams, the first to be considered is a solution with 1.50 metre-high beams and a 60 cm upper slab (VAR1 solution), and secondly another solution with 1.70 metre-high beams and a 30 cm upper slab (VAR2 solution) is considered.

Once the various alternative solutions have been established, the needed resources (machinery, materials and labour) are evaluated for each case. Subsequently, the methodology described in EHE (Annex 13) is applied to the four proposed scenarios. The Structure Contribution to Sustainability Index (ICSS) is obtained for each of the solutions studied (VAR1, VAR2 and HPO), obtaining similar results (Table 5) corresponding to the same sustainability interval (Level "D"). It should be noted that the methodology applied establishes five levels of sustainability, from Level "A" (more sustainable) to Level "E" (less sustainable), with regards to the structure's contribution to sustainability.

Table 5. Compared Index of Contribution of the Structure to Sustainability (ICSS) index for the foundation solutions considered.

Solution	ICSS	Level
VDT	0.37	Level D
VAR1	0.34	Level D
VAR2	0.35	Level D
HPO	0.24	Level D

As seen in the table, completely different solutions in terms of the type, operation and procedures for their implementation have obtained the same level of contribution to sustainability. This gives us reasons to define a model that allows for the complementing of the information obtained after determining the aforementioned index (ICSS), which should influence those variables not contemplated in the applied methodology, which could be constituted as an independent tool for the objective evaluation of the sustainability of a certain solution in the structures and foundations building phase.

Once the sub-models have been defined, corresponding to each of the non examined variables, the model is applied to the solution that was implemented in the case study: prestressed double "T" beams (VDT) as well as for the technically viable alternatives studied, which are VAR1, VAR2 and HPO.

In order to apply the described methodology and to calculate the level of emissions, we have taken into account the data from supplier companies and subcontractors who carried out the works with the VDT solution, hypothesising for rest of the solutions that the supplier companies and subcontractors are the same.

Table A1 of Appendix A includes the general summary of the estimated CO_2 emissions after the application of the sub-model corresponding to each of the variables not examined in the proposed disaggregated model. A summary of this table is shown in Table 13, with the global volume of emissions calculated for each of the technical solutions considered in the case study.

4.1. Transportation of Materials

Regarding the transportation of materials, and according to the volumes required in each of the analysed solutions, the transportation of fresh concrete from the manufacturing plant to the building site, the transportation of rebar (in 12 m bars) from the distribution centres, and the formwork systems from the storage facilities specialised in this type of material have been considered. Specifically, and depending on the case, the transportation of the prefabricated beams (VDT, VAR1 and VAR2) from the plant to the building site has been taken into account, and, in a similar way, in the case of the solution based on the use of post-tensioned concrete (HPO), the transportation of the active reinforcement from the same distribution centres as that of the passive reinforcement has been considered. The results obtained are shown in the attached table (Table 6).

Table 6. Calculation of CO_2 Emissions: Transportation of Materials (E_{TM}).

Transportation of Materials	VDT	VAR1	VAR2	HPO
Transportation of Concrete	4.549	6.334	4.491	6.334
Transportation of Prefabricated Beams	1.163	6.286	6.286	-
Transportation of B500S Steel	0.025	0.034	0.017	0.067
Transportation of Formwork	0.038	0.077	0.077	7.162
[T CO_2]	5.775	12.730	10.871	13.563

4.2. Workers' Transportation And Commute

Depending on the type of solution and building works to be carried out, the preparation and assembly of rebar, the formwork and pouring of concrete, the collection, preparation and placement of prefabricated beams and, in the case of the load assumption with post-tensioned concrete, the assembly work of post-tensioned reinforcement (laying of sheaths, routing and pulling of cables, placement of wedges and clamps, as well as stressing with a hydraulic multi-screw jack); we have determined the works for the execution period and the number of operators needed.

In accordance with the grouping of each subcontractor company and its different geographical locations, different hypotheses have been established to determine the degree of occupation and type of fuel used in the vehicles necessary for each solution studied; to determine the movements of the workers from their places of residence or accommodation to the work site.

The results obtained after the application of the corresponding sub-model are shown below (Table 7) and as can be seen, the lowest volume of CO_2 emissions corresponds to the VAR2 solution.

Table 7. CO_2 Emissions: Workers' commute (E_{DT}).

Workers' Commute	VDT	VAR1	VAR2	HPO
Gasoline Vehicles	0.385	0.355	0.337	0.296
Petrol Vehicles	0.423	0.391	0.370	0.938
[T CO_2]	0.808	0.746	0.707	1.234

4.3. Building process

The first thing to be taken into account were the permanent auxiliary means that were already in the building site, namely the tower cranes to lift loads and move materials. To determine the level of emissions associated with their use, the variables considered where the time spent unloading materials (passive and active reinforcement and formwork) and the time corresponding to the operations performed, depending on the type of solution considered, during the execution of the works.

The electric energy consumption has been estimated from the total usage time. The average duration of the usual manoeuvre has been estimated at 15 min, increased by 5% (due to possible unforeseen events or special difficulties in loading and/or unloading). Bearing in mind that other works are carried out during the building process which are not strictly related to the foundation, a usage coefficient of 70% of the total time has been considered in the case of the assembly process of passive reinforcement and formwork in solutions based on the use of pre-stressed prefabricated beams (VDT, VAR1 and VAR2) and a coefficient of 80% in post-tensioned concrete (HPO) works including the assembly of passive reinforcement and formwork as a result of its greater complexity in terms of geometry and functionality.

Secondly, regarding the load test assumptions based on the use of large-sized prefabricated beams (VDT, VAR1 and VAR2), the operations corresponding to the lifting and placement manoeuvre have been taken into account. In this sense, in the double "T" pre-stressed beams (VDT) solution, a 400-tonne crane was used on the capping slab of the subway tunnel, while the use of two 200-tonne cranes has been considered for the trough-beams of the VAR1 and VAR2 solutions, due to their dimensions and weight.

As for the consumption of fossil fuels (diesel) associated with the use of this type of large cranes, it has been estimated that for road travel, average consumption (by manufacturer, type of engine and speed on the road) is 1.5 L/km for the 400-tonne crane and 1.0 L/km for the 200-tonne crane. The fuel consumption associated with the movement of the cranes from the facilities of the subcontractors that carried out the work has also been considered, as well as that of the internal transfers during the manoeuvre, based on the data provided by the different manufacturers and suppliers, adopting a value of 5 L/h of consumption during the operation.

Thirdly, within the construction process, the placement of concrete has been taken into account. According to the current available data and taking into account the distances from the point where the pump is placed and the upper slab of the foundation, the use of 42-metre range pumps has been considered to pour the concrete, with a power of 400 CV (294 kW) and an average pumped concrete output of 80 m^3/h (according to the times for the changes of concrete and vibrated trucks), according to the data provided by the Spanish Society of Concrete Pumps (SEBHSA) [48].

To calculate the consumption of the pumps during concrete pumping, taking into account a power of 294 kW for a 42 m pump, a diesel consumption of 0.15 L/h per kW has been considered, that is, a consumption of 44.1 L/h. A percentage increase of 5% has also been considered (to account for possible divergence) due to possible unforeseen events during the pouring of the concrete. Similarly, fuel consumption has been determined as a result of the movement of the pumps from their base to the building site.

A differential aspect in contrast with other types of works is the on-site workshop. In general, it is becoming less common for iron to be produced in on-site facilities where the works are taking place, since its manufacture is usually sourced out to workshops when the works exceed a certain size. This way, the shaping and use of steel is optimised, the processing costs are reduced and so is the transport of the processed elements to the building site.

In this particular case study, the production workshop was located on site, so we have calculated the CO_2 emissions generated during the production of the necessary passive reinforcements in situ, for each of the hypotheses put forward. They are the following: bending machine, cutting machine and stirrup bar, whose data are obtained from the general attributes of the machines (as seen in the documentation provided by manufacturers and usual suppliers of companies dedicated to the production and assembly of corrugated steel in the region of Andalusia) [49].

Regarding the commission of the works, and with a lower comparative value, there is the use of concrete vibrators during the pouring of the concrete. In this case, we have considered the use of common single-phase universal electric needle vibrators with an average power of 2.3 kW which reach vibrating performances of up to 35 m^3/h, according to the technical data sheets of the different models.

Finally, regarding this specific construction process following the HPO solution compared to solutions using pre-stressed concrete beams (VDT, VAR1 and VAR2). The construction process requires the following machinery in general: laying of sheaths and trumpets, stringing of the strings, stressing of the tendon, cutting of the ends and sealing of boxes, and finally, the reinforcement of the foundation.

According to all these phases, we can summarise that the specific machinery needed during the execution of the specialised post-tensioning works are: cable straightening machine, hydraulic jack with hydraulic power unit (hydraulic pump with electric motor) for its operation, hand radial-type cutter and injection pump. In accordance with average characteristics for the execution of post-tensioned concrete in buildings, the following machines (all of them with three-phase current) have been considered based on the documentation provided by various manufacturers Hydraulic Cable Trimmer 0,5″–0,6″ (8 kW), Jack and Hydraulic (10 kW), Continuous Flow Injection Pump 12.5 L/min-8 bar (8 kW) and Radial Type Cutter (2.2 kW).

Applying the corresponding sub-model to calculate of emissions associated with the construction process, we can see how the solution with the lowest volume of emissions corresponds to the solution that was implemented in the actual case study: VDT (Table 8).

Table 8. CO_2 Emissions Calculation: Building Process (E_{PC}).

CONCRETE LAYING	VDT	VAR1	VAR2	HPO
Tower Crane				
Downloads / Transfers	0.095	0.124	0.095	0.666
Execution of works	0.998	1,198	0.998	3.728
Mobile Crane				
Displacement	1.483	1.977	1.977	-
Placement of Premanufactured Beams	1.297	2.076	2.076	-
Concrete Self-Pump				
Displacement	0.016	0.032	0.016	0.032
Execution of works	1.090	1.526	1.090	1.526
Iron Workshop				
Cutting machine	0.038	0.048	0.029	0.125
Bending machine	0.173	0.213	0.133	0.280
Stirrupper	0.008	0.008	0.004	0.033
Concrete Compression				
Concrete Vibrators	0.020	0.029	0.020	0.029
Post-tensioned concrete				
Stretcher	-	-	-	0.169
Jack	-	-	-	0.318
Injection pump	-	-	-	0.213
Cutter	-	-	-	0.028
[T CO_2]	5.219	7.230	6.438	7.145

4.4. Temporary Traffic Diversions

In order to determine the volume of traffic affected during the execution of the works, we took into account the traffic data corresponding to the year in which said works were carried out (2007) based on the data published in the city of Seville's urban traffic site [50], where the route used has an Annual Daily Traffic (ADT) of 12,974 vehicles/day for the route affected by the provisional detours put into place for the duration of the works. Traffic was dissected as follows: motorcycles (4.6%), cars including taxis (90.7%), trucks (1.8%) and buses (2.9%), with different hypotheses being established as to the type of fuel used by each type of vehicle.

The results of the implementation of the corresponding sub-model prove that VAR2 is the solution with the lowest volume of emissions (Table 9), followed very closely by VDT and VAR1, and with the HPO solution standing out negatively as a result of the longer runtime of the works.

Table 9. Calculation of CO_2 Emissions (T): Temporary Traffic Diversions.

Traffic Diversions	VDT	VAR1	VAR2	HPO
Motorbikes	0.228	0.232	0.212	0.374
Cars and Taxis	13.068	13.293	12.166	21.404
Trucks	1.127	1.146	1.049	1.846
Buses	2.263	2.302	2.107	3.707
[T CO_2]	16.686	16.973	15.535	27.330

It is worth noting that in the hypothetical case that the traffic diversion implied a reduction in the usual travel time of the vehicles in comparison to the original route, the runtime of the construction works would have played a fundamental role, and the importance of the solutions would have been reversed. If we had considered a reduced journey length equal to the one actually considered (hypothetically), the HPO solution would be the most efficient regarding the volume of emissions avoided, constituting the best solution compared to the rest.

4.5. Dust, Noise and Vibration Emissions

Firstly, the buildings that make up the immediate proximity ring have been assessed as potential receptors of the combined emissions of dust, noise and vibrations, which will act as a "screen" against the adjacent rings. Based on this assessment, we have first identified the buildings intended for residential use, taking into account the available cadastral data, along with surface area as well as the year of construction. Similarly, and with some consultation tools complementary to those used in the case of residential use, we have proceeded to assess the singular buildings of the first ring of affection, also specifying their address, as well as their main use.

Once an adequate assessment of the buildings that makes up the primary ring of potential impact or affection has been performed, the main variable to be taken into account according to the established hypotheses is the distance between the buildings and the point of the construction works (the source of the dust, noise and vibration emissions). To this end, we have used the catalogue of metadata provided by the Spatial Data Infrastructure (SDI) [51] to determine the distance between the geometric centres of the assessed buildings and the geometric centre of the propped-up foundation under study. Thus, we have considered a trapezoid-type weight distribution, in order for the minimum distance to be weighted with a percentage of effect of 100% and the maximum, with a degree of effect of 50%.

Following this, we determined the energy consumption attributed to air conditioning in each of the buildings based on the data provided by IDAE [38], which establishes the reference values for air conditioning, among others (heating and domestic hot water), which in the case of Seville amounts to 23.40 kW per m^2 per year. For the case in question (Seville), we have taken into account that the energy consumption of air conditioning systems occurs in the summer months and partially in late spring and early autumn. In operational terms, we have considered that this consumption occurs in the six months between May and October. Based on this distribution of consumption during "hot" months, we have established the "affected" demand, i.e., that part of the demand of energy during the period of execution of the works for each of the analysed building solutions.

Having determined the affected energy demand (energy consumption associated with the period of execution of the works) we have proposed a new variable to allow us to determine the increase in energy consumption caused by the need to switch on the air conditioning systems due to having closed the windows to avoid the effect of the aforementioned emissions, a need that would otherwise arise later in the day if it were not for these emissions. In our case, we have established an average daily increase of 4 h a day in the usage of air conditioning systems, and in days with an average use of 10 h, this represents an increase of 40% (the hypothesis raised is that the air conditioning systems are activated at 8:00 in the morning when in conditions of absence of dust, noise and vibrations, it would be activated on average at 12:00 noon, which is when the heat sensation at home and the workplace becomes uncomfortable.

In the case of the regarded areas and their types of activity and occupancy rates (homes, commercial spaces, hotel rooms, classrooms in learning centres and universities, premises in stations, etc.) different hypotheses have been established in accordance with the available data which affect the results obtained from different activity coefficients associated with the particularities of each case.

The only differential variable is the runtime of the building process. In the case of the study, the analysed solutions show different runtimes. However, if we consider the duration of the different hypotheses and the actual start date of the construction works, in every case the time period surpasses the following date: 31 October 2007, which is the deadline at which we have considered that energy consumption takes place, so the percentage of affected demand amounts to 32.61% (which corresponds to a proportion of 60 days, in all hypotheses, out of a total of 184 days corresponding to the six months considered from May to October: the hot months).

Therefore, when applying the proposed sub-model, the same level of emissions has been obtained for all the analysed solutions, as shown in the table below (Table 10).

As a hypothesis, we have obtained the possible result of the application of the sub-model in the case that the entire execution time of the works, for each analysed solution, was carried out within the yearly period affected by the higher demand for air conditioning (the hotter months). In this case, the period is the determining factor, and it can be seen that the solution with the lowest volume of emissions corresponds to the VAR2 solution, which is the solution with the shortest execution period (Table 11).

Table 10. Calculation of CO_2 Emissions: Dust, Noise and Vibrations.

Dust, Noise and Vibrations	VDT	VAR1	VAR2	HPO
Housing Buildings	29.913	29.913	29.913	29.913
Commercial Spaces	4.567	4.567	4.567	4.567
Singular Buildings	24.765	24.765	24.765	24.765
[T CO_2]	59.245	59.245	59.245	59.245

Table 11. Calculation of CO_2 Emissions (T): Dust, Noise and Vibrations. Hypothesis considering the deadline within the time most affected by the demand for Refrigeration.

Dust, Noise and Vibrations	VDT	VAR1	VAR2	HPO
Housing Buildings	38.887	39.884	36.394	63.814
Commercial Spaces	5.937	6.089	5.557	9.743
Singular Buildings	32.194	33.020	30.130	46.692
[T CO_2]	77.018	78.993	72.081	120.249

4.6. Load Test

With regard to the calculations performed, we have taken into account the movements of the trucks to the building sites for the load tests and the movement needed to perform the actual test (static and dynamic). Similarly, the data included in the results report of the loading test and an average speed of 50 km/h were considered for the calculation, in order to obtain an equivalent journey length that is accurately representative. Finally, the fuel consumption associated with the time used by a 75 kW backhoe loader during the loading of the trucks with soil until they reach the required weight for the different load tests has also been considered.

By applying the corresponding sub-model, it should be pointed out that the load test would have been the same for any of the technical building solutions proposed; therefore, the volume of emissions obtained is the same in all cases (Table 12).

Table 12. Calculation of CO_2 emissions: Load Test (E_{PPC}).

LOAD TEST	VDT	VAR1	VAR2	HPO
Transportation	0.816	0.816	0.816	0.816
Transit of Trucks (Load Test)	0.351	0.351	0.351	0.351
Loading of Trucks	0.001	0.001	0.001	0.001
[T CO_2]	1.168	1.168	1.168	1.168

As mentioned, IVMSA is the basis of the ICSS (value added), and this paper does not aim to take a deeper look at the sub-models associated with it. This paper focuses on the sub-models of the variables that are not taken into account in obtaining the ICSS. These are disaggregated models (the results are quantifiable in terms of CO_2 emissions, unlike the ICSS which gives us a level of contribution of the structure or foundation to sustainability, A, B, C, D or E).

5. Results

The obtained results are included in detail in Table A1 of Appendix A, which include the global volume of CO_2 emissions calculated for each of the building solutions considered in the case study, after the application of the sub-model corresponding to each of the non examined variables in the proposed disaggregated model. Table 13 shows its summary (in TCO_2): 88.9 for the implemented solution VDT, 98.092 for the alternative VAR1, 93.963 for the alternative VAR2 and 109.685 for the alternative HPO.

Table 13. Level of CO_2 Emissions. CO_2 total of non examined variables.

Results:	VDT	VAR1	VAR2	HPO
Transportation of Material	5.775	12.730	10.871	13.563
Worker's Commute	0.808	0.746	0.707	1.234
Laying of Concrete	5.219	7.230	6.438	7.145
Temporary Traffic Diversions	16.686	16.973	15.535	27.330
Dust, Noise and Vibration Emissions	59.245	59.245	59.245	59.245
Load Test	1.168	1.168	1.168	1.168
Total CO_2 Emissions (T)	88.900	98.092	93.963	109.685

Indeed, the significance of the research lies in the order of quantitative magnitude of the variables not contemplated in ICSS, not in the result of VDT, VAR1, VAR2 and HPO, hence the analysis of the representativeness of the obtained results. However, the results, according to the hypotheses raised, are the ones published. The discussion of results should emphasise the methodology on the one hand (already described) and the order of magnitude of the modelled aspects on the other hand.

6. Discussion

In view of the overall impact when considering all the non examined variables together, it appears that the actual solution implemented (VDT) has proved to be the most "sustainable", with a reduction in the volume of emissions of 9.4% compared to the solution based on the use of trough-beams of 1.50 m edge with 60 cm slab (VAR1), 5.4% compared to the solution also based on the use of trough-beams but 1.70 m edge with 30 cm slab (VAR2) and a considerable 18.9% in the case of the solution based on the use of post-tensioned concrete (HPO), as we can see in the following graph (Figure 5).

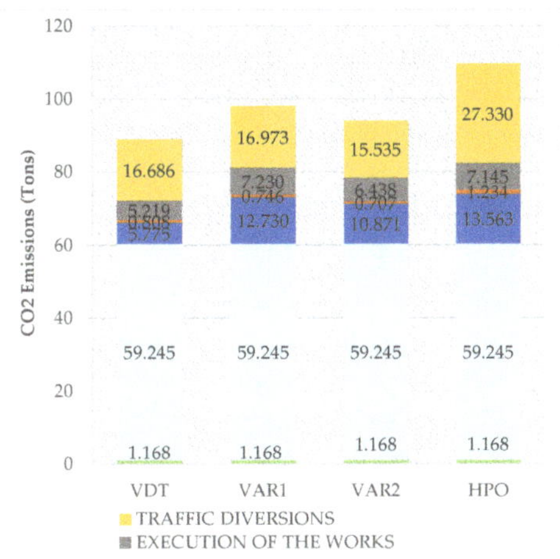

Figure 5. CO_2 Emissions Comparison: non examined variables.

In the case of the most sustainable solution, according to the results obtained (VDT), we can see that 67% of the emissions generated have their origin in the increases in electricity consumption associated with the combined emissions of dust, noise and vibrations, while the emissions derived from traffic diversions put into place in one of the lanes of Ramón y Cajal Avenue represent almost 20%

of the estimated total, with both aspects representing a volume of emissions close to 87% of the total. The rest of the emissions are distributed mainly between those derived from the transportation of material to the construction site (6%) and from the use of the necessary means for the laying of concrete (6%) as well as from the commute of workers and the required load test, in both cases, with percentages of around 1% of the total emissions, as shown in the following figure (Figure 6).

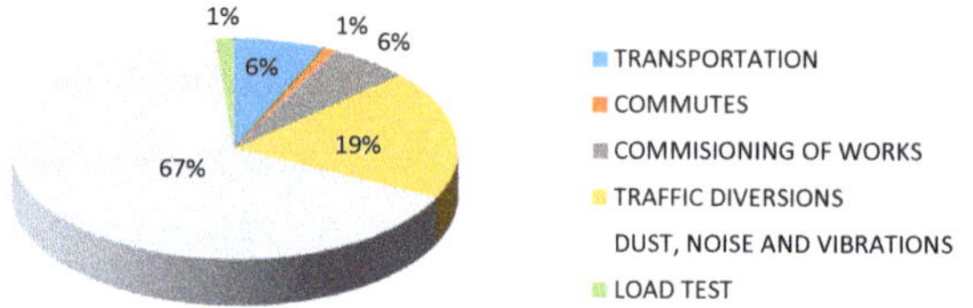

Figure 6. Origin of CO_2 Emissions in non examined variables. EHE-08 (VDT).

Similarly, it should be noted that out of all the variables not examined by EHE-08, the emissions of dust, noise and vibrations is the one that presents the highest uncertainty, unpredictability and difficulty when attempting to extrapolate it to other areas. If this variable where not taken into account, temporary traffic diversions would become the main source of CO_2 emissions (with a percentage of 56%), compared to the transportation of materials (19%) and the laying of concrete (18%), as shown in Figure 7.

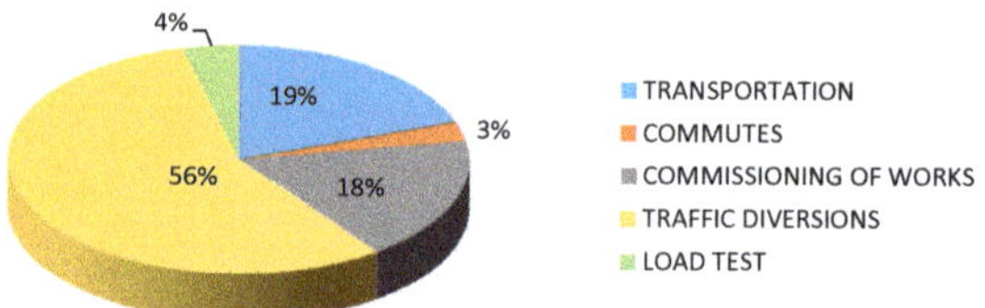

Figure 7. Origin of CO_2 Emissions in non examined variables. EHE-08 (VDT) without taking into account Dust, Noise and Vibration emissions.

In a similar way, temporary traffic diversions may or may not be necessary in construction works carried out in urban environments, as in fact is the case commonly. Regarding the building solution that was implemented in the case study (VDT), if the emissions of dust, noise and vibrations are not taken into account and traffic diversions are not necessary, we can see (as shown in Figure 8) that the main source of CO_2 emissions is the transportation of materials to the work site, with a percentage of 45%, compared to the implementation of the VDT building solution, which in the particular case of said solution, has a percentage of 40%. Similarly, the commute of workers from their places of residence or accommodation represents 6% of the overall emissions, while the load test raises its percentage to 9%, although the special circumstances surrounding its implementation in this type of unique foundation have already been mentioned.

In every work of research, highlighting its potential and its weaknesses becomes necessary. In this sense, regarding the limitations of the research, it is worth noting that they are closely linked to its level of scale (the execution of foundations and building structures, although they could be adapted, depending on the case, to structures and foundations of civil and/or industrial works), as well as to the starting methodology (the EHE Concrete Instruction), given that the CDM-SABSSF methodology focuses on the variables that are not explicitly considered by the aforementioned.

However, the greatest potential of this research is the evaluation of the sustainability levels of structures and foundations at this scale, regardless of other conditioning factors pertaining to the LCA of the building.

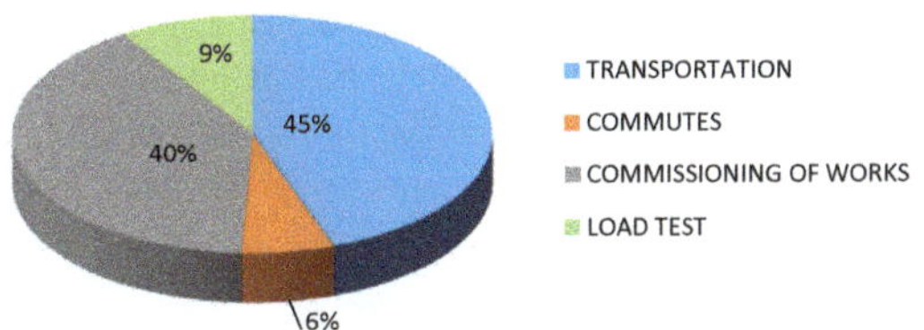

Figure 8. Origin of CO_2 Emissions in non examined variables. EHE-08 (VDT) without taking into account Dust, Noise and Vibration emissions and Temporary Traffic Diversions.

It is worth noting that most building projects are not subject to sustainability requirements (LEED type certification, or similar); however, a large percentage of these projects show relevant modifications at the level of structural and foundation solutions; in many cases, this is due to the technical and/or economic optimisation of the solutions included in the projects and in others to solve geological or geotechnical index problems that have not been taken into account for different reasons.

In both cases, the methodology proposed by the EHE (Concrete Instruction) in Spain, allows us to evaluate the sustainability levels of the different solutions and the proposed methodology (CDM-SABSSF), to take into account the variables that were previously unexamined, and that in certain situations can be differential when determining a building solution's contribution to sustainability at that scale (phase of execution of structures and foundations in projects without specific sustainability requirements).

From the performed analysis, when the proposed methodology is applied to the real case study, the most sustainable solution is the one established by ICSS. But this is a mere coincidence. In fact, in the sub-model with the greatest limitations, the same result is obtained by taking into account the real dates on which the works were carried out. If they had taken place on other (warmer) dates, the results would have been significantly different, taking into account the proportionality with the execution time of each of the solutions analysed. On the other hand, the load test was necessary in all the solutions analysed, but it would not have been necessary in a solution in situ that had not been tested, so that in quantitative terms we have obtained an order of magnitude of its representativeness in the decision-making process.

In conclusion, if certain conditions had changed, the most sustainable solution might have been a different one. In the case study, the choice was made without taking into account any requirements regarding sustainability. That the VDT solution was the most sustainable is a mere coincidence as shown in the research presented here, proving how the sustainability of a building solution can change in relation to the indicators or analytical tools used.

At this scale, this research is unprecedented in relation to certain variables (noise, dust, vibrations, temporary traffic diversions, load test, etc.), so it is not possible to mention a greater number of case studies, as this one is unique. Only a small amount of research has had an impact at this scale, on aspects such as the laying of concrete; but without taking into account, for instance, the transportation of workers to the building site [27]. It is to this very reason that this research owes its importance, to serve as a stepping stone for future cases of a similar kind. The selected case has allowed us to highlight and put into context the sustainability impact that the variables included in the proposed methodology (CDM-SABSSF) may have.

In relation to the sensitivity of the variables taken into account in the proposed methodology, it is worth highlighting:

- The most notable impact pertains to the influence that the generation of dust and noise can have in urban environments, and that in the case of the study, it does not have relative influence as the execution periods of the different alternatives are carried out outside the sensitivity period (according to the actual execution period of the works). This aspect is by far the most qualitatively relevant (potential for CO_2 emissions), and when it comes to the execution of building works in sensitive periods (hot weather) it can be decisive in relation to the decision making process. However, this sub-model is the one with the most limitations, according to the initial hypotheses put forward and the characterisation of the areas of influence, so this must be taken into account when extrapolating and adapting the methodology to other situations, geographical areas and countries.

- On the other hand, it is clear that the execution period is of vital importance in the case of temporary traffic diversions (in favour and against the sustainability of the solution) and yet its applicability is immediate and can be extrapolated to any situation based on the appropriate characterisation of the traffic affected.

- The construction process, and namely the transportation of materials and installation, depends mainly on the technical building solution put into place, and the proposed methodology allows for a quantification of the emissions generated to be obtained from a disaggregated model, so that its adaptation and applicability to other cases is immediate if the solutions are suitably characterised. These aspects are the most influential in the design and execution process and are the ones that are easiest for the agents involved to use to obtain reliable data.

- Finally, the transportation of workers and the execution of the load test are easily applicable to any situation, but the absolute results show a smaller overall impact in the contribution to the sustainability of the analysed building solutions.

7. Conclusions

The application of the methodology described in Annex 13 of EHE-08 to determine the Index of Contribution of the Structure to Sustainability (ICSS, ICES in Spanish), shows us how, in every case, the same level of "sustainability" (an equivalent ICSS index to level C) is obtained, both in the real case study and in the suggested alternatives. To evaluate the contribution of structures to sustainability, using the tools currently available seems like the sensible choice, as they provide the basis for a homogeneous comparison between the different agents involved in the construction process. However, as this research has shown, they must be complemented with other tools based on the disaggregation of the construction process, and the assessing of quantitative indicators (energy and fuel consumption, CO_2 emissions, etc.) so as to establish a suitable comparison in terms of sustainability, mainly in those cases where different solutions obtain similar sustainability rates (which means that a sufficiently representative comparison cannot be established). This is mainly due to the fact that some relevant variables have not been examined and accounted for. Among said variables that are worth noting: the transportation of materials, movement of workers, construction process or implementation, temporary traffic diversions, dust, noise and vibration emissions and the need to carry out a load test.

Based on this scenario, the main goal of this research has been developing a complementary disaggregated model for the evaluation of the sustainability of building solutions in the structure and foundation phase of the construction process (CDM-SABSSF). The reason why it was initially proposed as a complementary model is due to the aforementioned requirement to guarantee a single and homogeneous comparative framework within the reach of all the agents involved in the construction process, i.e., according to this premise, it is necessary to have the existing tools and to complement their use and the decision-making derived from their application with tools that allow us to take into account the variables that have otherwise been left out and unexamined.

On the other hand, based on the results obtained in the case study, and given the complexity of estimating the emissions derived from the overconsumption of energy associated with the emission of dust, noise and vibrations (increased use of air conditioning systems due to having the need to close the windows for longer than would be the case if the construction works had not taken place) as well as the need to carry out studies to contrast and calibrate the hypotheses put forward during the proposal of the corresponding sub-model, we have carried out a representativeness analysis, ignoring the results obtained with respect to the rest of the variables, and this has shown that the highest emission levels correspond to the temporary traffic diversions (varying from 56% in the VDT solution to 44% in the

VAR1 solution) followed by the transportation of materials (with percentages varying from 19% in the VDT solution to 33% in the VAR1 solution). In this analysis, the implementation shows less divergence between the different solutions, with levels of emissions varying from 14% for HPO solution to 19% for VAR2 solution. Finally, the joint percentage in relation to workers' commute and the load test shows some differences, varying between 7% of the VDT solution and 5% in the rest of the solutions analysed.

Similarly, and taking into account the exceptional nature of temporary traffic diversions (which may or may not occur at a given building site), we have performed a similar analysis, without considering the emissions for this aspect (as was done with the emissions derived from dust, noise and vibrations for the reasons already mentioned), and thus reaching important conclusions applicable to most actions at this level. From this analysis we can observe and conclude that, for the case study, the highest level of emissions corresponds to the transportation of materials to the worksite, with percentages of 45% in the case of the real case study (VDT) and percentages between 57% and 59% for the rest of the options considered. With regard to the works related to the laying of concrete, which is the second most important aspect, the percentages vary from 31% for the HPO solution to 40% for the VDT solution, while the combined percentage of emissions corresponding to the movement of workers and the load test vary from 9–10% for the VAR1, VAR2 and HPO solutions to 14% for the VDT solution.

Finally, in order to qualitatively scale the impact of the variables not examined by the currently available methodology and for which a disaggregated model has been developed in this paper, in relation to the level of overall emissions considering the steel and cement manufacturing process (as the main materials of reinforced concrete) we find that these variables can account for between 12% and 14% of the total volume of emissions depending on the particular solution considered in each case.

The results obtained in this research are a firm commitment to sustainability in foundations and building structures, taking into account that the unique cases are an opportunity in the design and execution phase to favour circular economy in all the construction systems and procedures that take place; the data we have obtained in this paper shows how foundation works executed without sustainability criteria could have been done if they had taken these variables into account, without modifying the required structural results and nevertheless contributing to mitigate climate change from the perspective of the inclusion of variables not hitherto examined at this level, with a high margin of action to reduce the consumption of material resources in the construction sector in particular, and in the field of civil engineering in general.

Author Contributions: Conceptualization, P.M.-M.; methodology, P.M.-M.; software, J.R.-P.; validation, J.R.-P.; formal analysis, J.R.-P. and P.M.-M.; investigation, J.R.-P. and P.M.-M.; resources, J.R.-P.; data curation, P.M.-M.; writing—original draft preparation, J.R.-P. and P.M.-M.; writing—review and editing, P.M.-M.; visualization, J.R.-P. and P.M.-M.; supervision, P.M.-M.; project administration, P.M.-M.; funding acquisition, P.M.-M. All authors have read and agreed to the published version of the manuscript.

Funding: The publication of this research paper is the result of the collaboration of the authors, carried out for the first phase of an Erasmus+ research project titled (CircularBIM) "Educational platform focused on advanced strategies of reinstatement of building materials in the industrial value chain to promote the transition to the circular economy through the use of BIM learning technologies" Code 2019-1-ES01-KA2013-065962, a project co-funded by the European Union and within the framework of an initiative of 2019 KA203, Strategic partnerships in the field of higher education), with the support of the "Servicio Español Para la Internacionalización de la Educación"(SEPIE, Spain). This paper would not have been possible without the results of the author's PhD dissertation and the knowledge acquired by the main author during her research fellowship at the High Technical School of Architecture of the University of Zaragoza.

Acknowledgments: To Antonio Ramírez-de-Arellano-Agudo for his invaluable help in this research, also to University of Seville and FERROVIAL AGROMAN S.A.

Appendix A

Table A1. Level of CO_2 Emissions. CO_2 total of non examined variables.

Transportation of Materials				
	VDT	**VAR1**	**VAR2**	**HPO**
Transportation of Concrete	4.549	6.334	4.491	6.334
Transportation of beams	1.163	6.286	6.286	0.000
Transportation of B500S Steel	0.025	0.034	0.017	0.067
Transportation of Formwork	0.038	0.077	0.077	7.162
[T CO_2]	5.775	12.730	10.871	13.563
Workers' commute				
	VDT	**VAR1**	**VAR2**	**HPO**
Petrol vehicles	0.385	0.355	0.337	0.296
Diesel Vehicles	0.423	0.391	0.370	0.938
[T CO_2]	0.808	0.746	0.707	1.234
Concrete Laying				
	VDT	**VAR1**	**VAR2**	**HPO**
Tower Crane				
Downloads / Transfers	0.095	0.124	0.095	0.666
Use	0.998	1.198	0.998	3.728
Mobile Crane				
Transportation	1.483	1.977	1.977	0.000
Placement of Prefabricated Beams	1.297	2.076	2.076	0.000
Concrete Pump				
Transportation	0.016	0.032	0.016	0.032
Use	1.090	1.526	1.090	1.526
Iron Workshop				
Cutting machine	0.038	0.048	0.029	0.125
Bending machine	0.173	0.213	0.133	0.280
Stirrupper	0.008	0.008	0.004	0.033
Concrete Compression				
Concrete Vibrator	0.020	0.029	0.020	0.029
Post-tensioned concrete				
Stretcher	0.000	0.000	0.000	0.169
Jack	0.000	0.000	0.000	0.318
Injection Pump	0.000	0.000	0.000	0.213
Cutter	0.000	0.000	0.000	0.028
[T CO_2]	5.219	7.230	6.438	7.145
Traffic Diversions				
		VDT	**VAR1**	**VAR2**
Motorbike	0.228	0.232	0.212	0.374
Cars & Taxis	13.068	13.293	12.166	21.404
Trucks	1.127	1.146	1.049	1.846
Buses	2.263	2.302	2.107	3.707
[T CO_2]	16.686	16.973	15.535	27.330
Dust, Noise and Vibrations				
	VDT	**VAR1**	**VAR2**	**HPO**
Housing Buildings	29.913	29.913	29.913	29.913
Commercial Spaces	4.567	4.567	4.567	4.567
Singular Buildings	24.765	24.765	24.765	24.765
[T CO_2]	59.245	59.245	59.245	59.245
Load Test				
	VDT	**VAR1**	**VAR2**	**HPO**
Transportation of Material	0.816	0.816	0.816	0.816
Transit of Trucks (Load Test)	0.351	0.351	0.351	0.351
Loading of Trucks	0.001	0.001	0.001	0.001
[T CO_2]	1.168	1.168	1.168	1.168
TOTAL [T CO_2]	88.900	98.092	93.963	109.685

References

1. Fuertes, A.; Casals, M.; Gangolells, M.; Forcada, N.; Macarulla, M.; Roca, X. An Environmental Impact Causal Model for improving the environmental performance of construction processes. *J. Clean. Prod.* **2013**, *52*, 425–437. [CrossRef]
2. Badi, S.; Murtagh, N. Green supply chain management in construction: A systematic literature review and future research agenda. *J. Clean. Prod.* **2019**, *223*, 312–322. [CrossRef]
3. Ahmad, M.; Zhao, Z.-Y.; Li, H. Revealing stylized empirical interactions among construction sector, urbanization, energy consumption, economic growth and CO_2 emissions in China. *Sci. Total Environ.* **2019**, *657*, 1085–1098. [CrossRef] [PubMed]
4. Zutshi, A.; Creed, A. An international review of environmental initiatives in the construction sector. *J. Clean. Prod.* **2015**, *98*, 92–106. [CrossRef]
5. Jiménez Herrera, L.M. *Desarrollo Sostenible—Transición Hacia la Coevolución Global*; Ediciones Pirámide: Madrid, Spain, 2000.
6. Fernández Latorre, F. *Indicadores de Sostenibilidad y Medioambiente: Métodos y Escala*; Consejería de Medio Ambiente: Junta de Andalucía, Sevilla, Spain, 2011. Available online: http://www.juntadeandalucia.es/medioambiente/web/Bloques_Tematicos/Sostenibilidad/Estrategia_andaluza_desarrollo_sostenible/libro_indicadores_sostenibilidad/3_cap_1_3.pdf (accessed on 23 June 2020).
7. Bell, S.; Morse, S. Sustainability Indicators Past and Present: What Next? *Sustainability* **2018**, *10*, 1688. [CrossRef]
8. Solís-Guzmán, J.; Rivero-Camacho, C.; Alba-Rodríguez, D.; Martínez-Rocamora, A. Carbon Footprint Estimation Tool for Residential Buildings for Non-Specialized Users: OERCO2 Project. *Sustainability* **2018**, *10*, 1359. [CrossRef]
9. Mercader, P. Cuantificación de los Recursos Consumidos y Emisiones de CO_2 Producidas en las Construcciones de Andalucía y sus Implicaciones en el Protocolo de Kioto. Ph.D. Thesis, Universidad de Sevilla, Seville, Spain, 2011. Available online: http://fondosdigitales.us.es/tesis/tesis/1256/cuantificacion-de-los-recursos-consumidos-y-emisiones-de-co2-producidas-en-las-construcciones-de-andalucia-y-sus-implicaciones-en-el-protocolo-de-kioto/ (accessed on 23 June 2020).
10. Sánchez, O.J.; Cardona, C.A. Análisis de ciclo de vida. *Rev. Univ. EAFIT* **2007**, *43*, 59–79.
11. Cabeza, L.F.; Rincón, L.; Vilariño, V.; Pérez, G.; Castell, A. Life cycle assessment (LCA) and life cycle energy analysis (LCEA) of buildings and the building sector: A review. *Renew. Sustain. Energy Rev.* **2014**, *29*, 394–416. [CrossRef]
12. Pilger, J.D.; Machado, Ê.L.; de Assis Lawisch-Rodriguez, A.; Zappe, A.L.; Rodriguez-Lopez, D.A. Environmental impacts and cost overrun derived from adjustments of a road construction project setting. *J. Clean. Prod.* **2020**, *256*, 120731. [CrossRef]
13. ISO. *Environmental Management: Life Cycle Assessment; Principles and Framework*; ISO: Geneva, Switzerland, 2006.
14. Simpson, B.; Tatsuoka, F. Geotechnics: The next 60 years. *Géotechnique* **2008**, *58*, 357–368. [CrossRef]
15. Brandl, H. Energy foundations and other thermo-active ground structures. *Géotechnique* **2006**, *56*, 81–122. [CrossRef]
16. DeJong, J.T.; Soga, K.; Kavazanjian, E.; Burns, S.; Van Paassen, L.A.; Al Qabany, A.; Chen, C.Y. Biogeochemical processes and geotechnical applications: Progress, opportunities and challenges. *Géotechnique* **2013**, *63*, 287–301. [CrossRef]
17. Camporeale, P.; Mercader-Moyano, P. Towards nearly Zero Energy Buildings: Shape optimization of typical housing typologies in Ibero-American temperate climate cities from a holistic perspective. *Sol. Energy* **2019**, *193*, 738–765. [CrossRef]
18. Shillaber, C.M.; Mitchell, J.K.; Dove, J.E. Energy and Carbon Assessment of Ground Improvement Works. I: Definitions and Background. *J. Geotech. Geoenviron. Eng.* **2016**, *142*. [CrossRef]
19. Shilaber, C.M.; Mitchell, J.K.; Dove, J.E. Energy and Carbon Assessment of Ground Improvement Works. II: Working Model and Example. *J. Geotech. Geoenviron. Eng.* **2016**, *142*. [CrossRef]
20. Ministerio de la Presidencia. REAL DECRETO 1247/2008, de 18 de Julio, por el que se Aprueba la Instrucción de Hormigón Estructural (EHE-08). *Boletín Oficial Estado* **2008**. Available online: http://www.fomento.es/nr/rdonlyres/e20dffb7-fd75-4803-8ca4-025064bb1c40/68186/1820103_2008.pdf (accessed on 23 June 2020).

21. Suárez, M.; Gómez-Baggethun, E.; Benayas, J.; Tilbury, D. Towards an Urban Resilience Index: A Case Study in 50 Spanish Cities. *Sustainability* **2016**, *8*, 774. [CrossRef]
22. Ariaratnam, S.T.; Piratla, K.; Cohen, A.; Olson, M. Quantification of Sustainability Index for Underground Utility Infrastructure Projects. *J. Constr. Eng. Manag.* **2013**, *139*, A4013002. [CrossRef]
23. Cianciarullo, M.I. Green construction—Reduction in environmental impact through alternative pipeline water crossing installation. *J. Clean. Prod.* **2019**, *223*, 1042–1049. [CrossRef]
24. Celauro, C.; Corriere, F.; Guerrieri, M.; Lo Casto, B.; Rizzo, A. Environmental analysis of different construction techniques and maintenance activities for a typical local road. *J. Clean. Prod.* **2017**, *142*, 3482–3489. [CrossRef]
25. Yang, S.-H.; Liu, J.Y.H.; Tran, N.H. Multi-Criteria Life Cycle Approach to Develop Weighting of Sustainability Indicators for Pavement. *Sustainability* **2018**, *10*, 2325. [CrossRef]
26. Paschalidou, A.K.; Kassomenos, P.A.; Kelessis, A. Tracking the association between metro-railway construction works and PM levels in an urban Mediterranean environment. *Sci. Total Environ.* **2016**, *568*, 1326–1332. [CrossRef] [PubMed]
27. Mel Fraga, J.; Del Caño Gochi, A.; De la Cruz López, M.P. Sustainability in the Preparation, Transport and Casting of the Concrete in Spain: Analysis of Energy Consumption and CO_2 Emissions. In Proceedings of the 18th International Congress on Project Management and Engineering, Alcañiz, Teruel, Spain, 16–18 July 2014. Available online: http://www.aeipro.com/files/congresos/2014alcaniz/CIDIP2014_0508_0520.4196.pdf (accessed on 23 June 2020).
28. Luo, W.; Sandanayake, M.; Zhang, G. Direct and indirect carbon emissions in foundation construction—Two case studies of driven precast and Cast-In-Situ piles. *J. Clean. Prod.* **2019**, *211*, 1517–1526. [CrossRef]
29. Li, L.; Li, Z.; Li, X.; Zhang, S.; Luo, X. A new framework of industrialized construction in China: Towards on-site industrialization. *J. Clean. Prod.* **2020**, *244*, 118469. [CrossRef]
30. Mercader Moyano, P.; Roldán Porras, J.; Ramirez de Arellano Agudo, A. Sustainability Assessment in Singular Structures, Foundations and Structural Rehabilitation in Spanish Legislation. *Open Constr. Build. Technol. J.* **2017**, *11* (Suppl. 1), 95–109. [CrossRef]
31. Hammad, A.W.A.; Akbarnezhad, A.; Wu, P.; Wang, X.; Haddad, A. Building information modelling-based framework to contrast conventional and modular construction methods through selected sustainability factors. *J. Clean. Prod.* **2019**, *228*, 1264–1281. [CrossRef]
32. Sandanayake, M.; Zhang, G.; Setunge, S. Environmental emissions at foundation construction stage of buildings—Two case studies. *Build. Environ.* **2016**, *95*, 189–198. [CrossRef]
33. Martínez, S.; González, C.; Hospitaler, A.; Albero, V. Sustainability Assessment of Constructive Solutions for Urban Spain: A Multi-Objective Combinatorial Optimization Problem. *Sustainability* **2019**, *11*, 839. [CrossRef]
34. FEMA. *Rapid Visual Screening of Buildings for Potencial Seismics Hazards, A Hand Book*, 2nd ed.; Federal Emergency Management Agency: Washington, DC, USA, 2002.
35. Zhang, H.; Xing, F.; Liu, J. Rehabilitation decision-making for buildings in the Wenchuan area. *Earthq. Constr. Manag. Econ.* **2011**, *29*, 569–578. [CrossRef]
36. Ewing, R.C. Foundation repairs due to expansive soils: Eudora Welty House, Jackson, Mississippi. *J. Perform. Constr. Facil.* **2011**, *25*, 50–55. [CrossRef]
37. Wilder, D.; Smith, G.C.G.; Gómez, J. Issues in design and evaluation of Compaction Grouting for foundation repair. In *Innovations in Grouting and Soil Improvement*; ASCE: Reston, VA, USA, 2005; pp. 1–12.
38. Seksinsky, E.J.; Qubain, B.S. An underpinning solution to a persistent foundation settlement problem. In *Soil Improvement for Big Digs*; ASCE: Reston, VA, USA, 1998; pp. 202–213.
39. Lutenegger, A.J.; Kemper, J.H. Preservation of historic structures using Screw-Pile foundations. In Proceedings of the 6th International Conference on Structural Analysis of Historic Construction, SAHCo8, Bath, UK, 2–4 July 2008; pp. 1079–1086. [CrossRef]
40. Da Casa, F.; Echevarría, E.; Celis, F. El movimiento en la ejecución de recalces con inyección armada. Análisis de tres casos con movimientos previos. *Inf. Construcción* **2012**, *64*, 507–518. [CrossRef]
41. Pineda, P.; García-Martínez, A.; Castizo-Morales, D. Environmental and structural analysis of cement-based vs. natural material-based grouting mortars. Results from the assessment of strengthening Works. *Constr. Build. Mater.* **2017**, *138*, 528–547. [CrossRef]
42. Ferreira Sánchez, A.; Alba Rodríguez, M.D.; de Ramirez Arellano Agudo, A.; Marrero, M. Recalce en condiciones de emergencia: 40 viviendas cercanas al río Guadalquivir, España. *Inf. Construcción* **2016**, *68*, 133. [CrossRef]

43. Oficina Española de Cambio Climático (OECC); Secretaría de Estado de Medio Ambiente del Ministerio de Agricultura; Alimentación y Medio Ambiente de España. *Guía Para el Cálculo de la Huella de Carbono y Para la Elaboración de un Plan de Mejora de una Organización*; OECC: Madrid, Spain, 2016. Available online: http://www.mapama.gob.es/es/cambio-climatico/temas/mitigacion-politicas-y-medidas/guia_huella_carbono_tcm7-379901.pdf (accessed on 23 June 2020).

44. Oficina Española de Cambio Climático (OECC); Secretaría de Estado de Medio Ambiente del Ministerio de Agricultura; Alimentación y Medio Ambiente de España. *Factores de Emisión. Registro de Huella de Carbono, Compensación y Proyectos de Absorción de Dióxido de Carbono*; OECC: Madrid, Spain, 2017. Available online: http://www.mapama.gob.es/es/cambio-climatico/temas/mitigacion-politicas-y-medidas/factores_emision_tcm7-359395.pdf (accessed on 23 June 2020).

45. De Fomento, M.; Consulting in Transportation and Logistics (SPIM). Resumen Ejecutivo. In *Ministry of Development: Study of the Costs of Goods Transportation by Road*; Ministry of Development, Government of Spain: Madrid, Spain, 2008. Available online: https://www.fomento.gob.es/NR/rdonlyres/D12A4405-3DE8-4D87-8F06-8CED0E11DD3E/40278/EstudioCostesMercanciasCarreteraoctubre2008.pdf (accessed on 23 June 2020).

46. Asociación de Empresas Constructoras y Concesionarias de Infraestructuras (SEOPAN). *Manual de Costes de Maquinaria de Construcción*; SEOPAN: Madrid, Spain, 2008.

47. Instituto para la Diversificación y Ahorro de la Energía (IDAE). Escala de la Calificación Energética. Edificios de Nueva Construcción. Madrid, Mayo de 2009. Available online: http://www.idae.es/uploads/documentos/documentos_CALENER_07_Escala_Calif_Energetica_A2009_A_5c0316ea.pdf (accessed on 23 June 2020).

48. Sociedad Española de Bombas de Hormigón (SEBHSA). Ficha Técnica Camión con Bomba Para Hormigón 42 m. BC-3907, BC-4147. PL-42/125-5X. (Spanish Society of Concrete Pumps- SEBHSA Technical Specifications for 42 m. BC-3907, BC-4147 and PL-42/125-5X Concrete Pump Trucks). Available online: http://www.sebhsa.com/_data/es/pl-42-125-5x-14.pdf (accessed on 23 June 2020).

49. ALBA-MACREL GROUP, S.L. Maquinas de Ferralla: Cortadora o Cizalla C32L-2,6 kW; Dobladora D36L-3,6 kW; Estribadora D24L-1,1 kW. (Iron Cutting Machines, Bending Machines, ALBA-MACREL GROUP). Available online: http://alba.es/productos/maquinas-de-ferralla (accessed on 23 June 2020).

50. De Sevilla, A. Centro de Gestión de la Movilidad. Delegación de Seguridad, Movilidad y Fiestas Mayores. Portal Tráfico Urbano. Intensidad Media Diaria: Seville, Spain. 2007. Available online: http://trafico.sevilla.org/pdf/ResumenIMD2007.pdf (accessed on 23 June 2020).

51. Infraestructura de Datos Espaciales (IDE). GeoPortal. (Spatial Data Infrastructure (SDI). GeoPortal). Available online: http://sig.urbanismosevilla.org/visorgis/geosevilla.aspx?Layers=IDES&Selected=01&xtheme=brown_12 (accessed on 23 June 2020).

Article

Carbon Footprint of Dwelling Construction in Romania and Spain. A Comparative Analysis with the OERCO2 Tool

Patricia González-Vallejo [1,*], Radu Muntean [2], Jaime Solís-Guzmán [1] and Madelyn Marrero [1]

[1] ArDiTec Research Group, Departament of Architectural Constructions II, Higher Technichal School of Building Engineering, Universidad de Sevilla, 41012 Seville, Spain; jaimesolis@us.es (J.S.-G.); madelyn@us.es (M.M.)

[2] Department of Civil Engineering, Transilvania University of Brasov, 500036 Brașov, Romania; radu.m@unitbv.ro

* Correspondence: pgonzalez1@us.es; Tel.: +34-954-556-697

Received: 13 July 2020; Accepted: 17 August 2020; Published: 20 August 2020

Abstract: CO_2 emissions due to the construction sector represent 40% of the total, either directly by the use of the building or indirectly by the emissions incorporated in construction materials and products. It is important to achieve a change in this sector to introduce these concepts in a simple way. There are various tools for evaluating emissions in construction projects. In the present work, the OERCO2 tool is used. This work studies housing projects in two European countries belonging to significantly different regions, Spain (Andalusia) and Romania (Bucharest and Transylvania). Although concrete or masonry structures are mainly used in Romania, due to an increased demand for residential buildings in recent years, a new niche has appeared in the construction sector: metallic and mixed (metal–concrete) structures for multi-storied buildings. For these reasons, a comparison between concrete and metallic buildings can be made in order to highlight their environmental impact. Twenty-four projects are selected from Romanian projects with metallic structures, and Spanish projects with concrete structures. They are also differentiated according to the type of foundation used. As expected, buildings with a metallic structure have more economic and environmental impact than reinforced concrete. The materials with greater impact are metal, concrete, cement, and ceramic products. The potential of the tool for the evaluation of various construction solutions, materials, and project phases is demonstrated.

Keywords: carbon footprint; assessment tool; dwelling construction; cost control

---

1. Introduction

Various environmental reports carried out in recent years highlight the construction sector as one of the main consumers of energy and generators of CO_2 emissions among the various industrial sectors, with estimates of 30–40% of the total environmental impact produced [1]. This concern has forced the appearance of different types of tools to assess these impacts: through certification and standardization, the promotion of international standards to use environmental labeling for construction products [2–5], the development and application of life cycle analysis (LCA) [6–8], and the environmental management of buildings from a life cycle perspective [9,10]. However, the implementation of these standards is not always easy to achieve, due to barriers of all kinds, economic, technical, practical, and cultural, which prevent professionals from selecting materials with less environmental impact [11,12].

If we focus on the analysis of the life cycle of buildings, the manufacturing and construction phase of the building life cycle, concentrated in a short period of time (1–2 years), causes the most intense environmental impact, mainly due to the consumption of concrete and steel for the structure, which represents a high percentage of the emissions produced during this phase [13,14]. This impact is diluted if the building's useful life is lengthened. The use and maintenance phase is generally responsible for 80–90% of the CO_2 emissions generated during the life cycle of the building [15], almost 60% of which is caused by the demand for energy for heating and cooling [16]. This implies that, in new standards such as zero energy buildings, emissions during the construction phase represent a higher percentage of the total emissions throughout the life cycle [17]. Therefore, once the energy consumption during the use phase is reduced, researchers' attention should be focused on materials that require less energy for their production [18].

Most of the recent studies that propose methodologies to estimate the environmental impact of buildings or the application of ecological indicators to the case studies of buildings have focused on aspects such as LCA [19], the analysis of energy consumption throughout of the life cycle [20], the carbon footprint of the life cycle [21], or a combination of these methods [22,23]. In recent years, these studies have been incorporated into more powerful computing tools, which is generating a new field of action for LCA, as is the case for building information modeling (BIM) platforms [24].

However, the LCA methodology and its derivatives are not always easy to implement by non-specialized users, and neither is their communication. For this reason, other methodologies have been implemented that have a smaller scope but are easier to use and implement by the agents involved in construction. Among the most employed, we find the ecological footprint (EF) or the carbon footprint (CF). The CF is an indicator of emissions of greenhouse gases generated by a given process [25], which stands out due to its simplicity and direct relationship with the main objectives of the Kyoto Protocol [26], along with its easy application in decision-making and environmental policy [27]. There are a large number of bibliographic reviews related to the use of the CF indicator in construction [28], however, the results are not always comparable, due to the absence of a methodology that follows international standards [29]. For this reason, studies have also been carried out in recent years to establish scales that allow for the definition of reasonable ranges of CO_2 emissions in construction processes [30].

Tools are in place that can ensure that new and already built buildings meet minimum requirements related to environmental sustainability. Most of these systems are currently developed by two international organizations [31]: the World Green Building Council (GBC), which develops tools from an international system to obtain sustainability data for buildings, adapting them to each country, and BRE Global is another independent organization that develops the BREEAM method.

In Spain, there is a variety of these tools that include the calculation of the CF of buildings in some way, for example, LEED or BREEAM, whose use has spread in the country thanks to national organizations such as the Spanish Green Building Council [32] and BREEAM Spain [33]. These tools include, among the categories evaluated, the CO_2 emissions due to the production of construction materials and the operational energy consumption; however, the final score does not reflect these CO_2 emissions, so it does not report each result separately for a better understanding and subsequent analysis of possible improvements.

However, other alternatives have emerged from various research projects in the last decade in Spain. For example, SpainGBC has presented VERDE tools [32], a set of environmental impact assessment tools for design assistance (HADES), new buildings (VERDE NE), rehabilitation (VERDE RH), and urban development (VERDE DU). In this set of tools, the CF obtains the highest percentage of the score, so it prevails over other sources of environmental impact. ECOMETRO is an open-source and online tool to measure the environmental impact of buildings [34]. The information generated is similar to an environmental product declaration (EPD), but it applies to entire buildings.

Highly specialized platforms such as the BEDEC cost database, SOFIAS tool [35], or E2CO2Cero [36] allow for the detailed calculation of CO_2 emissions according to the bill of quantities of a project. BEDEC was developed by the Institute of Construction Technology of Catalonia (ITeC), and uses environmental data of construction materials from the Ecoinvent database [37], known for being one of the most complete environmental databases at the European level [38] and for its integration with Simapro LCA software [39]. The SOFIAS tool uses data from the OpenDAP database [31,35]. An intermediate solution is E2CO2Cero [36], supported by the Basque Government, which is a software that estimates the embodied energy and CF of a building according to the materials consumed and the construction processes [36]. This tool also has two different versions: complete and simplified. The first one requires the presentation of the bill of quantities of the project, which is considered the appropriate way to reach the general public and create social awareness. A table with the scopes of each tool is included (Table 1).

Table 1. Comparison of assessment methods.

	Leed	Breeam	Verde	Ecometro	Sofias	Gabi	Green Homes
Site and construction	X	X	X	X	X	X	X
Transport	X	X		X	X	X	X
Energy management	X	X	X	X	X	X	X
Water management	X	X	X	X	X	X	X
Materials and resources management	X	X	X	X	X	X	X
Waste management	X	X	X	X	X	X	X
Comfort	X	X	X	X			X
Health	X	X	X	X			X
Environmental charges	X	X	X	X	X	X	X
Services	X	X	X				X
Economic aspects			X		X	X	X
Social aspects			X			X	X

In Romania, the awareness of green buildings and sustainable building materials has increased significantly over the past 10 years. The first LEED and BREEAM certifications appeared in 2008–2009. Currently in Romania, there are 39 buildings with BREEAM certificates, 21 with LEED, 3 with DGNB (German Sustainable Building Council) [40], and more than 4600 houses and apartments certified or in progress within the GREEN HOMES certification scheme developed by the Romania Green Building Council (GBC) [41].

The tools used by Romanian evaluators to calculate CO_2 emissions of buildings are One Click LCA, Integrated Environmental Solution (IES VE Pro) [42], GaBi [43], 360 Optimi [44], and others. All of these require data on the type of material, the name of the product, thickness, quantity, transport (distance and type), and durability.

Colliers International Romania used One Click LCA to calculate the entire building life cycle assessment for their first LEED v4 project, as part of LEED certification. The project aimed to achieve a Gold Level certification [45]. GaBi software [43] supported a study on the LCA methodology applied to optimize municipal solid waste management (MSW) systems in Cluj county, Romania.

Romania GBC has established procedures for EPDs to be easily integrated into environmental certification tools such as GREEN HOMES [41] and is promoting EPDs for the recognition of points in international LEED or BREEAM certification. In the case of the Living Building Challenge certification system, its materials category is designed to foster a successful materials economy that is non-toxic, transparent, and socially equitable [46].

In Romania, there is no accredited body to issue EPDs, and all declarations are issued by international entities. The National Institute for Research and Development in Buildings, Urban Planning and Regional Sustainable Development (URBAN INCERC), established in 2009, is the only recognized institution that tests materials and issues performance certifications.

Compared to Spain, in Romania, there are no tools such as the BEDEC cost database, SOFIAS, or the E2CO2Cero tool that can be used to calculate CO_2 emissions. There is cost-estimating software based on the quantities of materials, labor, transportation, and equipment used for buildings, but they do not include parameters such as energy consumption, CO_2 emissions, or other environmental data.

From this point of view, an instrument capable of estimating the CF of buildings, which is also available in Romania, is a necessity nowadays and may be important for the future development of the construction sector due to the possibility of increasing the awareness of all participants in the construction industry regarding environmental problems.

The experience of the authors in methodologies for calculating carbon footprints (CFs) is presented through an open-source software to estimate the CF of architectural projects from the design phase and the tool is part of the OERCO2 project [47]. It is developed for educational purposes and free access with an Erasmus Project granted by the European Union, and member countries include Spain and Romania. This research is part of the tool validation for the calculation of carbon emissions in the construction phase of the building life cycle [47,48]. In the present work, the OERCO2 tool evaluates the CO_2 emissions of the construction phase, including the extraction and manufacture of materials, as well as the management of construction and demolition waste (RCD) and the economic impact, and is compared to the rest of the tools in Table 1. Even though it does not cover all the aspects, it is easy to use and free accessible, making it an interesting teaching tool for college students and professionals [47]. The tool is valid to evaluate projects at the design stage.

The performance of this tool is explained through a comparative analysis of projects in both countries, Spain and Romania, which are part of the OERCO2 project. Representative typologies are assessed. In the particular case of studies in Romania, the CF of seismic reinforcements [49] is assessed. Additionally, a sensitivity analysis of the tool in these case studies is done.

A list with corresponding acronyms is included: Life cycle analysis (LCA); ecological footprint (EF); carbon footprint (CF), Green Building Council (GBC); Andalusian construction information classification system (ACICS); basic costs (BCs), auxiliary costs (ACs); simple unitary costs (SCs); life cycle inventory (LCI); metallic structure (MS).

2. Materials and Methods

The methodology for assessing the CF of the construction of residential building is based on the bill of the quantities of the project and a classification system for construction work that breaks down this information for materials, labor, and machinery (Figure 1). Budgets for 140 different projects are analyzed and classified; their budgets are reorganized into a construction breakdown system (CBS) that makes it easy to make comparisons. This organizing system has been successfully applied in previous research to assess the ecological footprint of buildings [50–56] and to estimate the generation of construction waste [45].

For the quantification of resources, the Andalusian construction cost database (ACCD) and its information classification system (ACICS) [57,58] are used, which estimate costs in the construction sector, and their use is mandatory in public works in Andalusia (Spain). The ACICS uses a hierarchical organization for work units. This is represented in Figure 2, with the example of a collector of the sewage system, showing categories called chapters, corresponding to the first two numbers of the code (for example, earthworks, foundations, installations, sewage system, etc.), which are subsequently divided into subcategories.

The base of the ACCD structure is formed by the basic costs (BCs), which are materials, machinery, or manpower, and those are added to form auxiliary costs (ACs) and simple unitary costs (SCs); the latter represent the various activities or work units (Figure 2). These three types of costs are those used by the OERCO2 tool for its calculations. The ASICS has been revised and adapted to construction in Romania through the analysis of housing construction projects facilitated by the project partners, but the costs refer to Spain to facilitate the comparison between countries.

Environmental data included in the OERCO2 software are obtained from the Ecoinvent database through Simapro, which was chosen to cover all the materials commonly used in the construction of buildings [57]. In order to obtain the CO_2 emissions embodied in construction materials, their life cycle inventory (LCI) is analyzed by applying the Intergovernmental Panel on Climate Change (IPCC) 100a methodology, which is used by the carbon footprint indicator to isolate CO_2 and other Greenhouse Gases (GHG) emissions from the LCI. A list of the main construction materials and their corresponding carbon footprints, as obtained from LCA data, is included in Table 2.

Table 2. Average values of environmental impacts by family of materials [59].

Material	Carbon Footprint (t CO_2 eq./t)
Soil	0.007
Wood	−0.992
Concrete	0.112
Asphalt	0.21
Ceramic	0.22
Aggregates and Stones	0.004
Metals	1.50
Plastics	3.25
Glass	0.669
Plaster and Pastes	0.002

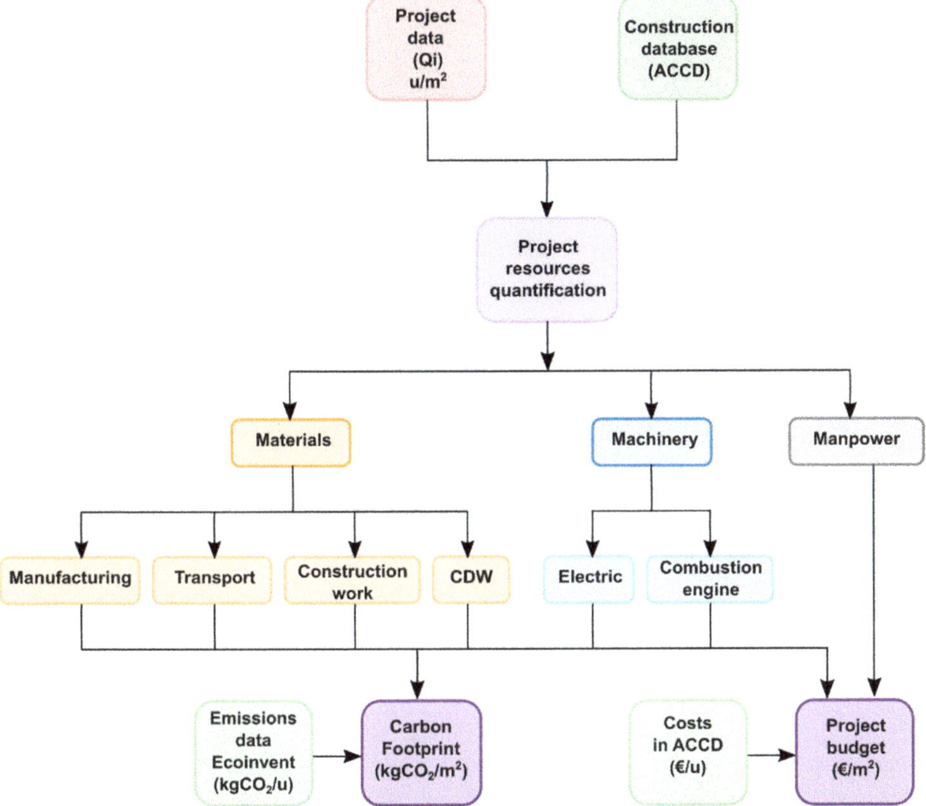

Figure 1. OERCO2 methodology.

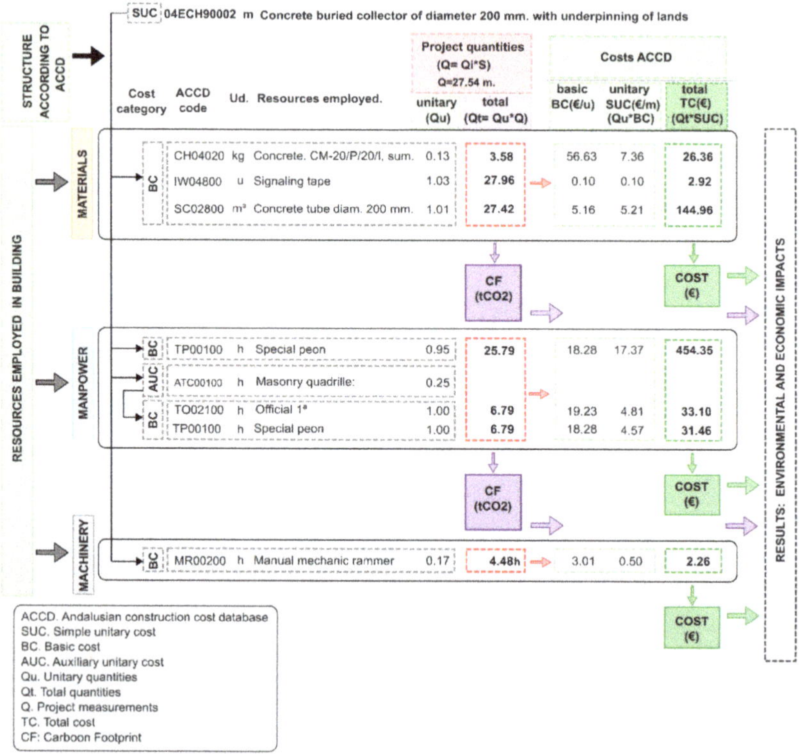

Figure 2. Cost and quantification of economic and Carbon Footprint (CF) model. Source: adapted from González-Vallejo et al. 2019 [56].

OERCO2 Tool

The OERCO2 tool [48] is an online application that allows for estimating the CF generated in the construction of residential buildings. It is derived from several previous research studies developed by the authors [50,51,54–56,60,61] and includes the evaluation of CO_2 emissions for the construction processes of 140 different types of residential buildings. The tool has been tested and evaluated by all the project partners and includes all the typologies and construction characteristics commonly used in Spain, Portugal, Italy, and Romania.

The data obtained from the bill of the quantities of each project are structured in accordance with the ACICS [57,58] mentioned above and are expressed in units per surface constructed (u/m²). The average quantity of each activity (Qi) of the 140 evaluated projects is obtained through a statistic process for each type of construction according to the model to assess the construction of buildings [50,62]. These average quantities are transformed into materials, labor, and machinery. The amounts of the various resources involved is evaluated using the CF methodology to obtain the emissions of CO_2 generated due to all the construction processes.

The OERCO2 tool has an initial screen where the user specifies general information about the project to be analyzed, such as number of floors, the type of structure, and floor area (Figure 3). The selection of these initial data allows for assigning one similar project from the database which collects the resource quantities needed in the building (Qi) [50,60].

In the following step, the user specifies data on the project related to construction solutions for each element (Figure 3). The OERCO2 tool uses this information to select which of the available SCs should be used for calculations. As mentioned above, SCs are made up of BCs and ACs. In this tool, the resources (BCs) contain not only economic information, but also environmental information (for example, CO_2 emission factor).

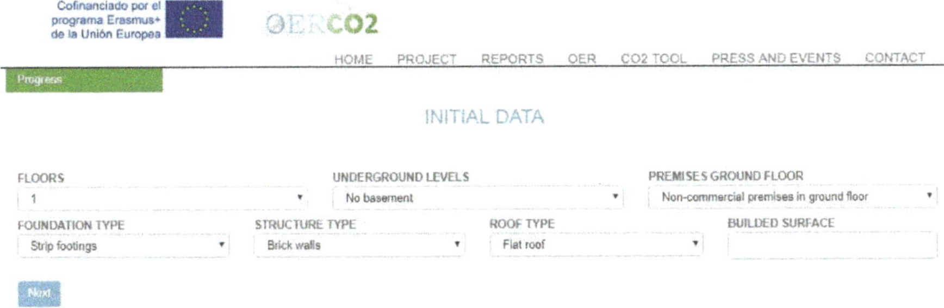

Figure 3. Selection interface for the initial data of the project in the OERCO2 tool [48].

Following this logic, when you select one specific work unit, the SC is assigned two factors: economic cost (€/unit of reference) and environmental cost (kg CO_2 eq/unit of reference), where the unit of reference is the selected SC (Figure 1). Initial data selection can meet the Qi of the work unit, which allows for obtaining the total cost and CF of this activity through Equations (1) to (4) whose factors are defined in Table 3.

$$CF_{MAT} = (\Sigma i \; Cmi \times UCF_{MAT}) + (UCF_{TRAN} \times Cmi) \tag{1}$$

$$CF_{MCOMB} = V \times UCF_{COMB} \tag{2}$$

$$V = (P \times T \times Ef) \tag{3}$$

$$CF_{MELEC} = (P \times T) \times UCF_{ELEC} \tag{4}$$

Table 3. Factors to calculate CF of the construction phase of materials and machinery.

Construction Material
E_{MAT}: emissions per material (kg CO_2/kg material)
Cmi: construction material i (kg)
UCF_{MAT}: unit carbon footprint per material (tCO_2eq/kg of material)
UCF_{TRAN}: unit carbon footprint of material transport (tCO_2eq/kg of material)
Machinery
CF_{MCOMB}: Combustion engine machinery (tCO_2eq)
UCF_{COMB}: unit impact of gasoline or gasoil (tCO_2eq/l) Data obtained from Ecoinvent [37]
V: gasoil consumption (liters)
P: power of the electric engine (kW)
T: working time (hours)
Ef: efficiency, liters of gasoil or gasoline consumed per engine power (l/kWh)
E_{COMB}: emission factor of the combustion engine of gasoil or gasoline 2.616 kg CO_2/l [63]
CF_{MELEC}: Electric engine machinery (tCO_2eq)
E_{ELEC}: emission factor of electric mix In Spain: 0.248 kg CO_2/kWh [64]/in Romania: 0.264 kg CO_2 [65]
UCF_{ELEC}: unit impact of the electric mix (tCO_2eq)

The transport of materials is included in the environmental impact by making approximations of the distance covered [66]. The transport is done by truck. As for the distance from the factory to the work site, most materials are manufactured close to the construction site, i.e., 250 km. In the specific case of concrete, a maximum distance of 20 km is considered before solidification starts [67]. The tons of CO_2 can be obtained with the data in Table 4. Material transport in Romania is similar to that in Andalusia, where pick-up points are in warehouses in peripheral zones, close to main cities. Fresh concrete has a similarly limited time for its transport.

Table 4. Data for calculating the impact of transport.

	Concrete	Other Materials
Truck load capacity (t)	24	2
Distance to factory (km)	20	250
Average diesel consumption (l/100 km)	26	26
Diesel emissions (tCO$_2$/l)	2.62×10^{-3}	2.62×10^{-3}
Diesel consumed water (m^3/l)	1.26	1.26
Diesel embodied energy (MJ/l)	57.70	57.70
Electricity embodied energy (MJ/kWh)	3.60	3.60

The tool displays the total cost of building construction (expressed in € and €/m^2) and the total CF (in kg CO_2 eq and kg CO_2 eq/m^2), the latter is also divided into materials and machinery (Figure 2).

In the OERCO2 tool, the users do not require specialized knowledge on the environmental evaluation of construction projects, such as the calculation of embodied energy or the CO_2 emission factor associated with construction processes. Instead, the users only need to know the solutions implemented in the building's construction. The OERCO2 tool has been verified by all the partners in the project to include variations between countries in processes and construction solutions [48,68].

3. Case Studies

Two project member countries are selected, for which building typologies are included in the OERCO2 tool, with Spain and Romania being two of the most remote countries and where more constructive differences can be found. The OERCO2 tool presents the alternative of concrete and metal structures, the latter option being included as it is a type of structure used in Romania, and therefore the analysis of these two countries and construction typologies is carried out.

Projects of Romanian (Bucharest and Transylvania) buildings with metal structures are studied and compared with the same typology of Spanish (Andalusia) buildings with a reinforced concrete structure. Although concrete or masonry structures are mainly used in Romania, due to an increased demand for residential buildings in recent years, the necessity to reduce the execution time, the necessity to improve quality and price control, the need for larger dimensions for open spaces, the development of automated execution technologies (welding), the increase in prices for formwork (especially labor), and the need for much better resilience, a new niche appeared in the construction sector: metallic and mixed (metal–concrete) structures for multi-storied buildings. A very important feature of metal structures is their multifunctionality and their ability to adapt, modify, and rebuild, as well as better seismic behavior, Romania being a country characterized by relatively strong earthquakes. Steel parts are prefabricated according to specific standards, are easy to transport, and their assembly is fast, with low costs and low dependence on environmental conditions [49].

Firstly, the current situation of construction in Romania is studied and the OERCO2 tool methodology is applied to residential buildings and compared with the study carried out in Spain. The chosen typology in both countries is the multi-family residential building, with the same number of floors below and above ground, with one of three different types of foundations, isolated footings, reinforced slab, and piles. Both countries use the most common constructive systems.

3.1. Statistical Data

From the statistical data, the most representative type of project in Romania is obtained. According to Table 5, dwellings represents nearly 80% of the buildings constructed since 2014, with an average floor area per apartment of 113 m^2 [69]. This percentage remains stable very year.

Data reflected in Table 5 justify the study of housing construction since it is an important sector and it maintains a stable rhythm of new buildings in 2015, slightly decreases in 2016, and goes up in 2017 and 2018, even exceeding previous years. The data are based on the total number of permits issued for residential buildings, with a large number of apartments/housing (units) finally authorized to be built, obtaining the total floor area in the last column.

Table 5. Statistical data of new buildings in Romania between 2014 and 2018.

Year	Number of Permits for Residential Buildings	Administrative Buildings	Others	Total Number of Permits	Apartments	Total Floor Area, Residential (m^2)
2014	37,672	234	7568	45,474	60,270	7,162,041
2015	39,112	237	7164	46,513	67,293	7,839,961
2016	38,653	196	6185	45,034	80,608	8,892,555
2017	41,603	204	6551	48,358	88,029	9,628,297
2018	42,694	255	7170	50,119	98,103	10,664,822

The dwelling characteristics, according to the report entitled "National Housing Strategy" and published by the Romanian Ministry of Environment, Water and Forests [70] in 2015, states that, in urban areas, approximately 70% of the housing units in Romania were multi-family buildings (blocks of flats/apartments) of 10 apartments or more, and the rest were individual dwellings. The building height was limited because elevators have high costs for construction and operation. The majority of multi-storied buildings were up to five floors, consisting in the ground floor and four upper floors [71]. Fewer floors were not economically feasible in terms of total floor area per lot.

In the case of Spain, according to data taken between the years 2007–2010 from the National Statistics Institute (INE) [72] and published studies [62], 78% of the constructed buildings were residential, of which 33% correspond to the type of building with four or five floors above ground.

3.2. Selected Projects

Once the statistical study was carried out, 24 types of projects were selected in Romania and 24 projects in Spain, in the OERCO2 tool, in Tables 6 and 7, Romanian projects have metallic structures, and in Spain, reinforced concrete structures. Both types of projects are also classified according to type of foundation, isolated footings, reinforced concrete slab, and piles. The roof could be inclined or horizontal and there are buildings with a basement (one or two floors) and without a basement. The use on the ground floor can be for premises or houses. Table 6 includes the general characteristics that are common in the 48 projects. The different construction characteristics between Romania and Spain are defined in Table 7. Data for Spain are based on the INE data that have already been analyzed [62].

Table 6. General information of selected buildings in OERCO2 tool.

	Number of Floors	Number of Underground Floors	Use of Ground Floor	
General information	5	0/1/2	Shops/dwelling	
	Foundation	**Structure**	**Roof Type**	**Floor Area (m^2)**
	Isolated Footings/Reinforced Slab/Piles	Metal/Reinforced Concrete	Flat/Sloping	72

Table 7. Selected data for the case studies from OERCO2 tool.

MS and CS Buildings (Romania and Spain)			

General information

Earthworks			
Excavations	**Fillings**	**Earth Transport**	
Excavator	Mechanic means	Mechanic means	
Sewer System			
Manholes	**Sewage pipes**	**Downpipes and roof sinks**	
In situ	PVC	Reinforced PVC	
Structure			
Formwork	**Floor slabs**	**Flat roof**	**Sloping roof**
Metallic	Waffle slab with recoverable caissons/One-way slab with concrete vaults	Passable and ventilated/Passable and inverted	Ceramic tiles
Masonry			
Façades	**Claddings**		**Partitions**
One-foot brick wall with chamber/0.5-foot brick wall with chamber	Single layer mortar		Double hollow brick 9 cm

Installations

Air-Conditioning System			
Air-Conditioning	**Terminal units**	**Ducts**	**Pipes**
None	None	None	None
Heating		**Insulation**	
Radiators	**Boiler**	**Pipe insulation**	
Classic steel	Gasoil	Applies	
Water supply and ventilation			
Cold-water pipes	**Hot-water pipes**	**Sinks**	**Ventilation**
Copper	Copper	PVC	Concrete
Domestic Hot water			
Heater	**Solar panels**		
Gas	Applies/Does not apply		
Accessibility			
Lift			
Applies			

Finishes

Insulation		**Vertical finishing**	
Thermal-acoustic		**Continuous**	
Polyurethane/Rock wool		Rendering	
Finishes			
Floors		**Ceilings**	
Ceramic		Continuous laminated gypsum	
Carpentry			
Windows	**Glazing**		**Doors**
Lacquered aluminum casement with thermal bridge break	Thermal-acoustic 6 + 12 + 6		Wood
Protection elements			
Blinds	**Protection grids**		**Railings**
Anodized aluminum	Hot-rolled steel		Steel

Table 7 includes the data selected in the tool for both cases according to the proposed alternatives, which are similar except for the type of slab, horizontal roof construction system, façade cladding, and type of insulation. Additionally, solar panels for domestic hot water are installed in Spain.

4. Results and Discussion

According to the characteristics of the projects proposed from the statistical analysis of each country, the 48 projects' data are classified and coded according to the OERCO2 tool information in Table 8 and average values are obtained, the numbers are generic for the software and do not have a specific meaning. The comparison is not only of Spanish and Romanian projects but also different constructive solutions are employed in each country that give rise to 24 combinations in each (Table 8). The codes are defined to identify the differences in the combinations as set in Table 8: column foundation, number underground floors, ground floor, and roof type. The MS and CS coding corresponds to the typical metal structure projects in Romania and the concrete structures of the projects in Spain, respectively. The numeric codes are those internally used by the software.

Table 8. Classification and codification of projects according to the OERCO2 tool, MS is metallic structure and CS is reinforced concrete structure.

Foundation	Number Underground Floors	Ground Floor	Roof Type	MS OERCO2 Code	CS OERCO2 Code
ISOLATED FOOTINGS	0	Dwelling	Flat	M012	C012
			Sloping	M013	C013
		Shops	Flat	M021	C021
			Sloping	M025	C025
ISOLATED FOOTINGS	1	Dwelling	Flat	M043	C043
			Sloping	M051	C051
		Shops	Flat	M040	C040
			Sloping	M048	C048
REINFORCED SLAB	0	Shops	Flat	M106	C106
			Sloping	M110	C110
REINFORCED SLAB	1	Dwelling	Flat	M090	C090
			Sloping	M098	C098
		Shops	Flat	M094	C094
			Sloping	M102	C102
REINFORCED SLAB	2	Dwelling	Flat	M147	C147
			Sloping	M148	C148
PILES	0	Dwelling	Flat	M062	C062
			Sloping	M054	C054
		Shops	Flat	M066	C066
			Sloping	M059	C059
PILES	0	Dwelling	Flat	M083	C083
			Sloping	M082	C082
		Shops	Flat	M081	C081
			Sloping	M080	C080

From the building typology and construction characteristics of the projects, the unit (Qi) and total (Qt) quantification of each project is obtained, from which the economic and CF results are obtained, and included in Supplementary Data Tables S1 and S2 (Romania and Spain, respectively).

The quantity of materials used in the projects is analyzed according to the weight, expressed in kg, in addition to the CF, and, finally, the CF is calculated for the construction phases of the projects.

Materials are grouped in families: concrete and cement, ceramics and bricks, aggregates and stones, and metals and alloys represent around 80% of the total weight.

In Figure 4, the materials are analyzed according to the weight of metallic structure (MS) buildings in Romania, and in Figure 5, concrete structure (CS) buildings in Spain. Buildings with an MS need a larger amount of materials, except in the case of aggregates, whose values are similar in both cases. Concrete and cement are the heaviest materials in both countries. Pile foundation buildings have the highest consumption, followed by those of isolated footings and finally of reinforced slab. Ceramics and bricks have a higher consumption in Romanian buildings, since façade cladding is thicker than in Spain. Aggregates and stones are used almost equally in the three types of foundations. Lastly, metals and alloys have a higher consumption, as expected, in MS buildings.

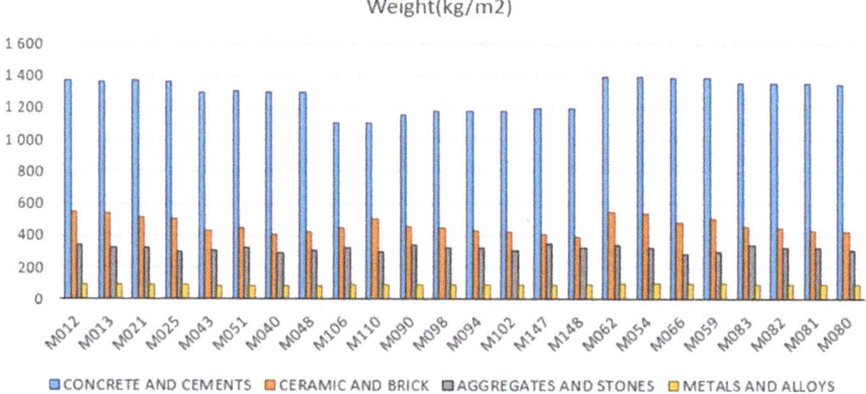

Figure 4. Weight per floor area (kg/m^2) of families of materials (Romania).

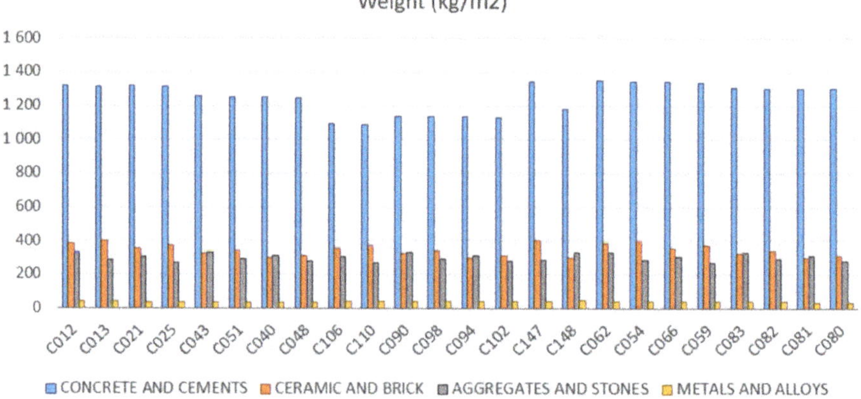

Figure 5. Weight per floor area (kg/m^2) of families of materials (Spain).

The analysis of the CF of materials introduces a new important family, plastics. Figure 6 shows that MS buildings in Romania produce a greater impact than those in Spain with CS (Figure 7) for all the families, except plastics, due to the use of projected polyurethane insulation in CS buildings. In MS projects (Figure 6), metals/alloys and concretes/cements produce the greatest CF, as the buildings with piles produce a slightly greater impact compared to the other two types. Ceramics/bricks also produce high CF values, due to their use in façades and interior cladding, and lastly plastics. However, in the case of CS buildings (Figure 7), the greatest impact is produced by concretes/cements, with those with piles and isolated footing being more polluting than those of reinforced slab. Ceramics/bricks have a much smaller impact, since Spanish buildings have less thick façades than Romanian ones, followed by metals and alloys, and then plastics. These results are similar to those of other studies [48].

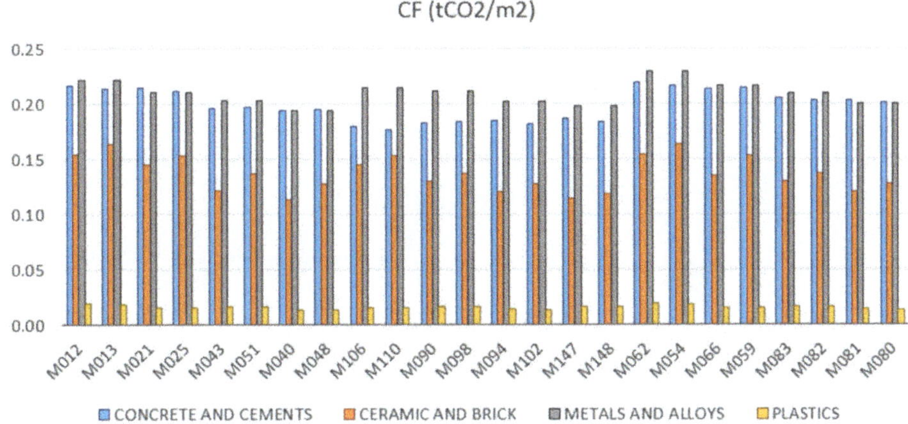

Figure 6. CF per area (tCO$_2$/m^2) by family of materials (Romania).

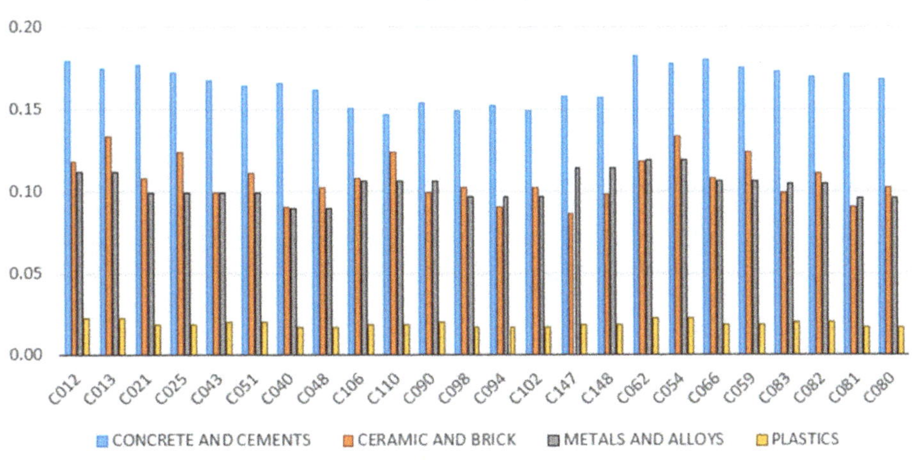

Figure 7. CF per area (tCO$_2$/m^2) by family of materials (Spain).

The patterns in Figures 4–7 are due to the project classification in Table 6, first fixing foundations, then the number of underground floors, followed by the type of construction, such as ground level use, and, finally, the roof type.

Simple costs included in the OERCO2 tool with the greatest impact (according to the ACCD) are 3HAL00002 (m^3 slab concrete); 03HMM00002 (m^3 mass concrete); 03CPS00007 (m pile on site); 05HHJ0010 (m^3 concrete assembled on beams); 05FBB00007 (m^2 waffle slab with concrete caissons); 05FUS00007 (m^2 one-way slab with concrete vaults); 05ACS00000 (kg steel in hot-rolled profiles); 05HAC00015 (kg corrugated steel in bars); 06LMM00101 (m^2 one-foot brick wall); 06LPC00001 (m^2 0.5-foot brick wall).

The following analysis is carried out by the construction phases or chapters of the project, as shown in Figure 8 (Romania) and Figure 9 (Spain). It is observed that the chapter with the greatest impact is structures in MS buildings and then the chapter of installations in CS buildings, due to solar panels, which have been mandatory in the construction of new housing in Spain since 2006. The next chapter that produces the greatest impact is masonry, in both countries, including ceramics/bricks for their use in façades, claddings, and partitions. The next important CF value is produced by the foundation phase, with the piles producing the greatest impact in both countries, followed by isolated footings and, lastly, with little difference, reinforced slabs. Of the five chapters, the one with the least impact is that of finishes, with similar values in all the cases analyzed since there are no variations in the construction systems and/or materials used.

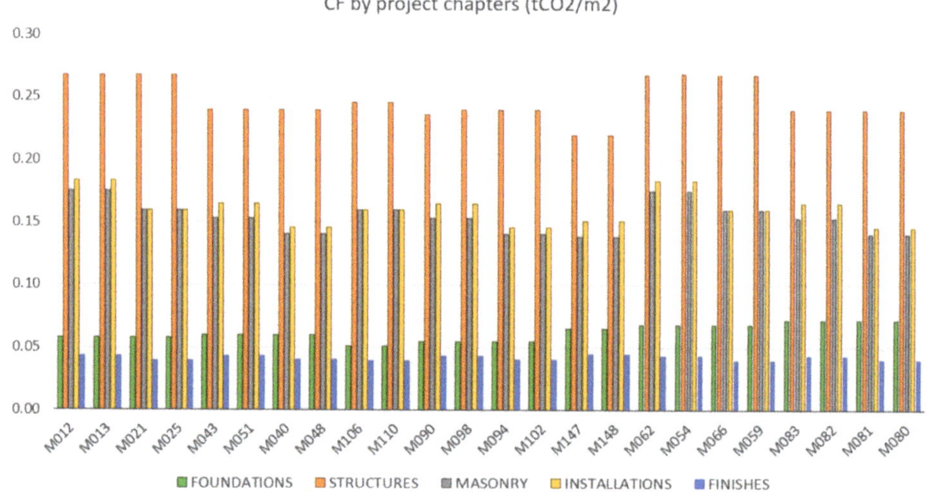

Figure 8. CF per area (tCO$_2$/m^2) according to chapter of project (Romania).

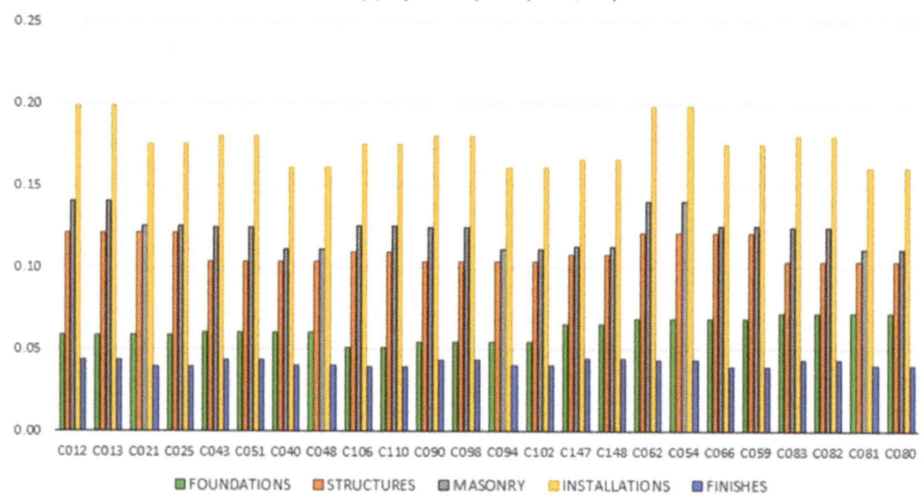

Figure 9. CF per area (tCO$_2$/m^2) according to chapter of project (Spain).

A sensitivity analysis of the CF produced by each of the different phases of the projects in each country is carried out, according to the percentage they represent of the total project, which is essential for decision-making at the design level in order to focus on the chapters with the greatest impact and to be able to reduce project emissions using more sustainable materials, based on the obtained results of the analysis.

Thanks to the data obtained from the OERCO2 tool, this analysis can be carried out, and it is presented in Tables 9 and 10, depending on whether they are from Spain or Romania, respectively, and grouped according to similarity in terms of building typology.

Table 9. Analysis of CF percentage by project chapters in Spain.

		Project of Spain						
	FOUNDATION	**ISOLATED FOOTINGS**		**REINFORCED SLAB**			**PILES**	
	UGF: Under Ground Floor	**W/O UGF**	**One UGF**	**W/O UGF**	**One UGF**	**Two UGF**	**W/O UGF**	**One UGF**
	CODE	C012/C013/ C021/C025	C040/C043/ C048/C051	C106/ C110	C090/C094/ C098/C102	C147/ C148	C054/C059/ C062/C066	C080/C081/ C082/C083
CHARTERS	FOUNDATION	9.79%	10.89%	9.16%	9.98%	11.47%	11.23%	12.73%
	STRUCTURES	20.29%	18.67%	19.61%	18.80%	19.05%	19.79%	18.28%
	MASONRY	21.81%	21.22%	22.36%	21.37%	19.93%	21.69%	20.78%
	INSTALLATIONS	30.63%	30.81%	31.28%	31.02%	29.29%	30.52%	30.17%
	FINISHES	6.79%	7.56%	6.99%	7.61%	7.87%	6.73%	7,40%
	% CF TOTAL	89.31%	89.15%	89.41%	88.78%	87.61%	89.97%	89.36%

Table 10. Analysis of CF percentage by project chapters in Romania.

		CF (%) of Romania Project Chapters						
	FOUNDATION	**ISOLATED FOOTINGS**		**REINFORCED SLAB**			**PILES**	
	UGF: Under Ground Floor	**W/O UGF**	**One UGF**	**W/O UGF**	**One UGF**	**Two UGF**	**W/O UGF**	**One UGF**
	CODE	M012/M013 M021/M025	M040/M043 M048/M051	M106/ M110	M090/M094 M098/M102	M147/ M148	M054/M059/ M062/M066	M080/M081/ M082/M083
CHARTERS	FOUNDATION	7.50%	8.56%	7.10%	7.77%	9.39%	8.74%	9.96%
	STRUCTURES	34.53%	34.12%	34.09%	33.90%	31.81%	34.18%	33.25%
	MASONRY	21.57%	20.88%	22.16%	20.82%	20.02%	21.34%	20.35%
	INSTALLATIONS	22.08%	22.10%	22.15%	22.04%	21.82%	21.84%	21.54%
	FINISHES	5.30%	5.95%	5.42%	5.93%	6.44%	5.24%	5.79%
	% CF TOTAL	90.97%	91.61%	90.92%	90.46%	89.48%	91.33%	90.89%

Thus, in the case of projects in Spain with a concrete structure (Table 9), it can be seen that the foundation for piles is the one with the highest CF, with values between 11.23% and 12.73% of the total project impact. The reinforced slab has the lowest CF, 3.57% less than piles.

In the chapter of structures, it is observed that buildings without underground floors have a higher CF than the rest, and in particular those with foundations with isolated footings, with a 2.01% difference from those with a lower CF, which are the buildings founded on piles and with a basement.

In the masonry and installation phases, it can be seen that the buildings that have commercial premises on the ground floor produce higher CFs, the difference being 2.43% for masonry and 1.99% for installations, between the highest and lowest CF projects.

Regarding cladding, the highest CF corresponds to buildings with the highest number of basement floors and without premises on the ground floor, with a difference of 1.08% between those with the highest and lowest value.

In the case of the projects in Romania with a metallic structure (Table 10), the projects with foundations of piles are those with the highest CF, with a difference of 2.86% with respect to the lowest CF, which are those with reinforced slab and without a basement. With respect to Spanish projects, they represent a lower percentage of the total, with an average difference of 2.50%.

The (metallic) structure in the Romanian buildings is the phase with the greatest impact on the project, and it is the foundation with footings that produce the highest CF, 2.72% higher with respect to the reinforced slab and two basement floors. With respect to Spanish concrete structures, they have a much higher CF in all cases, the average difference being 14.48%.

In masonry and installations, the situation is similar to Spanish buildings, those with premises on the ground floor and no basement have a higher CF than those without premises and consist of one or two basement floors, with a difference of 2.14% for masonry and 0.61% for installations between the highest and lowest with respect to the total CF of the project. The masonry values are very similar to those of the buildings in Spain, however, in the installations, by including the solar panels in Spain, the CF is higher by an average value of 8.59%.

Cladding has a similar impact in both countries, buildings without premises and with more basement floors produce a higher CF than those with premises on the ground floor and without a basement, with a difference of 1.20% between those of greater and lesser value. With respect to Spain, Romania has a lower CF of 1.56%.

Finally, an economic and environmental comparison is carried out (Table 11). The economic analysis highlights that MS buildings are more expensive per m^2 than CS buildings, and in both cases a pile foundation is the one with the greatest economic impact, followed by isolated footings and reinforced concrete slab. The OERCO2 costs are based on Spanish data.

Table 11. Economic and environmental comparison of MS and CS, according to type of foundation.

Foundation	Concrete Structure (CS)			Metallic Structure (MS)		
	Cost			Cost		
	Economic		Environmental	Economic		Environmental
	ACCD (Spain)	European	CF	ACCD (Spain)	European	CF
	($€/m^2$)	($€/m^2$)	($kgCO_2/m^2$)	($€/m^2$)	($€/m^2$)	($kgCO_2/m^2$)
Isolated footings	647.64	861.22	0.571	777.11	1033.39	0.740
Reinforced slab	639.05	849.80	0.556	754.80	1003.72	0.720
Piles	697.42	927.42	0.588	821.44	1092.34	0.753

The Romanian construction cost can be translated into Spanish cost by normalization tools such as European Construction Cost data [73,74], the coefficient for Spanish costs is 0.7052 and for Romanian costs, it is 0.464. In the present work the Spanish costs are used for both countries in order to facilitate the results comparison (Table 11).

Regarding the CF, MS buildings produce the highest emissions and pile foundations produce the greatest impact in both types of buildings, followed by isolated footings and, lastly, reinforced slab. These results are similar to others [68]. Materials (including their transport) are responsible for 95–97% of project emissions. The remaining percentage corresponds to machinery.

Muñoz et al. (2012) [75] carried out a study of the CF of social housing built in Chile, focusing on the LCA of construction materials, which included its implementation. The results showed that the energy of commissioning is negligible, while 35% corresponds to the extraction and manufacture of materials, and 65% to the use and maintenance. In addition, there are bibliographic reviews related to the use of the CF indicator in construction [28], however, the results are not always comparable due to the absence of a methodology that follows international standards [29]. For this reason, studies have also been carried out in recent years to establish scales that allow for defining reasonable ranges of CO_2 emissions in construction processes [30].

5. Conclusions

The OERCO2 tool is valid to compare constructive solutions between different project partner countries (in this case, Romania and Spain), because the tool includes the representative characteristics of the buildings in the partner countries. The tool allows for analyzing different constructive systems, in this case, the different types of foundations, structures, masonry, or installations that have been proposed. Therefore, it is possible to determine which constructive solution has the least impact from an economic and environmental point of view. According to the analyzed cases, the most efficient typologies are reinforced concrete buildings, with a significant difference with respect to metallic structures, both economically and environmentally.

For the 48 analyzed typologies, the families of materials that produce the most emissions are: metals/alloys, concretes/cements, and ceramics/bricks and, to a lesser extent, plastics. The structures, installations, foundations, and finishes produce the greatest impacts.

The OERCO2 tool evaluates a project's CF and economic impact simultaneously at the design stage and in detail, according to the project phases, such as earthwork, foundation, structure, etc. Decisions can be made regarding the construction systems and materials used in order to reduce emissions and economic impact, thus helping to understand and improve the eco-efficiency of projects. Therefore, OERCO2 can be a useful educational tool for architecture and engineering college students.

Of the types of foundation assessed both economically and environmentally, piles produce the highest emissions and cost, and the best option is reinforced slabs followed by insulated footings. The building ground floor use, as a dwelling or premises, also influences cost and emissions, especially in phases such as masonry, installations, or finishes.

All these analyses are important to decide how to design in the most environmentally friendly and economical way. The tool facilitates the decision-making of promoters and technicians, without needing prior knowledge about environmental indicators.

Supplementary Materials: The following are available online at http://www.mdpi.com/2071-1050/12/17/6745/s1, Table S1: Unit quantities (Qi) according to OERCO2 coding in Romanian projects based on building typology and construction characteristics, Table S2: Unit quantities (Qi) according to OERCO2 coding in Spanish Projects based on building typology and construction characteristics.

Author Contributions: All the authors have equally contributed to the final version of this paper. All authors have read and agreed to the published version of the manuscript.

Funding: This work has received funding from the University of Seville and the Junta de Andalucía for the consolidation of the research groups. Reference: TEP-172. 1821012924 (2020/595) and the VI Own Research and Transfer Plan of the University of Seville (VI PPIT-US). This research is also funded by the project: Andalusian Ecological Certificate for Building Construction According to Environmental Indicators Carbon, Ecological and Water Footprint (CEACE). Reference AT17_5913_USE. Financed within the PAIDI 2020 call: Knowledge Transfer Activities by the Ministry of Economy and Knowledge of the Andalusian Government (Spain).

Acknowledgments: This research has been carried out thanks to the OERCO2 project (code 2016-1-ES01-KA203-025422), an ERASMUS + project co-financed by the European Union and within the framework of an initiative in 2016 (KA2, strategic alliances in the field of education superior), with the support of the Spanish Service for the Internationalization of Education (SEPIE, Spain).

Conflicts of Interest: The authors declare no conflict of interest.

References

1. European Parliament—Council of the European Union DIRECTIVE (EU) 2018/844 of the European Parliament and of the Council of 30 May 2018 amending Directive 2010/31/EU on the Energy Performance of Buildings and Directive 2012/27/EU on Energy efficiency (Text with EEA Relevance). Available online: http://data.europa.eu/eli/dir/2018/844/oj (accessed on 1 August 2020).
2. UNE, Normalización Española. *UNE-EN ISO 14020 Environmental Labels and Declarations—General Principles*; UNE: Madrid, Spain, 2002.
3. UNE, Normalización Española. *UNE-EN ISO 14025 Environmental Labels and Declarations—Type III Environmental Declarations—Principles and Procedures*; UNE: Madrid, Spain, 2006.
4. UNE, Normalización Española. *UNE-EN 15804 Sustainability of Construction Works—Environmental Product Declarations—Core Rules for the Product Category of Construction Products*; UNE: Madrid, Spain, 2012.
5. UNE, Normalización Española. *UNE-EN ISO 14021 Environmental Labels and Declarations—Self-Declared Environmental Claims (Type II Environmental Labelling)*; UNE: Madrid, Spain, 2017.
6. UNE, Normalización Española. *UNE-EN ISO 14040 Environmental Management—Life Cycle Assessment—Principles and Framework*; UNE: Madrid, Spain, 2006.

7. UNE, Normalización Española. *UNE-EN ISO 14044 Environmental Management—Life Cycle Assessment—Requirements and Guidelines*; UNE: Madrid, Spain, 2006.

8. UNE, Normalización Española. *UNE-EN 15978 Sustainability of Construction Works. Assessment of Environmental Performance of Buildings. Calculation Methods*; UNE: Madrid, Spain, 2012.

9. UNE, Normalización Española. *UNE-EN ISO 14001 Environmental Management Systems—Requirements with Guidance for Use*; UNE: Madrid, Spain, 2015.

10. International Organization for Standardization (ISO). *ISO 15686-5 Buildings and Constructed Assets—Service Life Planning—Part 5: Life-Cycle Costing*; ISO: Geneva, Switzerland, 2017.

11. Giesekam, J.; Barrett, J.R.; Taylor, P. Construction sector views on low carbon building materials. *Build. Res. Inf.* **2016**, *44*, 423–444. [CrossRef]

12. Giesekam, J.; Barrett, J.; Taylor, P.; Owen, A. The greenhouse gas emissions and mitigation options for materials used in UK construction. *Energy Build.* **2014**, *78*, 202–214. [CrossRef]

13. Asif, M.; Muneer, T.; Kelley, R. Life cycle assessment: A case study of a dwelling home in Scotland. *Build. Environ.* **2007**, *42*, 1391–1394. [CrossRef]

14. Dimoudi, A.; Tompa, C. Energy and environmental indicators related to construction of office buildings. *Resour. Conserv. Recycl.* **2008**, *53*, 86–95. [CrossRef]

15. Radhi, H.; Sharples, S. Global warming implications of facade parameters: A life cycle assessment of residential buildings in Bahrain. *Environ. Impact Assess. Rev.* **2013**, *38*, 99–108. [CrossRef]

16. You, F.; Hu, D.; Zhang, H.; Guo, Z.; Zhao, Y.; Wang, B.; Yuan, Y. Carbon emissions in the life cycle of urban building system in China-A case study of residential buildings. *Ecol. Complex.* **2011**, *8*, 201–212. [CrossRef]

17. Cellura, M.; Guarino, F.; Longo, S.; Mistretta, M. Energy life-cycle approach in Net zero energy buildings balance: Operation and embodied energy of an Italian case study. *Energy Build.* **2014**, *72*, 371–381. [CrossRef]

18. Scheuer, C.; Keoleian, G.A.; Reppe, P. Life cycle energy and environmental performance of a new university building: Modeling challenges and design implications. *Energy Build.* **2003**, *35*, 1049–1064. [CrossRef]

19. Buyle, M.; Braet, J.; Audenaert, A. Life cycle assessment in the construction sector: A review. *Renew. Sustain. Energy Rev.* **2013**, *26*, 379–388. [CrossRef]

20. Ramesh, T.; Prakash, R.; Shukla, K.K. Life cycle energy analysis of buildings: An overview. *Energy Build.* **2010**, *42*, 1592–1600. [CrossRef]

21. Schwartz, Y.; Raslan, R.; Mumovic, D. The life cycle carbon footprint of refurbished and new buildings—A systematic review of case studies. *Renew. Sustain. Energy Rev.* **2018**, *81*, 231–241. [CrossRef]

22. Cabeza, L.F.; Rincón, L.; Vilariño, V.; Pérez, G.; Castell, A. Life cycle assessment (LCA) and life cycle energy analysis (LCEA) of buildings and the building sector: A review. *Renew. Sustain. Energy Rev.* **2014**, *29*, 394–416. [CrossRef]

23. Chau, C.K.; Leung, T.M.; Ng, W.Y. A review on life cycle assessment, life cycle energy assessment and life cycle carbon emissions assessment on buildings. *Appl. Energy* **2015**, *143*, 395–413. [CrossRef]

24. Marrero, M.; Wojtasiewicz, M.; Martínez-Rocamora, A.; Solís-Guzmán, J.; Alba-Rodríguez, M.D. BIM-LCA Integration for the Environmental Impact Assessment of the Urbanization Process. *Sustainability* **2020**, *12*, 4196. [CrossRef]

25. Weidema, B.P.; Thrane, M.; Christensen, P.; Schmidt, J.; Løkke, S. Carbon Footprint. A Catalyst for Life Cycle Assessment? *J. Ind. Ecol.* **2008**, *12*, 3–6. [CrossRef]

26. Cagiao, J.; Gómez, B.; Doménech, J.L.; Gutiérrez Mainar, S.; Gutiérrez Lanza, H. Calculation of the corporate carbon footprint of the cement industry by the application of MC3 methodology. *Ecol. Indic.* **2011**, *11*, 1526–1540. [CrossRef]

27. Bare, J.C.; Hofstetter, P.; Pennington, D.W.; Haes, H.A.U. Midpoints versus endpoints: The sacrifices and benefits. *Int. J. Life Cycle Assess.* **2000**, *5*, 319–326. [CrossRef]

28. Geng, S.; Wang, Y.; Zuo, J.; Zhou, Z.; Du, H.; Mao, G. Building life cycle assessment research: A review by bibliometric analysis. *Renew. Sustain. Energy Rev.* **2017**, *76*, 176–184. [CrossRef]

29. Dossche, C.; Boel, V.; De Corte, W. Use of Life Cycle Assessments in the Construction Sector: Critical Review. *Procedia Eng.* **2017**, *171*, 302–311. [CrossRef]

30. Chastas, P.; Theodosiou, T.; Kontoleon, K.J.; Bikas, D. Normalising and assessing carbon emissions in the building sector: A review on the embodied CO_2 emissions of residential buildings. *Build. Environ.* **2018**, *130*, 212–226. [CrossRef]

31. Oregi Isasi, X.; Tenorio, J.A.; Gazulla, C.; Zabalza, I.; Cambra, D.; Leao, S.O.; Mabe, L.; Otero, S.; Raigosa, J. SOFIAS—Software for life-cycle assessment and environmental rating of buildings. *Inf. De La Constr.* **2016**, *68*. [CrossRef]

32. SpainGBC VERDE Tool Website. Available online: http://www.gbce.es/es/pagina/herramientas-de-evaluacion-de-edificios (accessed on 22 November 2019).

33. BREEAM.ES Website. Available online: http://www.breeam.es/ (accessed on 9 November 2019).

34. Asociación Ecómetro Ecometro LCA Tool Website. Available online: http://acv.ecometro.org/ (accessed on 9 November 2019).

35. SOFIAS Project SOFIAS Project Website. Available online: http://www.sofiasproject.org/ (accessed on 9 November 2019).

36. e2CO2cero Tool Website. Available online: http://tienda.e2co2cero.com/ (accessed on 9 November 2019).

37. Ecoinvent Centre Ecoinvent Database. Available online: http://www.ecoinvent.org/database/ (accessed on 15 November 2019).

38. Martínez-Rocamora, A.; Solís-Guzmán, J.; Marrero, M. LCA databases focused on construction materials: A review. *Renew. Sustain. Energy Rev.* **2016**, *58*, 565–573. [CrossRef]

39. PRé Sustainability SimaPro 8. Available online: https://simapro.com/ (accessed on 9 November 2019).

40. Council, G.S.B. DGNB. German Sustainable Building Council. Available online: https://www.dgnb.de/en/index.php (accessed on 7 June 2020).

41. Romania Green Building Council GREEN HOMES Certification System. Available online: www.rogbc.org (accessed on 7 June 2020).

42. Integrated Environmental Solutions Limited Intelligent Communities Lifecycle. Available online: https://www.iesve.com/ (accessed on 26 November 2019).

43. Popiţ, G.-E.; Baciu, C.; Rédey, A.; Frunzeti, N.; Ionescu, A.; Yuzhakova, T.; Popovici, A. Life cycle assessment (LCA) of municipal solid waste management systems in Cluj County, Romania. *Environ. Eng. Manag. J.* **2017**, *16*, 47–57.

44. 360 Optimi. Available online: https://www.360optimi.com/app/sec/main/list (accessed on 5 July 2020).

45. Colliers International Romania. Available online: https://www.oneclicklca.com/case-study-warehouse-romania-leed/ (accessed on 4 July 2020).

46. Living Building Challenge Standard 3.0. Available online: https://living-future.org/ (accessed on 5 July 2020).

47. University of Seville and Partners OERCO2. Available online: http://oerco2.eu/es/ (accessed on 21 June 2020).

48. Solís-Guzmán, J.; Rivero-Camacho, C.; Alba-Rodríguez, D.; Martínez-Rocamora, A. Carbon footprint estimation tool for residential buildings for non-specialized users: OERCO2 project. *Sustainability* **2018**, *10*, 1359. [CrossRef]

49. Proiectare, P.D.E. *COD DE PROIECTARE SEISMICĂ—PARTEA I—PENTRU CLĂDIRI*; Ministerul Lucrărilor Publice, Dezvoltării și Administrației: Bucharest, Romania, 2013.

50. González-Vallejo, P.; Marrero, M.; Solís-Guzmán, J. The ecological footprint of dwelling construction in Spain. *Ecol. Indic.* **2015**, *52*, 75–84. [CrossRef]

51. Solís-Guzmán, J.; González-Vallejo, P.; Martínez-Rocamora, A.; Marrero, M. The Carbon Footprint of Dwelling Construction in Spain. In *The Carbon Footprint Handbook*; Taylor and Francis Group: Abingdon, UK, 2015; Volume 1, pp. 261–263. ISBN 978-1-4822-6222-3.

52. Solís-Guzmán, J.; Marrero, M.; Ramírez-de-Arellano, A. Methodology for determining the ecological footprint of the construction of residential buildings in Andalusia (Spain). *Ecol. Indic.* **2013**, *25*, 239–249. [CrossRef]

53. Solís-Guzmán, J.; Marrero, M.; Montes-Delgado, M.V.; Ramírez-de-Arellano, A. A Spanish model for quantification and management of construction waste. *Waste Manag.* **2009**, *29*, 2542–2548. [CrossRef]

54. Martínez-Rocamora, A.; Solís-Guzmán, J.; Marrero, M. Ecological footprint of the use and maintenance phase of buildings: Maintenance tasks and final results. *Energy Build.* **2017**, *155*, 339–351. [CrossRef]

55. Marrero, M.; Puerto, M.; Rivero Camacho, C.; Freire Guerrero, A.; Solís-Guzmán, J. "Assessing the economic impact and ecological footprint of construction and demolition waste during the urbanization of rural land". *Resour. Conserv. Recycling. Resour. Conserv. Recycl.* **2017**, *117*, 160–174. [CrossRef]

56. González-Vallejo, P.; Muñoz-Sanguinetti, C.; Marrero, M. Environmental and economic assessment of dwelling construction in Spain and Chile. A comparative analysis of two representative case studies. *J. Clean. Prod.* **2019**, *208*, 621–635. [CrossRef]

57. Marrero, M.; Ramírez de Arellano Agudo, A. The building cost system in Andalusia: Application to construction and demolition waste management. *Constr. Manag. Econ.* **2010**, *28*, 495–507. [CrossRef]
58. Consejería de Fomento y Vivienda Junta de Andalucía. Consejería de Fomento y Vivienda. Available online: https://www.juntadeandalucia.es/organismos/fomentoinfraestructurasyordenaciondelterritorio/areas/vivienda-rehabilitacion/planes-instrumentos/paginas/bcca-sept-2017.html (accessed on 3 July 2020).
59. Marrero, M.; Rivero-camacho, C.; Alba-rodríguez, M.D. What are we discarding during the life cycle of a building? Case studies of social housing in Andalusia, Spain. *Waste Manag.* **2020**, *102*, 391–403. [CrossRef] [PubMed]
60. González-Vallejo, P. Herramienta para la estimación de costes económicos y ambientales en el ciclo de vida de edificios residenciales. Fase de construcción. *Hábitat SustenTable* **2018**, *8*, 32–51. [CrossRef]
61. Alba Rodríguez, D.; Ferreira Sánchez, A.; Marrero, M.; Ramirez de Arellano Agudo, A. Ecological Footprint of Building Recovery: Foundation Pinning Under Emergency Conditions. In Proceedings of the II International Congress on Sustainable Construction and Eco-Efficient Solutions, Seville, Spain, 25–27 May 2015; pp. 35–45.
62. González-Vallejo, P.; Solís-Guzmán, J.; Llácer, R.; Marrero, M. La construcción de edificios residenciales en España en el período 2007-2010 y su impacto según el indicador Huella Ecológica. *Inf. De La Construcción* **2015**, *67*, e111. [CrossRef]
63. Instituto para la Diversificación y Ahorro de la Energía (IDAE) Conducción Eficiente de Vehículos Industriales. *Informe del Instituto para la Diversificación y Ahorro de la Energía*; IDAE: Madrid, Spain, 2012.
64. REE. *Red Eléctrica Española Avance*; Red Eléctrica de España: Madrid, Spain, 2015.
65. Autoritatea Nationala de Reglementare in Domeniul Energiei. Available online: https://www.anre.ro/en/about-anre/annual-reports-archive (accessed on 21 June 2020).
66. Freire-Guerrero, A.; Alba-Rodríguez, M.D.; Marrero, M. A budget for the ecological footprint of buildings is possible: A case study using the dwelling construction cost database of Andalusia. *Sustain. Cities Soc.* **2019**, *51*, 101737. [CrossRef]
67. Spain Government. *Instrucción de Hormigón Estructural. Structural Concrete Instruction, Normative Series 2011*; Spain Government: Madrid, Spain, 2008.
68. Solís-Guzmán, J.; Rivero-Camacho, C.; Tristancho, M.; Martínez-Rocamora, A.; Marrero, M. Software for Calculation of Carbon Footprint for Residential Buildings. In *Environmental Footprints and Eco-Design of Products and Processes*; Springer: Berlin, Germany, 2020; pp. 55–79.
69. National Institute of Statistics of Romania. Organization and Functioning of Official Statistics in Romania. Available online: http://www.insse.ro/cms/en/content/official-statistics-romania (accessed on 5 February 2019).
70. Romanian Government. *Strategia Națională a Locuirii 2017–2030*; Romanian Government: Bucharest, Romania, 2017.
71. Bruno, M.; Pleskovic, B. *Proceedings of the World Bank Annual Conference on Development Economics, 1994*; World Bank Group: Washington, DC, USA, 1995; ISBN 0821329081.
72. Instituto Nacional de Estadísticas (INE) de España. *Censo de Población y Viviendas*; INE: Madrid, Spain, 2001.
73. De Groot, I.B.; Bucknal, R.L. European Construction Costs. Available online: http://constructioncosts.eu/ (accessed on 3 August 2020).
74. Vázquez-López, E.; Garzia, F.; Pernetti, R.; Solís-Guzmán, J.; Marrero, M. Assessment model of end-of-life costs and waste quantification in selective demolitions: Case studies of nearly zero-energy buildings. *Sustainability* **2020**, *12*, 6255. [CrossRef]
75. Muñoz, C.; Zaror, C.; Saelzer, G.; Cuchí, A. Estudio del flujo energético en el ciclo de vida de una vivienda y su implicancia en las emisiones de gases de efecto invernadero, durante la fase de construcción Caso Estudio: Vivienda Tipología Social. Región del Biobío, Chile. *Rev. De La Construcción* **2012**, *11*, 125–145. [CrossRef]

 *sustainability*

Article

Energy and Economic Life Cycle Assessment of Cool Roofs Applied to the Refurbishment of Social Housing in Southern Spain

Antonio Dominguez-Delgado [1,*] and Helena Domínguez-Torres [2] and Carlos-Antonio Domínguez-Torres [3]

[1] Department of Applied Mathematics 1, Escuela Técnica Superior de Arquitectura, Universidad de Sevilla, Avda. Reina Mercedes 2, 41012 Sevilla, Spain
[2] Facultad de Económicas, Universidad de Sevilla, Avda. Ramón y Cajal, 41012 Sevilla, Spain; heldomtor@alum.us.es
[3] Escuela Técnica Superior de Arquitectura, Universidad de Sevilla, Avda. Reina Mercedes 2, 41012 Sevilla, Spain; cardomtor@alum.us.es
* Correspondence: domdel@us.es

Received: 8 June 2020; Accepted: 6 July 2020; Published: 12 July 2020

Abstract: Energy refurbishment of the housing stock is needed in order to reduce energy consumption and meet global climate goals. This is even more necessary for social housing built in Spain in the middle of the last century since its obsolete energy conditions lead to situations of indoor thermal discomfort and energy poverty. The present study carries out a life cycle assessment of the energy and economic performance of roofs after being retrofitted to become cool roofs for the promotion of social housing in Seville (Spain). Dynamic simulations are made in which the time dependent aging effect on the energy performance of the refurbished cool roofs is included for the whole lifespan. The influence of the time dependent aging effect on the results of the life cycle economic analysis is also assessed. A variety of scenarios are considered in order to account for the aging effect in the energy performance of the retrofitted cool roofs and its incidence while considering different energy prices and monetary discount rates on the life cycle assessment. This is made through a dynamic life cycle assessment in order to capture the impact of the aging dynamic behavior correctly. Results point out significant savings in the operational energy. However, important differences are found in the economic savings when the life cycle analysis is carried out since the source of energy and the efficiency of the equipment used for conditioning strongly impact the economic results.

Keywords: cool roof; energy efficiency; social housing; energy saving; dynamic numerical method; life cycle assessment

---

1. Introduction

Over the last few decades, there has been growing concern about global energy demand due to the exponential increase in its consumption. As stated in the document [1] from the European Commission, tackling energy consumption in European buildings is vital. According to this publication, nearly 40% of final energy consumption is attributable to buildings across the public and private sector. Because of this, extensive research in the field of energy efficiency has focused on the need to reduce energetic costs in the thermal conditioning of buildings in order to meet the directives set by the EU for H2030 [2], which establishes as a target for 2030 an improvement in energy efficiency of at least 32.5%.

In this research framework, the work focused on the use of techniques that take advantage of environmental and climatic resources in a suitable form is noteworthy. In warm climates, as in southern Spain, the high energy consumption needed to obtain internal comfort in buildings during the hot season poses a major problem. This is especially severe for the case of the social park housing built in southern Spain before the implementation of the first Spanish legislation aimed at regulating energy demand in buildings, NBE-CT-79 [3]. In this housing, the envelopes in general, and the roofs in particular, lack thermal insulation. Taking into account that in this region of Spain, the climate is dry and warm in the cooling season with usually clear skies at this time of year, it seems appropriate to take advantage of the radiative exchange with the sky by means of cold roof techniques to reduce the cooling load.

The European Cool Roofs Council [4] defines a "cool roof" as a roofing system able to reject solar heat and keep surfaces cooler under the sun. A "cool roof" is usually obtained by applying cold materials to the external surface of a roof. Its thermal performance is due to the properties of these materials, which are characterized by high values of solar reflectivity and thermal emissivity. Thereby, these kinds of materials are able to reduce the solar radiation absorption while releasing the heat absorbed by the roof. This way, during the day, the amount of absorbed solar radiation is reduced, and during the night, there is roof cooling because of the radiative exchange with the sky, which usually is colder. Thus cool roofs are considered a passive radiative cooling technique [5,6].

For cold materials, there are slightly differing definitions. Thereby, for ES-ENERGY STAR [7], cold materials must have a solar reflectance initially equal to or higher than 0.65, and after three years, it must be higher than 0.50, whilst APEC Energy Working Group [8] considered that they should have at least a reflectance of 0.70 and a thermal emissivity of 0.75.

The usefulness of cool roofs to enhance comfort conditions and reduce energy consumption in dwellings and more collective use buildings, schools, office buildings, libraries, etc., has been highlighted in a large number of experimental and theoretical studies. For example, Akbari et al. [9] monitored six types of commercial building roofs in California (USA) and found that increasing the solar reflectance of roofs by 0.33–0.60 reduced the peak temperatures by 33–42 °C and the daily cooling energy consumption by 4%, 18%, and 52% in a cold storage facility, a school building, and a department store building, respectively. Romeo and Zinzi [10] reported from field measurements that the application of a cool paint reduced the cooling load by 54% for a roof with an area of 700 m^2, while the peak temperatures of the external roof surface and the indoor air were reduced by 20 °C and 2.3 °C, respectively.

However, the extent of the benefits obtained by the use of cool roofs is found to depend on the local climatic conditions and on a number of other factors such as energy prices, the conditioning equipment, the building use, and the aging effect of the roof coating [11–22]. Likewise, benefits resulting from the decrease of the heat island effect together with an improvement of comfort in urban environments have been shown [13,14,17].

From an economic perspective, in [23], the usefulness of cool roofs was shown to produce significant savings when used to retrofit commercial buildings in the USA. Likewise, Reference [19] found that a high scale implementation of cool roofs in Andalusia, in the south of Spain, could potentially save, considering only residential buildings with flat roofs using electrical heating, 59 million euros annually in electricity costs, and the emission of 136,000 metric tons of CO_2 could be directly avoided every year from the production of electricity.

By contrast, only a limited number of studies have dealt with with the cost-effectiveness of cool roofs based on life cycle analysis. In [24], the cost-effectiveness of reflective white and colored roofs compared to conventional gray roofs in six selected cities in Mexico was analyzed. Based on the costs of electricity and reflective materials, a 10 year life cycle cost analysis showed that, in the absence of insulation, reflective white and colored roofs were more cost-effective than gray roofs for all locations. Zhang et al. [25] demonstrated the cost-effectiveness of using a cool paint in both unventilated and ventilated concrete roofs under the tropical climate in Singapore. Jo et al. [26] carried out research by including both on-site data and a building energy simulation model in order to quantify the electricity savings of a commercial building when replacing the existing dark roofing material with a reflective cool roof system. A 20 year cost benefit analysis, including the additional costs of cool roof retrofit and maintenance, showed that a 100% cool roof installation resulted in a savings of approximately $22,000 per year in energy costs and a consequent nine year payback period for the added cost. Yuan et al. [27] stove to find an optimal combination of the surface reflectivity and the insulation thickness of exterior walls based on a 10 year life cycle cost analysis. Optimum combinations of surface reflectivity and insulation thickness for six different regions in Japan were proposed.

However, the long-term performance of the reflective material was not taken into account in most of the papers referenced. In [28], a 20 year life cycle cost analysis proved that the cost-effectiveness of aged and restored cool roof when used to retrofit non-insulated roofs led to a net savings of up to 44.53 Tunisian dirhamsper m^2 and a payback period of 3.4 years. This work showed the interest in performing life cycle analysis in order to draw energy and economic conclusions. In [29], a new and structured approach was developed to carry out an uncertainty and sensitivity analysis in Life Cycle Assessment (LCA) in order to support the decision-making process in building renovation. Likewise, in [30], a life cycle assessment was used in order to make an economic-environmental valuation of a standard energy retrofit project for a public building in a Mediterranean area.

Moreover, the energy efficiency improvements that are achieved by means of the refurbishment of roofs when becoming cool roofs can be framed into sustainable energy transitions and the decarbonization process. However, the sustainability of an action whose aim is to improve energy efficiency must be ensured for its entire life cycle [31].

Considering that the energy refurbishment of the housing stock is needed in order to reduce energy consumption and that this is especially much more necessary for social housing built in Spain in the middle of the last century that has obsolete energy conditions, leading to situations of indoor thermal discomfort and energy poverty, the analysis carried out in this work tries to clarify the pertinence of using cool roofs as an efficient measure to reduce the energy consumption to obtain indoor comfort conditions for the considered housing park.

In the present paper, a study is proposed in order to appraise in a deeper way the effect of the aging on the energy performance of cool roofs, as well as the implications for the economic and energy life cycle assessment of this kind of roof when applied to the refurbishment of the social housing stock considered. This requires specific simulation codes that allow the incorporation of aging patterns to calculate the energy dynamic of the cool roof accurately when its solar reflectivity decreases because of the aging effect. Commercial packages for building energy simulation do not incorporate this ability, and this gap requires the development of specific simulation codes like the one presented here. This software limitation can be the cause of the absence of studies dealing with numerical computations of the time dependent aging effect for cool roofs. To the best of our knowledge, there exists a gap in the literature with regard to studies that perform dynamical life cycle assessment of cool roofs and analyze its energy and economic implications when used for the refurbishment of social dwelling roofs. Such a gap is particularly relevant in the geographical area that is analyzed here and that is characterized by high cooling demands.

This way, a comprehensive numerical study is done to simulate the energy performance of cool roofs when used to retrofit roofs from residential buildings belonging to the social park housing built in Seville, Spain, before the implementation of the first Spanish legislation aimed at regulating energy demand in buildings, NBE-CT-79 [3], in 1979. The main objective is to perform an energy and economic life cycle assessment of the cool roofs in a broad framework taking into account the effect on the LCA results of the real weather of Seville, the long-term performance of the coating layer solar reflectivities, the energy and equipment types used for conditioning the buildings, and the impact of economic parameters. Regarding the choice of the case study, it was taken into consideration that the weather of Sevilla makes this city a good place to analyze the potential advantages of the use of cool roofs in order to reduce energy consumption to achieve interior comfort conditions. Finally, it is necessary to point out that the findings of this study can be extended to other geographic scenarios as long as they are characterized by similar weather conditions to those existing in Sevilla. This would be the case of southern Spain and most of the meridional regions of Europe.

2. Methodology to Estimate the Roof Energy Performance

2.1. Physical Model

In this section, a generic description of the physical problem involving the heat transfer through the roof is set out. Specifically, the heat transfer in the roofs is determined by:

- The heat gain on the outer slab due to solar irradiation.
- The heat exchange by radiation between the outer surface and the sky.
- The heat exchange by convection between the outer surface and the ambient air.
- The heat transfer by conduction through the layers of the roof.
- The heat exchange by convection and radiation between the internal surface of the roof and the interior of the building.

Other possible factors, such as the radiative exchange between the roof and adjacent buildings or vegetable masses higher than the studied building, were not taken into account in this study for the sake of brevity.

In order to establish the physical model, the fact that the equations describing the heat diffusion equations through the different layers of the roof and the radiative exchanges must be computed in each time step in order to approximate the heat transfer through the roof adequately must be taken into account.

Heat exchanges between surfaces and air were calculated by using convective heat transfer correlations based on the use of suitable convective heat transfer coefficients together with the equations given by the energy balance for the external roof surface; see Section 2.2.2. Internal roof surface energy balance calculation was done by taking a fixed room temperature T_{room} constant for each season. Then, we calculated the exchange of heat by convection and radiation between the internal roof surface and the interior of the building by using a combined convective-radiative transfer coefficient and the energy balance for the internal roof surface; see Section 2.2.2.

2.2. Mathematical Formulation

In this section, the mathematical equations involved in the heat transfer through the roof are explained.

2.2.1. Thermal Conduction through the Roof

Heat conduction through the envelope is modeled by the equation:

$$\rho\, c_p\, \frac{\partial T}{\partial t} = \nabla \cdot (\kappa \nabla T), \tag{1}$$

where the density ρ, the specific heat c_p, and the conductivity κ take the value corresponding to each material of every layer of the envelope.

This equation is closed with boundary conditions for the external and internal surfaces of the roof that are given by the energy balance equation corresponding to each surface, as is explained in Section 2.2.2.

2.2.2. Energy Balance at the Roof Surfaces

For the external surface of the roof, the energy balance is given by:

$$\kappa\frac{\partial T}{\partial \vec{n}} + q_{c,ext} + q^{SW} + q^{LW} = 0, \tag{2}$$

where $q_{c,ext}$ is the intensity of the convective heat flux between the surface and the air flow that is given by:

$$q_{c,ext} = h_{c,ext}(T_a - T), \quad (W/m^2),$$

with T_a the outdoor air dry-bulb temperature, T the exterior roof surface temperature, $\vec{n}$ the outward normal vector to the exterior roof surface, and $h_{c,ext}$ the convective heat transfer coefficient described in Section 2.2.4. Finally, q^{SW} and q^{LW} are respectively the intensity of radiative flux of solar origin and the intensity of the balance of thermal long-wave radiation on the surfaces whose calculation is explained in Section 2.2.3.

For the internal surface of the roof the energy balance is given by:

$$\kappa \frac{\partial T}{\partial \vec{n}} + q_{cr,int} = 0, \tag{3}$$

where now, $\vec{n}$ is the outward normal vector to the interior roof surface, T is the interior roof surface temperature, and $q_{cr,int}$ is the intensity of the combined convective-radiative heat transfer between the internal roof surfaces and the interior temperature. This is given by:

$$q_{cr,int} = h_{conv,rad}(T_{room} - T), \quad (W/m^2), \tag{4}$$

where $h_{conv,rad}$ is the combined convective-radiative heat transfer coefficient explained in Section 2.2.4 and T_{room} is the interior air temperature.

2.2.3. Solar and Long-Wave Radiative Flux

For the external surface of a roof, the absorbed solar radiative flux is given by:

$$q^{SW} = \alpha^{SW} \cdot \left(I_b^{SW} \cdot cos(\theta) \cdot \frac{S_s}{S_{roof}} + I_{sd} \cdot F_{rs} + I_{gd} \cdot F_{rg} + I_{rn} \cdot F_{rn} \right), \quad (W/m^2), \tag{5}$$

where α^{SW} is the solar absorptance of the external roof surface, θ is the angle of incidence on the roof of the sun's rays, I_b^{SW} is the intensity of the beam solar radiation, S_s and S_{roof} are the sunlit and the entire roof areas, respectively, and I_{sd} and I_{gd} are the intensity of the diffuse radiation reflected by the sky and the ground, respectively, F_{rs} being the view factor between the roof and the sky and F_{rg} the view factor between the roof and the ground. Finally, I_{rn} is the intensity of the reflected radiation by the surrounding buildings, and F_{rn} is the the view factor between the roof and such buildings.

In the absence of buildings or shading elements higher than the roof under consideration, F_{rs} and F_{rg} are given by:

$$F_{rs} = \frac{1 + cos(\varphi)}{2}, \qquad F_{rg} = \frac{1 - cos(\varphi)}{2}$$

φ being the angle between the roof plane and the horizontal plane. For an unshaded flat roof, Expression (5) becomes:

$$q^{SW} = \alpha^{SW} \cdot I_H^{SW}, \quad (W/m^2), \tag{6}$$

where I_H^{SW} is the global horizontal solar radiation intensity given by $I_H^{SW} = I_b^{SW} \cdot cos(\theta) + I_{sd}$.

The long-wave radiation balance for the external roof surface is calculated as:

$$q^{LW} = Q_{sky}^{LW} - Q_w^{LW}$$

where Q_w^{LW} is the intensity of the long-wave radiation heat emitted by the external roof surface and Q_{sky}^{LW} is the sky downwelling long-wave radiation.

Q_w^{LW} is calculated using the law of Stefan–Boltzmann:

$$Q_w^{LW} = \epsilon \sigma T^4 \tag{7}$$

ϵ being the surface emissivity, T the temperature in Kelvin degrees of the surface, and $\sigma = 5.67 \cdot 10^{-8}$ [W/m$^2 \cdot$ K^4] the constant of Stefan–Boltzmann.

The sky downwelling long-wave radiation Q_{sky}^{LW} depends on several factors, but the most significant ones are outdoor temperature, the relative humidity of the environment, and cloud cover. Essentially, clouds absorb outgoing IR radiation and emit thermal IR radiation to a temperature higher than emitted by a clear sky. Thus, a cloudy day thermal downwelling sky irradiance can increase over 34% regarding the sky irradiance of a clear sky. Q_{sky}^{LW} can be computed as:

$$Q_{sky}^{LW} = \epsilon_{sky} \sigma T_a^4, \tag{8}$$

where ϵ_{sky} is the sky emissivity. Walton [32] and Clark et al. [33] estimated that ϵ_{sky} can be calculated as:

$$\epsilon_{sky} = (0.787 + 0.764 \ln(\frac{T_{dp}}{273}))(1 + \frac{224}{10^4}n - \frac{35}{10^4}n^2 + \frac{28}{10^5}n^3)$$

where T_{dp} is the absolute dew point temperature and n is the opaque sky cover in tenths. This is the correlation used in this work.

2.2.4. Convective Heat Transfer Coefficients

Suitable Convective Heat Transfer Coefficients (CHTC) for external building surfaces, $h_{c,ext}$, are essential in order to calculate accurately heat transfers between the roof surface and the ambiance air. However, while for vertical surfaces, there is a large number of correlations, for roofs, they are much scarcer. As stated by Mirsadeghi et al. [34], convective heat transfer calculations for roofs represent some of the most complex problems in wind flow around buildings due to the variety in their geometry and the complex flow patterns in the separated regions above the roofs.

In this work, the correlation used for the external surface of roof is $h_{c,ext} = 8.18 + 2.28\,V_R$ (W/m^{2}K), which was proposed by Hagishima and Tanimoto [35] based on experiments carried out on a roof. Here, V_R is the wind speed above the roof in m/s.

To calculate the heat exchange by convection and radiation between the internal roof surface and the interior of the building, a combined convective-radiative transfer coefficient given by $h_{conv,rad} = 8.3$ (W/m^{2}K) is used as recommended by ASHRAE [36]. This coefficient is often used in heat transfer calculation for buildings' interiors [37].

2.3. Aging Effect on the Cool Roof Absorptivity

This paper aims to carry out a long-term assessment of the energy behavior of the cool roofs in order to perform an LCA for a lifetime period of 20 years according to the service time of the roof coating. For such a long period, it is necessary to take into account the loss of the reflective properties of the roof coating over time due to its own aging or to the action of environmental agents as rain, dust, air particles, moisture, and sun.

According to [22], the aging of a paint depends on the characteristics of the paint itself (porosity, glass transition temperatures, water retention capacity). It also depends on the material on which the paint is applied (roughness) and on a number of environmental factors to which the paint is exposed. Mastrapostoli et al. [38], analyzing the weatherization of cool roofs in two Athenian schools, showed that the solar reflectance, after four years, decreased by around 25%. Xue et al. (2015) [39], by making use of a white roof coating based on styrene acrylate copolymer and cement, showed that after 400 h of artificial accelerated weathering, solar reflectance experienced a decrease of 11%.

In [40], it was found from a field survey that for all reflective roofing types, about 95% of aging occurred during the first two years, and 98% occurred within the first three years of installation. The conclusions for unwashed white roof coatings applied to a variety of substrates suggested a loss of solar initial reflectivity of about 20–25%, taking place in the first few months to one year, with little change beyond that and generally little variation by substrate.

According to Bretz and Akbary [41], most of the decrease in solar reflectivity occurred in the first year, possibly in the first few months. Then, values tended to stabilize. Eilert [42] also proved that the reduction in the solar reflectivity of white roofs was in the order of 10–30%, with most of the reduction occurring in the first year. In the field survey [40], it was concluded that the energy calculations should be based on at least one year aged reflectivity values, which were typically 75–80% of the initial values. The same was concluded in [22]. In this study, the results obtained from in-field measurements pointed to a reduction of the solar reflectance between 19 and 25% for different cool paints. In the same work, it was also observed that the aging of the cool coating layer did not significantly affect its thermal reflectivity.

On the other hand, in-field measurements [40,41] found that the original values for the reflectivity could be restored up to 90–100% by a periodic power washing. Nevertheless, the author of this field survey concluded that washing of cool coated roof surfaces as a maintenance measure did not constitute a cost-effective means of retaining or restoring solar-reflectance values, especially considering that most solar reflectance loss would reoccur in three to six months. While washing could usually restore most of the solar reflectivity, at least during the first several years of aging, the effects were short lived (typically three to six months) and yielded limited economic payback.

For the calculation of energy flows through the roofs, four scenarios for the cool coating were considered: roofs without the aging effect, roofs with the aging effect, but without any maintenance, roofs with the aging effect and a wash maintenance after a decade, and finally, aged roofs with quinquennial wash maintenance. Taking into account the social nature of the buildings in which the roof refurbishment was carried out, considering an annual roof washing seemed unlikely, it being more realistic to consider a maintenance wash with quinquennial or decennial periodicity or no maintenance.

Based on what was supported by most of the studies reviewed, the aging pattern considered in the present work for the cool roof coating was the following: throughout the first year, a loss of solar reflectivity equal to 20% of its initial value, concentrated mostly in the first months of the year, and a loss equal to 8.5% in the second year. For the third year, a loss equal to 0.9% was assumed. Then, the decrease in reflectance tended to become stabilized [41], until reaching a total loss of 30% for the whole life cycle span. This way, about 95% of the aging happened during the first two years, and 98% occurred within the first three years after the installation, as is stated in [40]. For the considered case of a wash a decade after the installation, the related literature assumed the solar reflectivity to be restored to 90% of the initial value. Afterwards, the same pattern as that taking place after the initial application of the cool coating was considered.

In Figure 1a, the monthly and accumulated percentages of solar reflectivity loss are shown, according to the aforementioned aging patterns. The evolution of the solar reflectivity for the aged case without maintenance and for the aged case with decennial washing are shown respectively in Figure 1b,c for a roof coating with an initial reflectivity equal to 0.9.

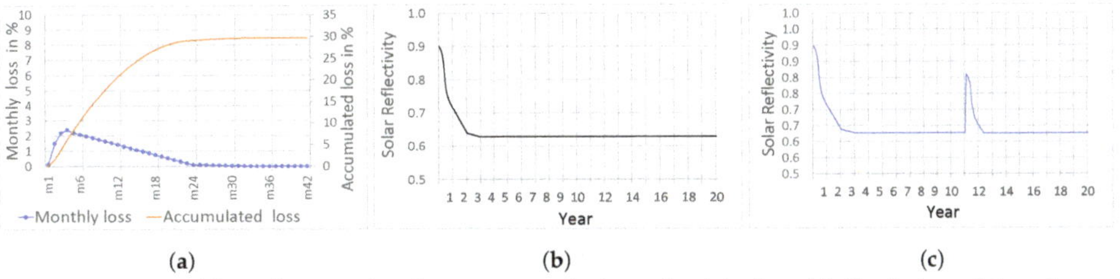

(a)　　　　　　　　　　(b)　　　　　　　　　　(c)

Figure 1. Monthly and accumulated percentage of solar reflectivity loss (**a**). Evolution of the solar reflectivity for the aged case (**b**) and for the aged case with decennial washing (**c**).

2.4. Case Buildings

For our study, a representative building of the social park housing built in Seville before 1979 was considered. This building belonged to the promotion of social housing called El Plantinar built in Seville in the 1960s. The constructive typology of this promotion was typical of social housing in this decade.

The roofs were flat type, and as can be seen in Figures 2 and 3, the promotion was made up of buildings of the same height, while there were no close buildings of higher height that could cast shadows over them. Therefore, the flux of the solar radiation incident on the roofs is given by Expression (6).

Figure 2. El Plantinar: aerial view.

Figure 3. El Plantinar: aerial view.

The roof configuration is described in Table 1, where dimensioning and thermophysical characteristics of the various components of the roofs are shown. The values shown in this table were obtained from [43,44].

The outer surface of the roof was considered covered by a layer of bituminous paint. Then, two absorptivity solar radiation coefficients of 0.8 and 0.9 were studied according to the usual values of the absorptivity for this kind of coating. These values represented the benchmark cases to be compared with the cool roofs.

In order to transform the roof to a cool roof, the application of a white elastomer layer on the surface of the outer layer of the roof was selected. Taking into account the wide variety of existing commercial cool paints and seeking to find their energy and economic performance in a broader way, cool roof coatings with absorptivities equal to 0.1, 0.2, 0.3, and 0.4, which could be easily found in commercial suppliers, were studied.

In Figure 4, the climatic chart of Seville is shown. As can be noted, winter could be characterized as mild and summers as hot and dry with high levels of solar radiation. Finally, autumns and springs were characterized by rainfall and moderate temperatures. According to the data from the Spanish State Meteorological Agency [45], the average annual temperature was 19.2 °C, with maximum average temperatures up to 40 °C in the months of July and August and a minimum average of 5.7 °C in January. The normal incident solar radiation value had a maximum average daily value of 8.3 kWh/m^2 in July and a minimum daily average value of 2.3 kWh/m^2 in December. In accordance with these characteristics, the climate of the area was classified as Mediterranean Csa according to the Köppen–Geiger climate classification.

Table 1. Thermophysical characteristics of the roof.

Layer	Description	Thickness (m)	Density (kg/m^3)	Specific Heat (J/kgK)	Conductivity (W/mK)
1 (Ext.)	Bituminous paint	0.0015	1150	1000	0.23
2	Ceramic tiles	0.005	2000	800	1.00
3	Mortar	0.01	2000	1000	1.40
4	Protective Layer	0.015	1150	1000	0.23
5	Mortar	0.01	2000	1000	1.40
6	Carbon cinders	0.1	640	657	1.40
7	Concrete vault	0.3	1330	1000	1.32
8 (Int.)	Gypsum plaster	0.01	1000	1000	0.32

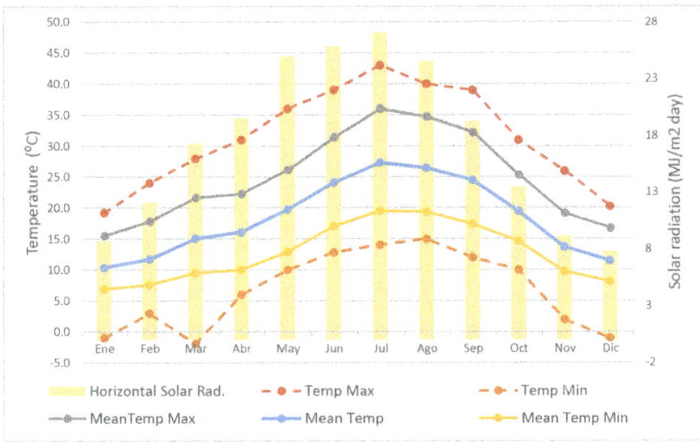

Figure 4. Climatic chart of Seville.

2.5. Numerical Resolution

Equation (1) for the heat transfer through the roof was solved by using a one-dimensional finite difference approximation in space and a semi-implicit Euler method for time. The entire computation process was executed developing a code based on the described numerical method using the open source FreeFem++ software [46]. For the temperature on the solid surfaces, the border conditions were chosen from the energy balance equations as described in Section 2.2.2.

In order to ensure the reliability of the proposed numerical method, it was first validated through an inter-model comparison with the Energy Plus software [47]. This was made by considering short time intervals for the calculation, typically months, due to the fact that Energy Plus does not implement time dependent absorptance since it is only able to deal with constant values of solar absorptance. This feature was implemented in our numerical model by means of allowing the simulation of the aging effect on cool roof absorptance in long time intervals, typically decades, as was done in our LCA calculations. Likewise, from a purely numerical point of view, the method was checked through the analytical calculation of the evolution of the temperatures on the nodes used for the difference finite method calculations by using a low number of nodes for the discretization of the roof [48].

The general calculation process had the following steps:

- Phase I. Preprocess:

  I.A   Input of geometry and materials' data: in this phase, the geometry of the roof was entered and stored, as well as the data of the materials involved in the simulations.

  I.B   Input of meteorological data: wind speed, ambient air temperature, ambient relative humidity, and solar radiation were stored. These values could be taken from standard weather files as the Energy Plus Weather (EPW) files, or from a local meteorological station, or derived from some model.

I.C   The downwelling long-wave radiation incident on the roof was calculated for every time considered in computations. These calculations were performed once and could be incorporated as data in the simulation process.

- Phase II. Time iterations:

For every time step, heat transfer coefficients and radiative heat exchange were upgraded, and the border conditions given by surface energy balances on every surface, Equations (2) and (3), were updated. Finally, the heat transfer Equation (1) for the roof was solved at the current time.

For the interior zone under the roof, we considered a test zone inspired by [28,49,50] assuming the envelope was adiabatic except for the roof. The thermal behavior of the roof depended on the operating mode of the conditioning system. Different modes can be found in the literature: continuous air-conditioning [27,28,51–53], intermittent air-conditioning with a fixed indoor set-point temperature [54], and no air-conditioning with fluctuating indoor temperature [55].

In order to evaluate the energy behavior of the roof and its impact on the energy loads, we assumed that the ambient interior temperature was kept constant for every season. We assumed a continuous mode of the conditioning system with an indoor set-point temperature for the cooling season set at 24 °C, for the heating season set at 20 ° C, and for the intermediate seasons set at 22.5 °C. These values were selected by taking into consideration the comfort temperature intervals established in the Spanish regulation of thermal installations in buildings (RITE) [56]. On the other hand, these values were used in previous research on cool roofs under climatic conditions similar to those of the present study [28].

3. Life Cycle Cost Analysis

An accurate approach to the economics of the cool roof is the assessment of its cost-effectiveness through an LCA that takes into account future costs [57].

This method allows a comparison of future costs with today's costs while evaluating throughout the life cycle period the savings obtained when cool roofs are used.

To make the life cycle cost analysis, all anticipated costs were calculated and discounted to their present worth. The usual procedure is to compute the costs for each future year taking into account the future value of the variables that generate the costs and then to discount each annual cost to its present worth. Finally, the sum of all the present values constituted the life cycle cost.

When the life cycle cost was determined for each of the roofing systems under consideration, including different cool and non-cool roof options and different scenarios for aging, the cost-effectiveness of each system was appraised by its life cycle costs; the lowest was the life cycle cost, the most cost effective was the system, and equivalently, the systems with highest life cycle savings were selected as the most cost effective.

3.1. Methodology

For the LCA, we considered a lifetime of $n = 20$ years, that is the average life service of the considered white reflective coating [58]. Then, the life cycle cost and saving together with the payback period for each of the considered cool roof scenarios were determined. In this analysis, the costs of energy, the performance of the equipment used for air-conditioning, the initial cost of application of the reflective coating, and the maintenance costs derived from the power washing of the cool roofs were considered, as well as the economic variables that drove the economic process: the increase of energy costs i and the monetary discount rate d.

In the LCA analysis, the method P1-P2introduced in [57] was used with some small differences due to the dynamic character of the yearly energy operating costs induced by the aging and maintenance of the cool roofs. To determine the present worth of one monetary unit of the future time period k

(usually expressed in years), with a market discount rate d (fraction per time period), the relationship to be used is:

$$PW_k = 1/(1+d)^k.$$

Then, if $P_{e,C}$ and $P_{e,H}$ are the current prices of energy for cooling and heating, respectively, and if i is the inflation rate for energy costs, the present worth of any future energy payment $C_e(k)$ in the period k is given [57] by:

$$PW_k(C_e(k)) = C_e(k)/(1+d)^k$$

where:

$$C_e(k) = \left(\frac{Q_c(k) \times P_{e,c}}{SEER \times (3.6 \times 10^6)} + \frac{Q_h(k) \times P_{e,h}}{SCOP \times (3.6 \times 10^6)} \right) (1+i)^{k-1} \tag{9}$$

where SEER is the Seasonal Energy Efficiency Ratio and SCOP is the Seasonal Coefficient Of Performance of the equipment used for conditioning. Equation (9) takes into account the fact that the energy loads $Q_c(k)$ and $Q_h(k)$ for the cool roofs varied annually because of the aging effect and the maintenance to which the cool coating was subjected. In the calculation of $C_e(k)$, it was considered that the energy used for cooling was always electricity, but for heating, the costs were evaluated both considering the use of electricity and gas.

Likewise, if i_M is the inflation rate for the costs of maintenance, the present worth of any future maintenance payment $C_M(k)$ in the period k is given by:

$$PW_k(C_M(k)) = C_M(k)/(1+d)^k = \delta(k)\,C_M(1+i_M)^{k-1}/(1+d)^k$$

where $\delta(k)$ is equal to 1 if the maintenance is done in year k or equal to 0 if not and C_M is the current price of maintenance per unit area of roof surface.

Then, the present worth of the life cycle total cost per unit area of the roof surface is given by:

$$C_t = C_e + C_M + C_I$$

where C_e and C_M are the present worth of energy and maintenance life cycle total cost, respectively, calculated as:

$$C_e = \sum_{k=1}^{n} PW_k(C_e(k)) \quad \text{and} \quad C_M = \sum_{k=1}^{n} PW_k(C_M(k)),$$

and C_I is the cost of the initial investment of installing the cool roof coating.

When compared with the initial reference roof, due to the refurbishment, there will be annual operational changes in terms of energy together with the corresponding changes in economic terms related mainly to the energy costs. In order to calculate the difference between the use of the cool roofs and the reference case, we calculated the Net Savings (NS) for the whole life cycle period through:

$$NS = \sum_{k=1}^{n} \left[PW_k(C_e^{(ref)}(k)) - PW_k(C_e(k)) \right] - C_M - C_I. \tag{10}$$

Here, $PW_k(C_e^{(ref)}(k))$ is the present worth value of the energy costs for the reference case in the period k given by:

$$PW_k(C_e^{(ref)}(k)) = C_e^{(ref)}(k)/(1+d)^k,$$

being:

$$C_e^{(ref)}(k) = \left(\frac{Q_c^{(ref)}(k) \times P_{e,c}}{SEER \times (3.6 \times 10^6)} + \frac{Q_h^{(ref)}(k) \times P_{e,h}}{SCOP \times (3.6 \times 10^6)} \right) (1+i)^{k-1} \tag{11}$$

where now $Q_c^{(ref)}(k)$ and $Q_h^{(ref)}(k)$ are the cooling and heating loads, respectively, for the reference roofs in the time period k.

Then, $PW_k(C_e^{(ref)}(k)) - PW_k(C_e(k))$ is the difference in the period k between the energy costs for the reference roof and the cool roof under analysis. A positive value of NS means that the use of the cool coating produces savings with respect to the reference case. It is worth highlighting the fact that energy consumption varies annually due to the aging effect, and therefore, Expression (10) cannot be reduced to a single analytical expression.

The Payback Period (PB) is the time horizon t for which the value of net savings is equal to zero [59]. Because the series coefficients in (10) are variable for the time step k, the value t giving the payback period is obtained through the computations of the net savings accumulated for every horizon k_0:

$$NS(k_0) = \sum_{k=1}^{k_0} \left[PW_k(C_e^{(ref)}(k)) - PW_k(C_e(k)) \right] - C_M - C_I. \tag{12}$$

Then, if $NS(k_0) < 0$ and $NS(k_0 + 1) > 0$, the value of the payback period t is calculated through the expression:

$$t = \frac{-NS(k_0)}{NS(k_0 + 1) - NS(k_0)} + k_0. \tag{13}$$

The value of t provided by (13) is consistent with the usual value found in the literature [28–57] when energy consumption is assumed to be constant for all years.

3.2. Economic Indicators

The variables considered to perform the economic calculations involved in the LCA are listed in Table 2.

Table 2. Economic variables used in the LCA.

Variable	Value	
Cool paint application	9.45	€/m^2
Washing cost	1.63	€/m^2
Electricity cost	0.2403	kWh
Gas cost	0.0736	kWh
Energy inflation rate	0%, 3%, 6%	
Discount rate	0.5%, 1.5%, 3%	
Lifetime	20	years

The costs of the cool paint application and its maintenance were taken from [58]. The electricity and natural gas prices were the prices in Spain (including taxes) for household consumers [60]. Following the approach of [61], three possible values for the energy inflation and the discount rates were considered, as shown in Table 2. Finally, the maintenance discount rate i_M was taken equal to d under the supposition that the evolutions of both indexes were close to each other.

For the conditioning equipment, we considered that cooling was always done by the air-conditioning machinery that used electricity as the only energy source. For heating, the following cases were considered:

- Heating by natural gas.
- Heating by electricity radiators.
- Heating by air (heat pumps).

Since the purchasing power of the inhabitants of these homes was not high, it was considered that the air-conditioning devices were mid-range with an A++ energy rating according to the European regulations, under the assumption that these devices were usually less expensive than those with the energy certification A+++. The efficiency values of the conditioning equipment are shown in Table 3.

Table 3. Conditioning equipment efficiency. SEER, Seasonal Energy Efficiency Ratio; SCOP, Seasonal Coefficient Of Performance.

Device	Efficiency		
Cooling pump	SEER	=	7.3
Heating pump	SCOP	=	4.85
Gas natural heating	η_s	=	0.8
Electrical radiator heating	η_s	=	1

The values shown in Table 3 for the pump were the means of the values established in the EU Regulation 626/2011 [62] for an air-conditioning equipment labeled A++.

4. Results

In this section, the results from the energy and economic LCA analysis are presented.

4.1. Energy Results

4.1.1. Cool Roof Energy Performance

In this section, an analysis of the energy performance of the cool roof is done by using the introduced numerical method applied to the case study with the non-insulated roof described in Section 2.4. In the first stage, the energy performance of the roof was assessed with solar reflectivity covering a range from 0.1 to 0.9. Then, transmission loads per unit area of the roof surface were calculated for the whole climatic year.

In Figure 5, the behavior of the temperatures over a week that covered the last days of July and the first ones of August is shown for a reference roof and a cool roof. For the reference roof, a solar absorptivity equal to 0.9 and a thermal emissivity of 0.85 were considered, while for the cool roof, a solar absorptivity equal to 0.1 and a thermal emissivity of 0.9 were assumed.

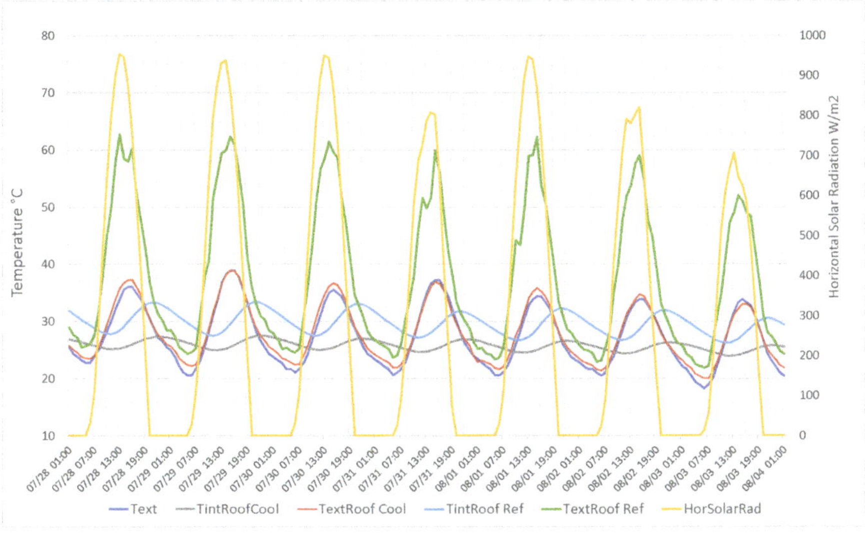

Figure 5. Temperatures for the reference and cool roofs during a week of summer.

In this figure, the typical behavior of the cool roof is observed. Most of the time, the external layer of the cool roof had a temperature close to the ambient temperature, whereas that of the reference roof had temperatures much higher, especially in the central hours of the day, where the solar irradiation was more intense. At night, there was a significant drop in surface temperatures caused by the absence of solar irradiation and by the cooling caused by the radiative exchange with the sky vault. However, the thermal energy accumulated in the mass of the roofs made the temperatures of the surface layers in general higher than the environmental ones, a fact that was much more evident for the reference roof. Moreover, the daily fluctuations of the temperatures were much smaller for the cool roof. This way, cool roofs were better protected against thermal fatigue. This resulted in a longer service life for the roof.

Another important fact was the remarkable difference observed between the temperatures of the interior surfaces of the two types of roofs. Firstly, the temperature of the internal surface of the reference roof was several degrees higher than that of the cool roof, but it also had much stronger oscillations. Both events strongly affected the energy consumption to achieve interior comfort conditions and the sensation of thermal stress.

In Figure 6, the monthly heat flux through the roof is shown, considering inwards heat flux as positive and outwards heat flux as negative. It can be seen that the effect of low absorptivity values was very pronounced in summer. Thus, for an absorptivity of 0.1, the cooling load was significant only for the months of July and August. On the other hand, this low absorptivity value introduced a considerable penalty in wintertime, which resulted in a higher value of the heating load. This combined effect was the basis for for the estimation of the energy savings achieved by the use of cool roofs. In general, it was observed that as the absorptivity value increased, a gradual increase occurred of the cooling load in the warm season and a decrease of the heating load in the heating season.

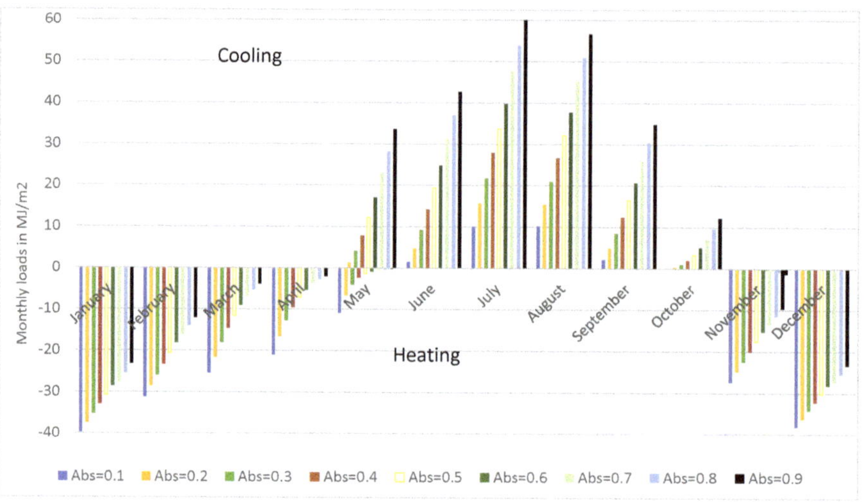

Figure 6. Effect of the solar absorptivity on monthly heat flux through the roof.

Figure 7 shows for the different values of the absorptivity the annual transmission loads per unit area of the roof surface for the case study. From this graph, it can be concluded that the lowest energy consumption values were obtained for the lowest values of solar absorptivity. Specifically, for the values 0.1 and 0.2 of the absorptivity, while the values of total consumption were practically the same, the difference lied in the values of loads for heating and cooling. Such values showed a notable difference according to what was said before: higher levels of solar reflectivity corresponded to lower values of cooling load and higher values of heating load.

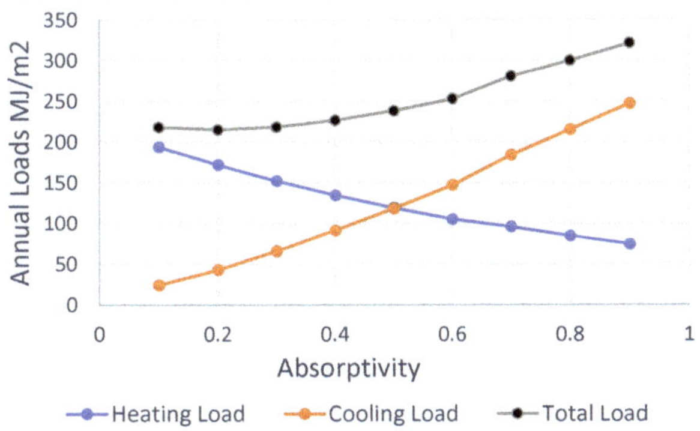

Figure 7. Effect of the solar absorptivity on annual heat flux through the roof.

4.1.2. Aging Effect on the Energy Loads

In Figure 8a–c, the evolutions of the heating, cooling, and total loads are shown for the aged case with decennial wash maintenance for the four absorptivity scenarios studied. It can be observed that the progressive loss of solar reflectivity in the first years produced an increase of the cooling loads and a decrease of the heating for all the cases. After four years, the loads tended to become constant until the decennial wash maintenance was done. At this moment, the initial behavior for the loads was repeated, although a variation of the initial values of the loads was observed according to the fact that the wash recovered only 90% of the initial reflectivity. For the total loads, it was observed that for the value of the absorptivity equal to 0.1, the aging effect produced a decay of the total load. This was due to the fact that, in spite of the cooling load increasing during the summer, such an increase was advantageously offset by the reduction of the winter penalty.

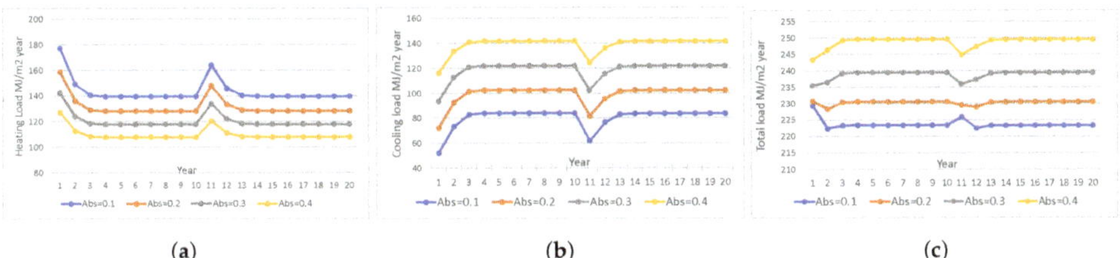

(a)   (b)   (c)

Figure 8. Aged case with decennial washing. Yearly evolutions of the: heating loads (a), cooling loads (b), and total load (c).

For the value of the absorptivity equal to 0.2, the aging effect also produced an initial decrease in the second year for the total load, although it rapidly recovered the initial load values. However, after the ten year maintenance, a somewhat longer further decrease in total loads than at the beginning of the cycle was observed. For the higher absorptivity values, that is 0.3 and 0.4, it was observed that the aging effect, both initially and after the ten year maintenance, resulted in an increase in total loads since in these cases, the increases in cooling loads were not compensated by the reductions in heating loads.

In Figure 9a–c, the evolutions of the heating, cooling, and total loads are shown for the aged case with a quinquennial washing. In these figures, the same patterns as for the decennial washing can be observed although with a five year periodicity, that is, an increase of the heating loads and a decrease of the cooling loads in the first years after every washing together with an initial decrease of the total loads after washing for all the roofs' absorptivities except for the case of the roof absorptivity equal to 0.1, where the aging effect gave rise to the same decrease of the total roof as that of the case of the decennial maintenance.

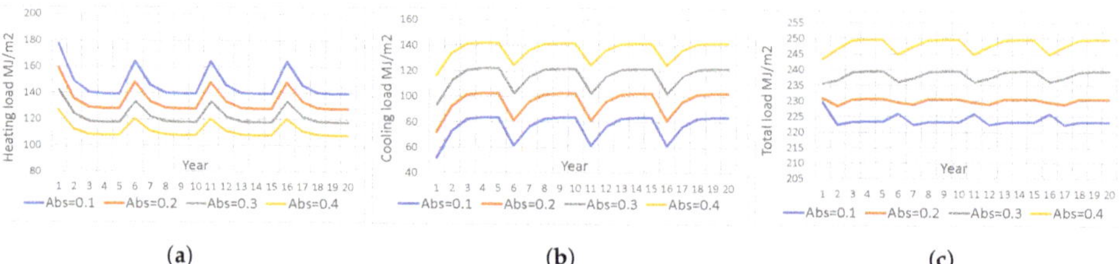

(a)   (b)   (c)

Figure 9. Aged case with quinquennial washing. Yearly evolutions of the: heating loads (**a**), cooling loads (**b**), and total loads (**c**).

Finally, in Figure 10a–c, the evolutions of the heating, cooling, and total loads are shown for the aged case without any maintenance along the entire LCA lifetime period. In this case, the aging effect reproduced the initial pattern observed in the previous cases, that is a decrease of the heating loads and an increase of the cooling loads for all the cool roof scenarios, a decrease of the total loads for the case of absorptivity equal to 0.1 due to the increase of the cooling load being offset by a decrease in the heating load, an initial decrease and a later recovery of the initial load values for the absorptivity equal to 0.2, and finally, for the remaining cases, a net increase of the total loads for the first years, which remained stabilize from the fourth year until the end of the lifetime period.

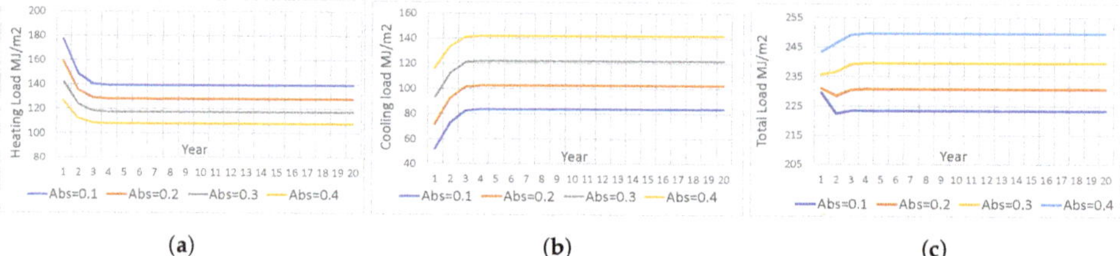

(a)   (b)   (c)

Figure 10. Aged case without maintenance. Yearly evolutions of the: heating loads (**a**), cooling loads (**b**), and total loads (**c**).

In Figure 11, the range of variation of the loads related to the range of variation of the roof absorptivities for the three studied aged cases is shown. The heating variation range curves show the difference between the heating load for the absorptivity equal to 0.1 minus the heating load for an absorptivity equal to 0.4, and the cooling range curve shows the difference between the cooling load for the roof with an absorptivity equal to 0.4 minus the cooling load for an absorptivity equal to 0.1. As can be seen, for all the cases, the effect of the change in absorptivity values was greater on the cooling loads than on the heating ones. In fact, initially, this range of variation was about 65 MJ/m^2 year for the cooling range and of 50 MJ/m^2 year, that is a difference of 15 MJ/m^2 year. However, once the aged effect stabilized the solar absorption values, the difference was close to 30 MJ/m^2 year , that is almost double, and the same pattern was observed after the washing maintenance. In short, the effect of changes in absorptivity produced a noticeably greater impact on the cooling loads than on the heating ones.

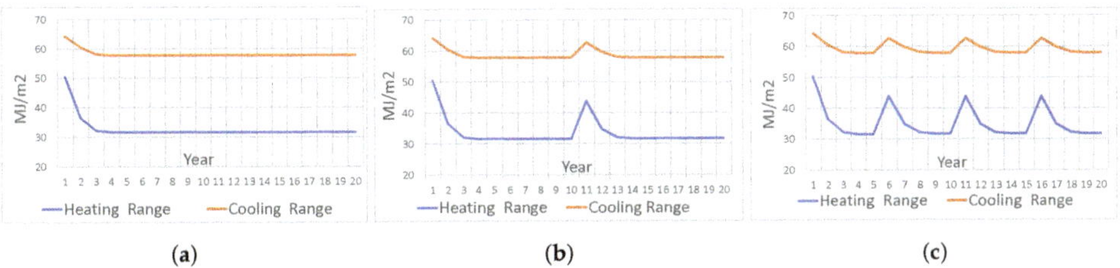

Figure 11. Yearly range variation of the heating and cooling loads for the aged case: without maintenance wash (**a**), with decennial wash (**b**), and with quinquennial wash (**c**).

The total loads in GJ/m^2 along the entire lifetime period for the different cool roof and maintenance scenarios are shown in Figure 12 together with the two reference roofs without retrofitting. It can be noted that for all cool roof cases and scenarios, the total loads were lower than those of the reference roofs. Moreover, it was also observed that the lower the absorptivity, the lower was the total load. Regarding the maintenance regime, no significant differences were found among the total loads, except for the unrealistic case of no aged roof that was only included as an indicator of the need to include the aging effect.

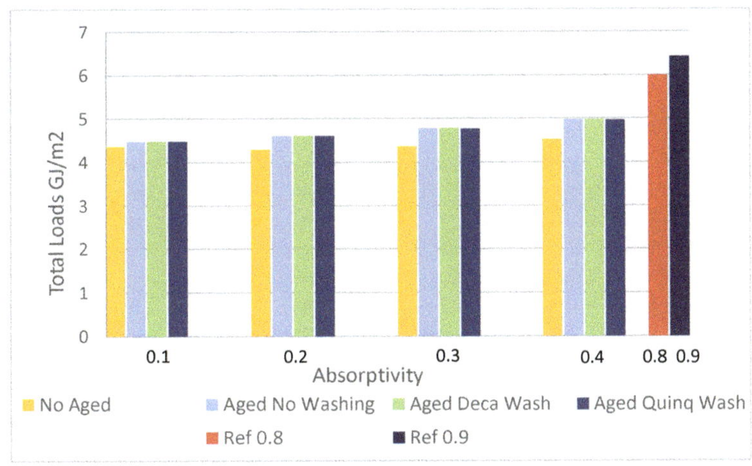

Figure 12. Total loads for the entire LC period.

In Figure 13a,b and Table 4, the total load savings are shown for the entire LC period. In these, it can be observed that the decreases of the total load for the reference roof with $abs_{ref} = 0.9$ were higher than for the reference roof with $abs_{ref} = 0.8$ according to the idea that for warm climates' cool coatings were more efficient in order to reduce the loads the higher was the absorptivity of the roof retrofitted with the cool coating. Likewise, the growing effect on the total loads' reduction that resulted from the decrease in absorptivity was observed. On the other hand, no significant differences were appreciated among the different maintenance scenarios, as stated previously.

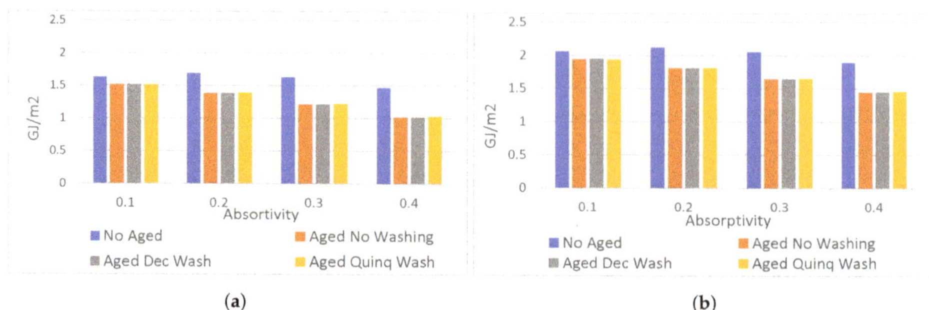

Figure 13. Life cycle period total load decreases (GJ/m^2) for: roof reference absorptivity $= 0.8$ (**a**); roof reference absorptivity $= 0.9$ (**b**).

Table 4. Life cycle period total load decreases in GJ/m^2.

Cool Roof Absorptivity	Reference Absorp. = 0.8				Reference Absorp. = 0.9			
	No Aged	Aged No Wash	Aged Dec.Wash	Aged Quinq.Wash	No Aged	Aged No Wash	Aged Dec. Wash	Aged Quinq. Wash
0.1	1.638	1.515	1.519	1.515	2.065	1.942	1.946	1.942
0.2	1.699	1.384	1.385	1.391	2.126	1.811	1.812	1.818
0.3	1.632	1.217	1.216	1.228	2.060	1.644	1.644	1.655
0.4	1.471	1.020	1.018	1.032	1.898	1.447	1.445	1.460

4.2. LCA Economic Results

In this section, the LCA results of the cool roof refurbishment are presented using the NS and PB as performance indicators. The net savings and payback periods owing to refurbishment investment with the four cool roofs' absorptivities $abs_{cool} = 0.1, 0.2, 0.3, 0.4$ when compared with the reference roofs with absorptivities $abs_{ref} = 0.8, 0.9$ were analyzed under the economic framework described in Section 3 and for all the aging scenarios described in Section 2.3. Positive values for NS mean real savings, while negative values imply a higher LCA cost for the cool roof and, therefore, no real savings.

Two cases were taken into account for the LCA:

- The cool retrofit installation was not necessarily required, and the decision about its installation was driven by energy savings considerations. In this case, the installation costs were considered in full.
- The roof needed renovation, and the installation of a cool roof was opted for instead of implementing a classic system, which in our study was considered to be made with bituminous insulating paint. In this case, the installation costs were considered marginal, since they were computed as the difference between the cost of installation of the cool roof minus the installation cost for the bituminous paint.

In Tables 5 and 6, the maximum total cost savings are summarized for the whole life cycle period of time. Details about these results are analyzed in the next sections.

Table 5. Full costs: life cycle period maximum total savings in €/m^2.

Cool Roof Absorptivity	Reference Absorp. = 0.8				Reference Absorp. = 0.9			
	No Aged	Aged No Wash	Aged Dec. Wash	Aged Quinq. Wash	No Aged	Aged No Wash	Aged Dec. Wash	Aged Quinq. Wash
0.1	−0.77	5.09	3.23	−0.79	4.46	10.32	8.47	4.43
0.2	3.76	4.76	3.05	−0.71	9.00	9.99	8.28	4.51
0.3	14.61	3.96	2.29	−1.31	11.13	9.19	7.52	3.92
0.4	6.15	2.52	0.87	−2.62	11.39	7.76	6.11	2.61

Table 6. Marginal costs: life cycle period maximum total savings in €/m^2.

Cool Roof Absorptivity	Reference Absorp. = 0.8				Reference Absorp. = 0.9			
	No Aged	Aged No Wash	Aged Dec. Wash	Aged Quinq. Wash	No Aged	Aged No Wash	Aged Dec. Wash	Aged Quinq. Wash
0.1	7.23	13.10	11.24	8.4	12.47	18.33	16.48	13.64
0.2	11.77	12.77	11.06	8.48	17.01	18.01	16.29	13.72
0.3	13.9	11.97	10.3	7.89	19.14	17.20	15.53	13.12
0.4	14.16	10.53	8.88	6.58	19.4	15.77	14.12	11.82

The results are detailed in Sections 4.2.1–4.2.8, which describe the results obtained for each scenario. The tables that are referenced in these sections can be found in Appendix A to the paper.

4.2.1. Aged Cool Roof without Maintenance Full Costs

The NS and PB results for the aged roof without any maintenance are presented in Tables A1 and A2. For both reference roofs, only when heat pumps were used for heating, some savings were found. Likewise, it was observed that the smaller was the absorptivity of the cool roof, the greater was the savings.

For the reference roof with $abs_{ref} = 0.8$, savings only occurred for a differential $i - d$ greater than or equal to 2.7% when the cool roofs' absorptivities were 0.1, 0.2, and 0.3, while for $abs_{cool} = 0.4$, savings were found only if the differential $i - d$ was greater than 3%. The best result in this case was a savings of 5.09 €/m^2 with a payback of 16.43 years that was obtained for the roof with $abs_{cool} = 0.1$ and for the economic indicators $d = 0.3\%$, $i = 6\%$.

For the reference roof with $abs_{ref} = 0.9$, savings were higher than for the case of $abs_{ref} = 0.8$. Now, for the cool roof with $abs_{cool} = 0.1$, savings happened for all the economic scenarios except when $d = 3\%$ and $i = 0\%$, that is for a negative differential $i - d$ equal to -3%. For cool roofs with $abs_{cool} = 0.2$ and 0.3, savings required values of the differential $i - d$ equal to or greater than -1.5%. Finally, for $abs_{cool} = 0.4$, savings required a strictly positive differential $i - d$. The best result in this case was a savings of 10.32 €/m^2 with a payback of 12.33 years, which was obtained for the cool roof $abs_{cool} = 0.1$ and the economic indicators $d = 0.3\%$, $i = 6\%$.

4.2.2. Aged Cool Roof without Maintenance Marginal Costs

The results of the case of the aged roof without maintenance considering marginal costs are presented in Tables A3 and A4. In this case, the savings were called marginal savings. They showed once again a tendency to take higher savings as the absorptivity of the cool roof decreased, the savings attained for the reference roof $abs_{ref} = 0.9$ being higher.

The best results were a savings of 13.10 €/m^2 with a payback of 3.73 years and a savings of 18.33 €/m^2 with a payback of 2.83 years for the reference roofs $abs_{ref} = 0.8$ and $abs_{ref} = 0.9$, respectively, which were both obtained for the cool roof with $abs_{cool} = 0.1$ and $d = 0.3\%$, $i = 6\%$.

It is worth noting that for economic scenarios with a lower inflation differential, good results were also achieved. Therefore, for the case of $d = 0.3\%$, $i = 0\%$, which gave a negative differential, savings of 6.4, 6.27, 5.86, and 5.10 €/m^2 and payback periods of 4.05, 3.91, 3.97, and 4.32 years were obtained for cool roofs with absorptivities 0.1, 0.2, 0.3, and 0.4, respectively, when the reference roof had $abs_{ref} = 0.8$. For the same economic scenario for $abs_{ref} = 0.9$, the marginal savings were greater, being 9.26, 9.13, 8.72, and 7.96 €/m^2, and the payback periods were 2.99, 2.86, 2.85, and 3.01 years for the same cool roof absorptivities.

In this case, a small amount of savings was also obtained when gas was used for heating, but only for $abs_{cool} = 0.4$.

4.2.3. Aged Cool Roof with Decennial Maintenance Full Costs

The results for the aged cool roof with decennial maintenance full costs are shown in Tables A5 and A6. It can be observed that including the maintenance costs penalized the LCA savings of refurbishing with cool paint when compared with the previous case of no maintenance.

Again, only heating with a pump gave savings. The biggest savings were 3.23 €/m² with a payback of 14.07 years and a savings of 8.47 €/m² with a payback of 2.83 years for the reference roofs $abs_{ref} = 0.8$ and $abs_{ref} = 0.9$, respectively. Both results were obtained for $abs_{cool} = 0.1$ and $d = 0.3\%$, $i = 6\%$.

As in the previous cases, the decrease in absorptivity was associated with an increase in savings and higher savings for $abs_{ref} = 0.9$ than for $abs_{ref} = 0.8$.

Now, the inflation differential required to get savings for $abs_{ref} = 0.8$ was greater than 4.5% for $abs_{cool} = 0.1\%$, 0.2%, and 0.3% and equal to 5.7% for $abs_{cool} = 0.4$. For $abs_{ref} = 0.9$, savings were obtained if the differential was greater than or equal to 1.5% for $abs_{cool} = 0.1\%$, 0.2%, and 0.3% and greater than 2.7% for $abs_{cool} = 0.4$.

4.2.4. Aged Cool Roof with Decennial Maintenance Marginal Costs

The results for the aged cool roof with decennial maintenance and marginal costs are shown in Tables A7 and A8.

Once again, savings were found only if the heating worked through the heat pump, which led to savings for all the economic scenarios considered. Likewise, smaller values of absorptivity produced greater savings, which took the larger values for the reference roof with $abs_{ref} = 0.9$.

The best result for the reference roof $abs_{ref} = 0.9$ was a savings of 16.48 €/m² with a payback of 2.13 obtained for $d = 0.3\%$, $i = 6\%$ and for the reference roof $abs_{ref} = 0.8$, a saving of 11.24 €/m² with a payback of 3.73 and the same economic indicators.

Now, the inflation differential required to get savings for $abs_{ref} = 0.8$ was greater than 4.5% for $abs_{cool} = 0.1\%$, 0.2%, and 0.3% and equal to 5.7% for $abs_{cool} = 0.4$. For $abs_{ref} = 0.9$, savings were obtained if the differential was greater than or equal to 1.5% for $abs_{cool} = 0.1\%$, 0.2%, and 0.3% and greater than 2.7% for $abs_{cool} = 0.4$.

4.2.5. Aged Cool Roof with Quinquennial Maintenance Full Costs

The results for the aged cool roof with quinquennial maintenance and full costs are shown in Tables A9 and A10. The most relevant fact of this case was that for almost all the studied scenarios, there were no savings. Specifically, for the reference roof with $abs_{ref} = 0.8$, there were no savings in any case, while for the reference roof with $abs_{ref} = 0.9$, only for the differential $i - d$ equal to or greater than 4.5%, some savings were found. The best results in this case were obtained for $abs_{cool} = 0.2$, the greater savings equal to 4.51 €/m² with a payback of 17.01 obtained for $d = 0.3\%$, $i = 6\%$.

4.2.6. Aged Cool Roof with Quinquennial Maintenance Marginal Costs

The results for the aged cool roof with quinquennial maintenance and marginal costs are shown in Tables A11 and A12.

As in the previous cases, savings were found only for heating using heat pumps and existed for all absorptivity and economic scenarios considered. The greatest savings were given for $abs_{cool} = 0.2$ for both reference roof absorptivities, the savings for $abs_{ref} = 0.9$ being greater than the ones for $abs_{ref} = 0.8$ for equivalent economic indicators.

For $abs_{ref} = 0.8$, the greatest value was a savings of 8.48 €/m² with a payback of 3.73 years, and for $abs_{ref} = 0.9$, the greatest savings was 13.72 €/m² with a payback of 2.71 both for $d = 0.3\%$ and $i = 6\%$.

4.2.7. No Aged Cool Roof Full Costs

In Tables A13–A16, results are presented for the case of the non-aged roof with full costs. Although this case may be considered unrealistic, in the literature, it is a frequently treated case. For this reason and because it can give information on the importance of the aging effect when comparing both cases, its results are included here. Maintenance was not considered in the LCA of this case because the non-aged effect made it unnecessary.

For the case of full costs (Tables A13 and A14), savings were found only when heat pumps were used for heating.

Savings were greater for $abs_{ref} = 0.9$ than for $abs_{ref} = 0.8$ under the same economic scenarios. When existing, savings increased as the absorptivity of the cool roofs grew, giving $abs_{cool} = 0.4$ the highest savings both for $abs_{ref} = 0.8$ and $abs_{ref} = 0.9$.

Specifically, when $abs_{ref} = 0.8$, there were no savings for $abs_{cool} = 0.1$; there were savings for $abs_{cool} = 0.2$ if the economic differential $i - d$ was equal to or greater than 2.7%; and there were savings as well for $abs_{cool} = 0.3$ and $abs_{cool} = 0.4$ if the inflation differential was equal to or greater than 1.5%.

When $abs_{ref} = 0.9$, there were savings for $abs_{cool} = 0.1$ if the inflation differential was equal to or greater than 2.7%, for $abs_{cool} = 0.2$ if the inflation differential was equal to or greater than −0.3%, while for $abs_{cool} = 0.3$ and $abs_{cool} = 0.4$, the only scenario for which there were no savings was when $d = 0.3\%$ and $i = 6\%$, which involved a strong negative economic differential $i - d = -3\%$.

The highest savings for $abs_{ref} = 0.8$ was 6.15 €/m² with a payback of 14.45 years obtained when $abs_{cool} = 0.4$, $d = 0.3\%$, and $i = 6\%$. For $abs_{ref} = 0.9$, the greatest savings was 11.39 €/m² with a payback of 11.76 years and the same economic parameters.

4.2.8. No Aged Cool Roof Marginal Costs

In Tables A15 and A16, marginal savings are presented. It can be pointed out again that savings were found only if heat pumps were used for heating. In this case, savings existed in all the absorptivity and economic scenarios analyzed.

The savings took an increasing value from $abs_{cool} = 0.1$ to $abs_{cool} = 0.4$. This way, the greatest values of the savings were found for $abs_{cool} = 0.4$ for both reference roofs.

For $abs_{ref} = 0.8$, the greatest savings was 14.16 €/m² with a payback of 3.09 years, and for $abs_{ref} = 0.9$, the greatest savings was 17.01 €/m² with a payback of 2.64 years both for $d = 0.3\%$ and $i = 6\%$.

5. Discussion

For the climatic zone under consideration and the building case study, the results presented in Section 4.1 indicated a decrease in the annual total load for roofs when the absorptivity of the external coating of the roof decreased. This way, a decrease close to 32% when comparing a reference roof of $abs_{ref} = 0.9$ to a cool roof of $abs_{cool} = 0.1$ was found, as is shown in Figure 7. As can be observed in this figure, the pattern found was the following: the less absorptive of the roof, the lower was the cooling load and the higher the heating load, in such a way that the increase in the heating load was clearly compensated by the decrease in the cooling load as the absorptivity decreased and fell within the range of what was a cool roof, that is for absorptivity values less than or equal to five [7].

Furthermore, it can be established from Figure 11 that the effect of the change in absorptivity values was greater on cooling loads than on heating ones. That is, when absorptivity decreased, the decrease of cooling loads was proportionally much greater than the increase of heating loads. This meant that the effect of changes in absorptivity produced a noticeably greater impact on cooling loads than on heating ones.

The energy analysis showed that for the different cases of cool roofs under consideration, that is aged without washing, aged with decennial and quinquennial washing, and non-aged, for the entire LC time period of twenty years, the total loads had very similar results for each value of the cool roof absorptivity; see Figure 13a,b and Table 4. This meant that maintenance through power washing did not have a significant effect on the energy behavior of the cool roof, a fact that has been mentioned in the literature [40]. For the case of no aging effect, smaller values of the total loads were found; this suggested the need to consider the aging effect to have a realistic energy analysis, as was stated in the previous literature [28,40].

For the life cycle's total period, energy savings were found for the four cool roofs' absorptivities and for the two reference roofs studied for all the aged cases considered; see Table 4. For the reference roof $abs_{ref} = 0.9$, the savings ranged between 1.946 GJ/m^2 obtained for $abs_{cool} = 0.1$ and 1.445 GJ/m^2 obtained for $abs_{cool} = 0.4$. For the reference roof $abs_{ref} = 0.8$, the savings were less than for $abs_{ref} = 0.9$, and they were in a range between 1.519 GJ/m^2 obtained for $abs_{cool} = 0.1$ and 1.018 GJ/m^2 obtained for $abs_{cool} = 0.1$.

As can be noted in Table 4, the differences in the energy savings induced by the washing of the cool roofs were very small for the LCA total period, which cast doubt on its advisability, especially when taking into account possible damage to the cool coating when performing maintenance [40].

For the non-aged case, as observed in Figure 13a,b and Table 4, the total loads were the smallest, and consequently, the savings obtained in this case were the greatest. However, as mentioned before, this case could be considered unrealistic, so its results should be taken into account only as a reference by means of which both scenarios (aged and non-aged roofs) could be contrasted while determining the influence of the aging effect on the energy behavior of cool roofs during the LCA period of time.

Regarding the LCA economic analysis, the first issue to highlight was that, generally, economic savings were only obtained when the heat pump was used for heating. The other heating systems considered did not provide savings, except in the specific case of aged roofs without maintenance, in which for the case of $abs_{cool} = 0.4$, some marginal savings were obtained for gas heating; see Tables A3 and A4. This was due to the fact that, in spite of the considerable decrease in total loads pointed out in the energy analysis, the effects of the cost of energy for heating gave rise to an increase in the heat load, which from an economic point of view, was not compensated with the decrease in the refrigeration load, the only exception to this being the case of an even system of energy efficiency for heating and cooling as provided by heat pumps.

On the other hand, whether rehabilitation was necessary or not also had a considerable impact on the results of the LCA. If the refurbishment is done because it is necessary, the costs of applying the cool coating layer should be compared with those of applying another type of coating which in the case of the present study was the bituminous paint layer, while the additional cost of the cool roof represented only the marginal cost. On the contrary, if the application of the cool roof was done without the need for rehabilitation, the initial costs were complete when the LCA was performed.

When marginal savings were considered, for all the cases except one, significant economic savings existed as long as the heat pump was used for heating. These marginal savings were greater for the case of no maintenance, while the marginal savings from a ten year maintenance were greater than those obtained from the five year maintenance; see Tables A3, A4, A7, A8, A11 and A12.

The savings obtained in the case of complete costs were logically lower. In this case, the biggest savings were also obtained in the case that maintenance was not performed, while again, the savings from a ten year maintenance gave rise to better results than those obtained from the five year savings; see Tables A1, A2, A5, A6, A9 and A10. Now, when maintenance was considered, there was a noticeable decrease in the number of cases that resulted in savings. These were mainly concentrated in values of the roof reference with absorptivity equal to 0.9 and high values of the economic differential $i - d$. Even for the quinquennial maintenance, there were no savings for any case when the reference absorptivity was 0.8.

Note also that in all scenarios considered for the aged roofs, the best savings results were obtained for the reference roof with $abs_{ref} = 0.9$ and for the lowest values of the cool roof absorptivity.

On the other hand, as can be seen in Tables 5, 6, and A1–A10, the punctual improvement that was obtained after washing in terms of energy loads had a small reflecting in the economic savings.

This was mainly due to the fact that maintenance costs were not counterbalanced enough so as to obtain relevant energy costs savings. A similar result was suggested in [40].

It is also important to point out the role of economic indicators. As evidenced by the LCA savings, the higher the values of the economic differential $i - d$, the higher the values for savings. This was closely related to an increase in the economic benefits provided by cool roofs in the face of high energy prices.

Finally, for the non-aged case, the savings obtained were close to those for the aged case with no maintenance; see Tables A5 and A6. However, now, the different heat dynamics throughout the period of time considered implied that the highest savings rates were obtained for cool roofs with the highest values. Specifically, the greatest savings were obtained for $abs_{cool} = 0.4$, while for the aged case, they were obtained for $abs_{cool} = 0.1$. The same can be stated for the non-aged case marginal savings: values close to the marginal savings from the aged case with no maintenance, but with the same behavior related to the absorptivities.

Although the performed analysis seemed to point out the existence of benefits when using cool roofs to refurbish the obsolete energy conditions of the social housing stock under study, some extra analysis must be done in order to assess its social impact and to identify the steps that the involved parties, that is tenants, social services, municipality, etc., should take in order to exploit the advantages of this kind of energy improvement. Some research about the topic of social impacts and stakeholder analysis as that made in [31] could help to clarify the role of social agents regarding the implementation of the proposed refurbishment measures.

Thus, taking into account the social implications of the analysis made in this work, it could be useful in order to shed light on the social implications of the present research to perform a Life Cycle Sustainability Assessment (LCSA) [31], which considers environmental LCA, life cycle costing, and Social Life Cycle Assessment (SLCA) [63,64].

6. Conclusions

The energy and economic performances of residential social building roofs retrofitted to become cool roofs were analyzed under the southern Spain climate. The thermal dynamic of the roofs was simulated through a computational code using a finite difference approach in order to assess the impact of different cool coatings when used to retrofit the roofs of dwellings belonging to a promotion of social housing in Seville (Spain) built in the 1960s.

Four different cases of cool coating with values of solar absorptivity equal to 0.1, 0.2, 0.3, and 0.4 when applied to two reference roofs with solar absorptivity equal to 0.8 and 0.9 were analyzed. In order to get realistic simulation conditions, the aging effect of the cool coatings was taken into account by means of designing a pattern of the incidence of the aging effect over the solar absorptivities throughout the service lifetime of the cool roof.

This gave rise to different scenarios regarding the energy behavior of the cool roofs according to the temporary evolution of the cover resulting from the aging effect and the maintenance type considered for the cool coatings.

The results obtained for the different combinations of cool coatings and aging effect pointed to significant savings in the operational energy. This way, a decrease in the annual total loads for roofs was found. Such decreases were larger as the absorptivity of the external coating of the retrofitted roof was reduced. The maximum decrease found was close to 32%, and it was obtained when a roof with solar absorptivity equal to 0.9 was retrofitted with a cool paint with solar absorptivity equal to 0.1.

Moreover, the limited role played by the cool roofs' maintenance over the total loads during the LCA period of time was revealed, since the values of such loads were similar to each of the cool absorptivities considered. However, some variations in the evolution of the heating and cooling loads were observed.

In order to evaluate the economic value of using cool roofs for social housing rehabilitation, a 20 year life cycle analysis was conducted. To carry out a comprehensive LCA analysis, a variety of values for the variables of solar absorptivity, cool roof maintenance, energy and maintenance costs, monetary discount rate, conditioning equipment, and installation cost were taken into account.

The LCA analysis pointed out that initial costs had a strong impact on the economic results. This way, if the costs of the cool coating were not compared to any other roof retrofit, savings were only obtained when, in the case of heating, the equipment used was a heat pump and the economic differential between the energy inflation costs and the monetary discount rate was high. Namely, the best result was found when an aged cool roof without maintenance was used and the cool absorptivity was equal to 0.1, while the reference absorptivity equaled 0.9, and the economic indicators were $d = 0.3\%$ and $i = 6\%$. Following this, the economic LCA savings was 10.32 €/m^2, and the payback period was 12.33 years.

If marginal costs were considered, the LCA analysis reported savings for all the scenarios studied as long as heating was carried out by means of pump heating. In this case, the best result was found again for the case of an aged cool roof without maintenance when the cool absorptivity was equal to 0.1, the reference absorptivity was equal to 0.9, and the economic indicators were $d = 0.3\%$ and $i = 6\%$. The economic LCA savings was now 18.33 €/m^2, and the payback period was 2.83 years.

Another result confirmed in the LCA analysis was the impact of the economic differential $i - d$ on the savings. High values of such a differential yielded higher economic savings and consequently shorter payback periods. This result, which could be considered as expected, was ratified by means of the implemented cost-effectiveness analysis and the fact that investments in refurbishment seeking to reduce energy consumption were from a purely economic standpoint more profitable within raising energy price scenarios.

Furthermore, its use gave rise to economic savings when adequate equipment to condition dwellings was used. Finally, other results described in the literature such as the scarce economic gains stemming from a power washing of this kind of cool coating were revealed for the time period in which the LCA was carried out.

According to the obtained results, in terms of energy consumption and environmental benefits, the use of cool coatings with a solar absorptivity as low as possible is recommended for the rehabilitation of cool roofs in the social building case study in the climatic zone under consideration.

The main limitation of this study rested on the lack of social information on the valuation of all the involved parties of the society, that is tenants, social services, municipality etc., of the application of the present research results. This could be a venue of future research together with the extension of the analysis implemented with the aim of broadening such analysis along the lines of an LCSA considering environmental LCA, life cycle costing, and SLCA.

Author Contributions: Conceptualization, all the authors; methodology, all the authors; software, all the authors; formal analysis, all the authors; investigation, all the authors; writing, original draft preparation, H.D.-T. and C.-A.D.-T.; writing, review and editing, all the authors. All authors read and agreed to the published version of the manuscript.

Conflicts of Interest: The authors declare no conflict of interest.

Appendix A. LCA Savings

Table A1. LCA savings for the aged roof without maintenance vs. the reference roof with absorptivity = 0.8.

		Absorptivity Solar: 0.1 vs. 0.8						Absorptivity Solar: 0.2 vs. 0.8					
		Gas		Electr.		Pump		Gas		Electr.		Pump	
		Savings	PayB	Savings	PayB	Savings	PayB	Savings	PayB	Savings	PayB	Savings	PayB
d=0.3%	*i* = 0%	−15.15	-	−62.48	-	−1.6	-	−12.55	-	−50.33	-	−1.73	-
	i = 3%	−16.82	-	−79.59	-	1.14	19.43	−13.41	-	−63.51	-	0.92	19.45
	i = 6%	−19.24	-	−104.24	-	5.09	16.43	−14.65	-	−82.49	-	4.76	16.43
d=1.5%	*i* = 0%	14.58	-	−56.69	-	−2.52	-	−12.25	-	−45.87	-	−2.63	-
	i = 3%	−15.99	-	−71.1	-	−0.2	-	−12.98	-	−57.01	-	−0.38	-
	i = 6%	−18.02	-	−91.85	-	3.11	16.48	−14.02	-	−72.94	-	2.83	18.47
d=3%	*i* = 0%	−13.98	-	−50.69	-	−3.47	-	−11.94	-	−41.25	-	−3.55	-
	i = 3%	−15.14	-	−62.5	-	−1.58	-	−12.54	-	−50.34	-	−1.72	-
	i = 6%	−16.78	-	−79.24	-	1.1	19.45	−13.38	-	−63.23	-	0.88	19.46

		Absorptivity Solar: 0.3 vs. 0.8						Absorptivity Solar: 0.4 vs. 0.8					
		Gas		Electr.		Pump		Gas		Electr.		Pump	
		Savings	PayB	Savings	PayB	Savings	PayB	Savings	PayB	Savings	PayB	Savings	PayB
d=0.3%	*i* = 0%	−10.34	-	−38.98	-	−2.14	-	−8.7	-	−28.97	-	−2.9	-
	i = 3%	−10.5	-	−48.46	-	0.35	19.57	−8.35	-	−35.18	-	−0.67	-
	i = 6%	−10.74	-	−62.11	-	3.96	16.5	−7.85	-	−44.13	-	2.52	18.529
d=1.5%	*i* = 0%	−10.28	-	−35.77	-	−2.98	-	−8.82	-	−26.87	-	−3.65	-
	i = 3%	−10.42	-	−43.78	-	−0.87	-	−8.52	-	−32.12	-	−1.77	-
	i = 6%	−10.62	-	−55.25	-	2.15	18.37	−8.1	-	−39.63	-	0.91	19.43
d=3%	*i* = 0%	−10.21	-	−32.44	-	−3.85	-	−8.93	-	−24.68	-	−4.42	-
	i = 3%	−10.33	-	−38.98	-	−2.12	-	−8.69	-	−28.97	-	−2.89	-
	i = 6%	−10.49	-	−48.26	-	0.31	19.58	−8.35	-	−35.05	-	−0.71	-

Table A2. LCA savings for the aged roof without maintenance vs. the reference roof with absorptivity = 0.9.

		Absorptivity Solar: 0.1 vs. 0.9						Absorptivity Solar: 0.2 vs. 0.9					
		Gas		Electr.		Pump		Gas		Electr.		Pump	
		Savings	PayB	Savings	PayB	Savings	PayB	Savings	PayB	Savings	PayB	Savings	PayB
d=0.3%	*i* = 0%	−14.64	-	−70.21	-	1.25	17.639	−12.04	-	−58.06	-	1.12	17.84
	i = 3%	−16.15	-	−89.95	-	4.97	14.327	−12.73	-	−73.86	-	4.75	14.46
	i = 6%	−18.31	-	−118.38	-	10.32	12.33	−13.73	-	−96.63	-	9.99	12.43
d=1.5%	*i* = 0%	−14.13	-	−63.53	-	0.01	19.97	−11.8	-	−52.71	-	−0.09	-
	i = 3%	−15.40	-	−80.22	-	3.15	15.6	−12.39	-	−66.07	-	2.97	15.76
	i = 6%	−17.22	-	−104.09	-	7.64	13.18	−13.22	-	−85.19	-	7.37	13.29
d=3%	*i* = 0%	−13.59	-	−56.62	-	−1.27	-	−11.55	-	−47.18	-	−1.36	-
	i = 3%	−14.63	-	−70.24	-	1.28	17.65	−12.03	-	−58.08	-	1.14	17.86
	i = 6%	−16.11	-	−89.55	-	4.91	14.45	−12.71	-	−73.55	-	4.70	14.58

		Absorptivity Solar: 0.3 vs. 0.9						Absorptivity Solar: 0.4 vs. 0.9					
		Gas		Electr.		Pump		Gas		Electr.		Pump	
		Savings	PayB	Savings	PayB	Savings	PayB	Savings	PayB	Savings	PayB	Savings	PayB
d=0.3%	*i* = 0%	−9.83	-	−46.71	-	0.71	18.55	−8.2	-	−36.7	-	−0.04	-
	i = 3%	−9.83	-	−58.81	-	4.19	14.92	−7.68	-	−45.54	-	3.15	15.903
	i = 6%	−9.82	-	−76.25	-	9.19	12.78	−6.93	-	−58.27	-	7.76	13.5
d=1.5%	*i* = 0%	−9.83	-	−42.61	-	−0.44	-	−8.3	-	−33.71	-	−1.11	-
	i = 3%	−9.82	-	−52.85	-	2.48	18.32	−7.3	-	−41.18	-	1.58	17.46
	i = 6%	−9.82	-	−67.49	-	6.68	16.3	−13.36	-	−51.87	-	5.45	14.49
d = 3%	*i* = 0%	−9.83	-	−38.38	-	−1.65	-	−8.55	-	−30.62	-	−2.23	-
	i = 3%	−9.82	-	−46.73	-	0.73	18.55	−8.19	-	−36.71	-	−0.02	-
	i = 6%	−9.82	-	−58.57	-	4.13	15.05	−7.68	-	−45.36	-	3.1	16.02

Table A3. LCA marginal savings for the aged cool roof without maintenance vs. the reference roof with absorptivity = 0.8.

		Absorptivity Solar: 0.1 vs. 0.8						Absorptivity Solar: 0.2 vs. 0.8					
		Gas		Electr.		Pump		Gas		Electr.		Pump	
		Savings	PayB	Savings	PayB	Savings	PayB	Savings	PayB	Savings	PayB	Savings	PayB
$d=0.3\%$	$i=0\%$	−7.14	-	−54.47	-	6.4	4.05	−4.54	-	−42.32	-	6.27	3.91
	$i=3\%$	−8.81	-	−71.58	-	9.14	3.88	−5.4	-	−55.501	-	8.93	3.75
	$i=6\%$	−11.23	-	−96.23	-	13.1	3.73	−6.64	-	−74.48	-	12.77	3.61
$d=1.5\%$	$i=0\%$	−6.57	-	−48.68	-	5.48	4.17	−4.24	-	−37.86	-	5.37	4.03
	$i=3\%$	−7.98	-	−63.14	-	7.8	3.991	−4.97	-	−49.00	-	7.62	3.859
	$i=6\%$	−10.01	-	−83.84	-	11.12	3.832	−6.01	-	−63.15	-	10.84	3.70
$d=3\%$	$i=0\%$	−5.97	-	−42.68	-	4.53	4.33	−3.93	-	−33.24	-	4.45	4.18
	$i=3\%$	−7.13	-	−54.492	-	6.42	4.13	−4.53	-	−42.33	-	6.28	3.99
	$i=6\%$	−8.77	-	−71.23	-	9.11	3.959	−5.377	-	−55.22	-	8.89	3.83

		Absorptivity Solar: 0.3 vs. 0.8						Absorptivity Solar: 0.4 vs. 0.8					
		Gas		Electr.		Pump		Gas		Electr.		Pump	
		Savings	PayB	Savings	PayB	Savings	PayB	Savings	PayB	Savings	PayB	Savings	PayB
$d=0.3\%$	$i=0\%$	−2.33	-	−30.97	-	5.86	3.97	−0.69	-	20.96	-	5.104	4.32
	$i=3\%$	−2.49	-	−40.45	-	8.36	3.81	−0.34	-	−27.17	-	7.33	4.12
	$i=6\%$	−2.73	-	−54.1	-	11.97	3.66	0.15	19.6	−36.12	-	10.53	3.95
$d=1.5\%$	$i=0\%$	−2.27	-	−27.76	-	5.02	4.09	−0.81	-	−18.86	-	4.35	4.46
	$i=3\%$	−2.41	-	−36.12	-	7.13	3.919	−0.51	-	−24.11	-	6.23	4.24
	$i=6\%$	−2.61	-	−47.24	-	10.16	3.76	−0.09	-	−31.62	-	8.92	4.06
$d=3\%$	$i=0\%$	−2.2	-	−24.43	-	4.15	4.26	−0.92	-	−16.67	-	3.58	4.66
	$i=3\%$	−2.32	-	−30.9	-	5.88	4.06	−0.68	-	−20.96	-	5.11	4.41
	$i=6\%$	−2.48	-	−40.25	-	8.32	3.89	−0.34	-	−27.04	-	7.29	4.21

Table A4. LCA marginal savings for the aged cool roof without maintenance vs. the reference roof with absorptivity = 0.9.

		Absorptivity Solar: 0.1 vs. 0.9						Absorptivity Solar: 0.2 vs. 0.9					
		Gas		Electr.		Pump		Gas		Electr.		Pump	
		Savings	PayB	Savings	PayB	Savings	PayB	Savings	PayB	Savings	PayB	Savings	PayB
$d=0.3\%$	$i=0\%$	−6.63	-	−62.2	-	9.26	2.99	−4.03	-	−50.05	-	9.13	2.86
	$i=3\%$	−8.14	-	−81.94	-	12.98	2.9	−4.72	-	−65.85	-	12.76	2.78
	$i=6\%$	−10.3	-	110.37	-	18.33	2.83	−5.72	-	−88.62	-	18.01	2.716
$d=1.5\%$	$i=0\%$	−6.12	-	−55.52	-	8.02	3.05	−3.79	-	−44.7	-	7.91	2.92
	$i=3\%$	−7.39	-	−72.21	-	11.16	2.96	−4.38	-	−58.06	-	10.98	2.84
	$i=6\%$	−9.21	-	−96.08	-	15.6	2.88	−5.21	-	−77.18	-	15.38	2.77
$d=3\%$	$i=0\%$	−5.58	-	−48.61	-	6.73	3.14	−3.54	-	−39.17	-	6.64	3.01
	$i=3\%$	−6.62	-	−62.23	-	9.29	3.05	−4.02	-	−50.07	-	9.15	2.92
	$i=6\%$	−8.09	-	−81.54	-	12.92	2.96	−4.70	-	−65.54	-	12.71	2.84

		Absorptivity Solar: 0.3 vs. 0.9						Absorptivity Solar: 0.4 vs. 0.9					
		Gas		Electr.		Pump		Gas		Electr.		Pump	
		Savings	PayB	Savings	PayB	Savings	PayB	Savings	PayB	Savings	PayB	Savings	PayB
$d=0.3\%$	$i=0\%$	−1.82	-	−38.70	-	8.72	2.85	−0.19	-	−28.69	-	7.96	3.01
	$i=3\%$	−1.81	-	50.80	-	12.20	2.78	0.32	17.44	−37.53	-	11.16	2.91
	$i=6\%$	−1.80	-	−68.24	-	17.20	2.71	1.07	14.63	50.26	-	15.77	2.83
$d=1.5\%$	$i=0\%$	−1.82	-	−34.60	-	7.56	2.92	−0.36	-	−25.70	-	6.89	3.07
	$i=3\%$	−1.81	-	−44.84	-	10.49	2.84	0.07	19.27	−33.17	-	9.59	2.98
	$i=6\%$	−1.81	-	−59.48	-	14.69	2.77	0.70	15.75	−43.86	-	13.46	2.90
$d=3\%$	$i=0\%$	−1.82	-	−30.37	-	6.35	3.01	−0.54	-	−22.6	-	5.77	3.17
	$i=3\%$	−1.81	-	−38.72	-	8.75	2.92	−0.182	-	−28.7	-	7.98	3.07
	$i=6\%$	−1.81	-	−50.56	-	12.14	2.84	0.33	17.48	−37.35	-	11.11	2.98

Table A5. LCA savings of the aged roof with 10 years maintenance vs. the reference roof with absorptivity = 0.8.

		Absorptivity Solar: 0.1 vs. 0.8						Absorptivity Solar: 0.2 vs. 0.8					
		Gas		Electr.		Pump		Gas		Electr.		Pump	
		Savings	PayB	Savings	PayB	Savings	PayB	Savings	PayB	Savings	PayB	Savings	PayB
	$i=0\%$	−17.28	-	−65.64	-	−3.43	-	−14.53	-	−53.08	-	−3.49	-
$d=0.3\%$	$i=3\%$	−19.06	-	−83.2	-	−0.7	-	−15.44	-	−66.54	-	−0.81	-
	$i=6\%$	−21.6	-	−108.4	-	3.23	17.18	−16.73	-	−85.86	-	3.05	17.28
	$i=0\%$	−16.64	-	−59.67	-	−4.33	-	−14.19	-	−48.49	-	−4.37	-
$d=1.5\%$	$i=3\%$	−18.16	-	74.53	-	−2.02	-	−14.96	-	−59.89	-	−2.1	-
	$i=6\%$	−20.3	-	−95.72	-	1.28	18.64	−16.06	-	−76.13	-	1.13	18.76
	$i=0\%$	−15.98	-	−53.47	-	−5.25	-	−13.83	-	−43.72	-	−5.27	-
$d=3\%$	$i=3\%$	−17.22	-	−65.62	-	−3.37	-	−14.47	-	−53.04	-	−3.43	-
	$i=6\%$	−18.97	-	−82.79	-	−0.69	-	−15.36	-	−66.21	-	−0.8	-

		Absorptivity Solar: 0.3 vs. 0.8						Absorptivity Solar: 0.4 vs. 0.8					
		Gas		Electr.		Pump		Gas		Electr.		Pump	
		Savings	PayB	Savings	PayB	Savings	PayB	Savings	PayB	Savings	PayB	Savings	PayB
	$i=0\%$	−12.25	-	−41.52	-	−3.87	-	−10.56	-	−31.35	-	−4.61	-
$d=0.3\%$	$i=3\%$	−12.45	-	−51.23	-	−1.3476	-	−10.24	-	−37.75	-	−2.36	-
	$i=6\%$	−12.71	-	−65.16	-	2.29	17.85	−9.76	-	−46.91	-	0.87	19.1
	$i=0\%$	−12.15	-	−38.2	-	−4.69	-	−5.35	-	−29.15	-	−5.35	-
$d=1.5\%$	$i=3\%$	−12.32	-	−46.43	-	−2.56	-	−10.37	-	−34.57	-	−3.45	-
	$i=6\%$	−12.55	-	−58.14	-	0.48	19.43	−9.97	-	−42.28	-	−0.73	-
	$i=0\%$	−12.05	-	−34.762	-	−5.54	-	−10.73	-	−26.87	-	−6.11	-
$d=3\%$	$i=3\%$	−12.19	-	−41.48	-	−3.81	-	−10.51	-	−31.3	-	−4.56	-
	$i=6\%$	−12.38	-	−50.98	-	−1.34	-	−10.19	-	−37.56	-	−2.35	-

Table A6. LCA savings of the aged roof with 10 years maintenance vs. the reference roof with absorptivity = 0.9.

		Absorptivity Solar: 0.1 vs. 0.9						Absorptivity Solar: 0.2 vs. 0.9					
		Gas		Electr.		Pump		Gas		Electr.		Pump	
		Savings	PayB	Savings	PayB	Savings	PayB	Savings	PayB	Savings	PayB	Savings	PayB
	$i=0\%$	−16.77	-	−73.37	-	−0.57	-	−14.02	-	−60.81	-	−0.63	-
$d=0.3\%$	$i=3\%$	−18.39	-	−93.56	-	3.12	16.58	−14.76	-	−76.9	-	3.02	16.63
	$i=6\%$	−20.68	-	−122.54	-	8.47	14.07	−15.81	-	−100	-	8.28	14.09
	$i=0\%$	−16.2	-	−66.52	-	−1.79	-	−13.74	-	−55.33	-	−1.83	-
$d=1.5\%$	$i=3\%$	−17.57	-	−83.6	-	1.33	18.02	14.37	-	−68.95	-	1.24	18.28
	$i=6\%$	−19.5	-	−107.96	-	5.81	15.08	−15.26	-	−88.37	-	5.66	15.11
	$i=0\%$	−15.59	-	−59.4	-	−3.05	-	−13.44	-	−49.66	-	−3.07	-
$d=3\%$	$i=3\%$	−16.72	-	−73.36	-	−0.5	-	−13.96	-	−60.79	-	−0.56	-
	$i=6\%$	−18.29	-	−93.1	-	3.12	16.63	−14.69	-	−76.52	-	3.01	16.68

		Absorptivity Solar: 0.3 vs. 0.9						Absorptivity Solar: 0.4 vs. 0.9					
		Gas		Electr.		Pump		Gas		Electr.		Pump	
		Savings	PayB	Savings	PayB	Savings	PayB	Savings	PayB	Savings	PayB	Savings	PayB
	$i=0\%$	−11.74	-	−49.25	-	−1.01	-	−10.06	-	−39.08	-	−1.755	-
$d=0.3\%$	$i=3\%$	−11.77	-	−61.59	-	2.48	17.11	−9.56	-	−48.1	-	1.46	18.16
	$i=6\%$	11.79	-	−79.29	-	7.52	14.40	−8.83	-	−61.05	-	6.11	15.15
	$i=0\%$	−11.7	-	−45.05	-	−2.16	-	−10.2	-	−36	-	−2.81	-
$d=1.5\%$	$i=3\%$	−11.73	-	−55.49	-	0.79	18.86	−9.78	-	−43.641	-	−0.09	-
	$i=6\%$	−11.76	-	−70.38	-	5.02	15.49	−9.17	-	−54.52	-	3.8	16.34
	$i=0\%$	−11.66	-	−40.696	-	−3.35	-	−10.34	-	−32.8	-	−3.91	-
$d=3\%$	$i=3\%$	−11.68	-	−49.23	-	−0.941	-	−10	-	−39.05	-	−1.69	-
	$i=6\%$	−11.71	-	−61.29	-	2.47	17.15	−9.51	-	−47.87	-	1.46	18.20

Table A7. LCA marginal savings of the aged roof with 10 years maintenance vs. the reference roof with absorptivity = 0.8.

		Absorptivity Solar: 0.1 vs. 0.8						Absorptivity Solar: 0.2 vs. 0.8					
		Gas		Electr.		Pump		Gas		Electr.		Pump	
		Savings	PayB	Savings	PayB	Savings	PayB	Savings	PayB	Savings	PayB	Savings	PayB
d=0.3%	$i = 0\%$	−9.27	-	−57.63	-	4.571	4.05	−6.52	-	−45.07	-	4.51	3.91
	$i = 3\%$	−11.05	-	−75.19	-	7.3	3.88	−7.43	-	−58.53	-	7.19	3.75
	$i = 6\%$	−13.59	-	−100.39	-	11.24	3.73	−8.72	-	−77.85	-	11.06	3.61
d=1.5%	$i = 0\%$	−8.63	-	−51.66	-	3.67	4.17	−6.18	-	−40.48	-	3.63	4.03
	$i = 3\%$	−10.15	-	−66.52	-	5.98	3.99	−6.96	-	−51.88	-	5.9	3.85
	$i = 6\%$	−12.29	-	−87.71	-	9.29	3.83	−8.051	-	−68.12	-	9.145	3.70
d=3%	$i = 0\%$	−12.29	-	−45.46	-	2.75	4.33	−5.82	-	−35.71	-	2.73	4.18
	$i = 3\%$	−7.97	-	−57.61	-	4.63	4.13	−6.46	-	−45.03	-	4.57	3.99
	$i = 6\%$	−10.96	-	−74.78	-	7.31	3.96	−7.35	-	−58.2	-	7.2	3.83

		Absorptivity Solar: 0.3 vs. 0.8						Absorptivity Solar: 0.4 vs. 0.8					
		Gas		Electr.		Pump		Gas		Electr.		Pump	
		Savings	PayB	Savings	PayB	Savings	PayB	Savings	PayB	Savings	PayB	Savings	PayB
d=0.3%	$i = 0\%$	−4.24	-	−33.51	-	4.13	3.97	−2.55	-	−23.34	-	3.39	4.32
	$i = 3\%$	−4.44	-	−43.22	-	6.66	3.81	−2.23	-	−29.74	-	5.64	4.12
	$i = 6\%$	−4.7	-	−57.15	-	10.3	3.66	−1.75	-	−38.9	-	8.88	3.95
d=1.5%	$i = 0\%$	−4.14	-	−30.19	-	3.31	4.09	−2.64	-	−21.14	-	2.65	4.46
	$i = 3\%$	−4.32	-	−38.42	-	5.44	3.91	−2.36	-	−26.56	-	4.55	4.24
	$i = 6\%$	−4.54	-	−50.13	-	8.49	3.76	−1.96	-	−34.27	-	7.27	4.06
d=3%	$i = 0\%$	−4.04	-	−26.75	-	2.46	4.26	−2.72	-	−18.86	-	1.89	4.66
	$i = 3\%$	−4.18	-	−33.47	-	4.19	4.06	−2.5	-	−23.29	-	3.44	4.41
	$i = 6\%$	−4.37	-	−42.97	-	6.66	3.89	−2.18	-	−29.55	-	5.65	4.21

Table A8. LCA marginal savings of the aged roof with 10 years maintenance vs. the reference roof with absorptivity = 0.9.

		Absorptivity Solar: 0.1 vs. 0.9						Absorptivity Solar: 0.2 vs. 0.9					
		Gas		Electr.		Pump		Gas		Electr.		Pump	
		Savings	PayB	Savings	PayB	Savings	PayB	Savings	PayB	Savings	PayB	Savings	PayB
d=0.3%	$i = 0\%$	−8.76	-	−65.36	-	7.434	2.99	−6.01	-	−52.8	-	7.37	2.86
	$i = 3\%$	−10.38	-	−85.55	-	11.13	2.9	−6.75	-	−68.89	-	11.03	2.78
	$i = 6\%$	−12.67	-	−114.53	-	16.48	2.13	−7.8	-	−91.99	-	16.29	2.71
d=1.5%	$i = 0\%$	−8.19	-	−58.51	-	6.21	3.05	−5.73	-	−47.32	-	6.17	2.92
	$i = 3\%$	−9.56	-	−75.59	-	9.34	2.96	−6.36	-	−60.94	-	9.25	2.84
	$i = 6\%$	−11.49	-	−99.952	-	13.82	2.88	−7.25	-	−80.36	-	13.67	2.77
d=3%	$i = 0\%$	−7.58	-	−51.39	-	4.95	3.14	−5.43	-	−41.65	-	4.93	3.01
	$i = 3\%$	−8.71	-	−65.35	-	7.5	3.05	−5.95	-	−52.78	-	7.44	2.92
	$i = 6\%$	−10.28	-	−85.09	-	11.13	2.96	−6.68	-	−68.51	-	11.02	2.84

		Absorptivity Solar: 0.3 vs. 0.9						Absorptivity Solar: 0.4 vs. 0.9					
		Gas		Electr.		Pump		Gas		Electr.		Pump	
		Savings	PayB	Savings	PayB	Savings	PayB	Savings	PayB	Savings	PayB	Savings	PayB
d=0.3%	$i = 0\%$	−3.73	-	−41.24	-	7.01	2.85	−2.05	-	−31.07	-	6.25	3.02
	$i = 3\%$	−3.76	-	−53.58	-	10.49	2.78	−1.55	-	−40.09	-	9.47	2.91
	$i = 6\%$	−3.78	-	−71.28	-	15.53	2.71	−0.82	-	−53.04	-	14.12	2.83
d=1.5%	$i = 0\%$	−3.69	-	−37.04	-	5.84	2.92	−2.19	-	−27.99	-	5.191	3.07
	$i = 3\%$	−3.72	-	−47.48	-	8.8	2.84	−1.77	-	−35.63	-	7.91	2.98
	$i = 6\%$	−3.74	-	−62.37	-	13.03	2.76	−1.16	-	−46.51	-	11.81	2.9
d=3%	$i = 0\%$	−3.65	-	−32.68	-	4.65	3.01	−2.33	-	−24.79	-	4.096	3.17
	$i = 3\%$	−3.67	-	−41.22	-	7.06	2.92	−1.99	-	−31.04	-	6.31	3.07
	$i = 6\%$	−3.70	-	−53.28	-	10.49	2.84	−1.50	-	−39.86	-	9.47	2.98

Table A9. LCA savings of the aged roof with 5 years maintenance vs. the reference roof with absorptivity = 0.8.

		Absorptivity Solar: 0.1 vs. 0.8						Absorptivity Solar: 0.2 vs. 0.8					
		Gas		Electr.		Pump		Gas		Electr.		Pump	
		Savings	PayB	Savings	PayB	Savings	PayB	Savings	PayB	Savings	PayB	Savings	PayB
d=0.3%	i = 0%	−21.73	-	−72.44	-	−7.22	-	−18.7	-	−59.12	-	−7.13	-
	i = 3%	−26.92	-	−91.22	-	−4.59	-	−19.87	-	−73.53	-	−4.5	-
	i = 6%	−21.6	-	−118.16	-	−0.79	-	−21.52	-	−94.21	-	−0.71	-
d=1.5%	i = 0%	−20.94	-	66.01	-	−8.03	-	−18.24	-	−54.17	-	−7.95	-
	i = 3%	−22.75	-	−81.91	-	−5.82	-	−19.23	-	−66.37	-	−5.73	-
	i = 6%	−25.32	-	−104.57	-	−2.63	-	−20.62	-	−83.76	-	−2.55	-
d=3%	i = 0%	−20.1	-	−59.34	-	−8.87	-	−17.74	-	−49.03	-	−8.78	-
	i = 3%	−21.59	-	−72.33	-	−7.06	-	−18.55	-	−59	-	−6.97	-
	i = 6%	−23.68	-	−90.7	-	−4.49	-	−19.69	-	−73.1	-	−4.41	-
		Absorptivity Solar: 0.3 vs. 0.8						Absorptivity Solar: 0.4 vs. 0.8					
		Gas		Electr.		Pump		Gas		Electr.		Pump	
		Savings	PayB	Savings	PayB	Savings	PayB	Savings	PayB	Savings	PayB	Savings	PayB
d=0.3%	i = 0%	16.23	-	−47.01	-	−7.42	-	−14.4	-	−36.38	-	−8.11	-
	i = 3%	−16.61	-	−57.47	-	−4.92	-	−14.21	-	−43.35	-	−5.86	-
	i = 6%	−17.14	-	−72.46	-	−1.31	-	−13.91	-	−53.34	-	−2.62	-
d=1.5%	i = 0%	16.04	-	−43.41	-	−8.2	-	−14.4	-	−33.96	-	−8.81	-
	i = 3%	−16.36	-	−52.26	-	−6.08	-	−14.24	-	−39.86	-	−6.91	-
	i = 6%	−17.14	-	−72.46	-	−1.31	-	−13.99	-	−48.26	-	−4.18	-
d=1.5%	i = 0%	−15.82	-	−39.65	-	−8.99	-	−14.39	-	−31.43	-	−9.51	-
	i = 3%	−16.09	-	−46.89	-	−7.27	-	−14.26	-	−36.25	-	−7.96	-
	i = 6%	−16.46	-	−57.11	-	−4.82	-	−14.06	-	−43.07	-	−5.76	-

Table A10. LCA savings of the aged roof with 5 years maintenance vs. the reference roof with absorptivity = 0.9.

		Absorptivity Solar: 0.1 vs. 0.9						Absorptivity Solar: 0.2 vs. 0.9					
		Gas		Electr.		Pump		Gas		Electr.		Pump	
		Savings	PayB	Savings	PayB	Savings	PayB	Savings	PayB	Savings	PayB	Savings	PayB
d=0.3%	i = 0%	−21.23	-	−80.17	-	−4.35	-	−18.19	-	−66.85	-	−4.27	-
	i = 3%	−23.19	-	−101.57	-	−0.75	-	19.87	-	−73.53	-	−4.5	-
	i = 6%	−25.99	-	−132.3	-	4.43	17.11	−20.6	-	−108.35	-	4.51	17.01
d=1.5%	i = 0%	−20.49	-	−72.86	-	−5.5	-	−17.79	-	−61.02	-	−5.41	-
	i = 3%	−22.16	-	−90.97	-	−2.46	-	−18.63	-	−75.44	-	−2.37	-
	i = 6%	−24.52	-	−116.81	-	1.89	18.5	−19.82	-	−96	-	1.97	18.4
d=3%	i = 0%	19.71	-	−65.27	-	−6.67	-	−17.35	-	−54.96	-	−6.59	-
	i = 3%	−21.08	-	−80.08	-	−4.19	-	−18.05	-	−66.75	-	−4.11	-
	i = 6%	−23.01	-	−101.01	-	−0.67	-	−19.02	-	−83.41	-	−0.59	-
		Absorptivity Solar: 0.3 vs. 0.9						Absorptivity Solar: 0.4 vs. 0.9					
		Gas		Electr.		Pump		Gas		Electr.		Pump	
		Savings	PayB	Savings	PayB	Savings	PayB	Savings	PayB	Savings	PayB	Savings	PayB
d=0.3%	i = 0%	−15.73	-	−54.74	-	−4.56	-	−13.9	-	−44.11	-	−5.25	-
	i = 3%	−15.94	-	−67.82	-	−1.08	-	−13.53	-	−53.71	-	−2.03	-
	i = 6%	−16.22	-	−86.6	-	3.92	17.29	−12.98	-	−67.48	-	2.61	18.08
d=1.5%	i = 0%	−15.59	-	−50.25	-	−5.66	-	−13.96	-	−40.8	-	−6.27	-
	i = 3%	−15.77	-	−61.32	-	−2.73	-	−13.65	-	−48.92	-	−3.55	-
	i = 6%	−16.01	-	−77.1	-	1 1.47	18.76	−13.19	-	−60.5	-	0.34	19.68
d=3%	i = 0%	−15.43	-	−45.59	-	−6.8	-	−14.39	-	−31.43	-	−9.51	-
	i = 3%	−15.58	-	−54.64	-	−4.4	-	−14.26	-	−36.25	-	−7.96	-
	i = 6%	−15.78	-	−67.43	-	−1.01	-	−13.39	-	−53.38	-	−1.94	-

Table A11. LCA marginal savings of the aged roof with 5 years maintenance vs. the reference roof with absorptivity = 0.8.

		Absorptivity Solar: 0.1 vs. 0.8						Absorptivity Solar: 0.2 vs. 0.8					
		Gas		Electr.		Pump		Gas		Electr.		Pump	
		Savings	PayB	Savings	PayB	Savings	PayB	Savings	PayB	Savings	PayB	Savings	PayB
	$i = 0\%$	−12.53	-	−63.23	-	1.984	4.056	−9.49	-	−49.91	-	2.071	3.91
d=0,3%	$i = 3\%$	−14.6675	-	81.52	-	4.61	3.88	−10.66	-	−64.33	-	4.69	3.75
	$i = 6\%$	−17.71	-	−108.95	-	8.4	3.73	−12.32	-	−85.01	-	8.48	3.61
	$i = 0\%$	−11.74	-	−56.82	-	1.15	4.17	−9.047	-	−44.98	-	1.23	4.03
d=1.5%	$i = 3\%$	−13.56	-	−72.72	-	3.37	3.99	−10.03	-	−57.18	-	3.45	3.85
	$i = 6\%$	−16.13	-	−95.37	-	6.55	3.83	−11.43	-	−74.57	-	6.63	3.7
	$i = 0\%$	−10.92	-	−50.166	-	0.304	4.33	−8.56	-	−39.85	-	0.38	4.18
d=3%	$i = 3\%$	−12.41	-	−63.16	-	2.11	4.131	−9.38	-	−49.83	-	2.19	3.99
	$i = 6\%$	14.5	-	−81.52	-	4.68	3.95	−10.52	-	−63.93	-	4.76	3.833

		Absorptivity Solar: 0.3 vs. 0.8						Absorptivity Solar: 0.4 vs. 0.8					
		Gas		Electr.		Pump		Gas		Electr.		Pump	
		Savings	PayB	Savings	PayB	Savings	PayB	Savings	PayB	Savings	PayB	Savings	PayB
	$i = 0\%$	−7.03	-	37.812	-	1.78	3.97	−5.2	-	−27.18	-	1.08	4.32
d=0.3%	$i = 3\%$	−7.41	-	−48.26	-	4.28	3.811	−5	-	−34.15	-	3.33	4.12
	$i = 6\%$	−7.94	-	−63.25	-	7.89	3.665	−4.7	-	−44.137	-	6.58	3.95
	$i = 0\%$	−6.84	-	−34.21	-	0.98	4.09	−5.21	-	24.76	-	0.37	4.46
d=1.5%	$i = 3\%$	−7.17	-	−43.06	-	3.1	3.91	−5.05	-	−30.66	-	2.282	4.24
	$i = 6\%$	−7.62	-	−55.67	-	6.13	3.76	−4.8	-	−39.07	-	5.005	4.06
	$i = 0\%$	−6.64	-	−30.48	-	0.17	4.26	−5.22	-	−22.255	-	−0.34	-
d=3%	$i = 3\%$	−6.91	-	−37.72	-	1.9	4.06	−5.08	-	−27.08	-	1.2	4.41
	$i = 6\%$	−7.28	-	−47.94	-	4.35	3.89	−4.89	-	−33.89	-	3.409	4.21

Table A12. LCA marginal savings of the aged roof with 5 years maintenance vs. the reference roof with absorptivity = 0.9.

		Absorptivity Solar: 0.1 vs. 0.9						Absorptivity Solar: 0.2 vs. 0.9					
		Gas		Electr.		Pump		Gas		Electr.		Pump	
		Savings	PayB	Savings	PayB	Savings	PayB	Savings	PayB	Savings	PayB	Savings	PayB
	$i = 0\%$	−12.02	-	−70.96	-	4.84	2.99	−8.99	-	−57.64	-	4.93	2.86
d=0.3%	$i = 3\%$	−13.99	-	−92.36	-	8.44	2.9	−9.98	-	−74.68	-	8.53	2.78
	$i = 6\%$	−16.78	-	−123.097	-	13.64	2.83	−11.39	-	−99.14	-	13.72	2.71
	$i = 0\%$	−11.3	-	−63.66	-	3.68	3.05	−8.6	-	−51.82	-	3.77	2.92
d=1.5%	$i = 3\%$	−12.97	-	−81.78	-	6.72	2.96	−9.44	-	−66.25	-	6.81	2.84
	$i = 6\%$	−15.33	-	−107.61	-	11.08	2.88	−10.63	-	−86.81	-	11.17	2.77
	$i = 0\%$	10.54	-	−56.09	-	2.5	3.14	8.18	-	−45.78	-	2.58	3.013
d=3%	$i = 3\%$	−11.9	-	−70.9	-	4.97	3.05	−8.87	-	−57.57	-	5.06	2.92
	$i = 6\%$	13.83	-	−91.84	-	8.5	2.96	−9.84	-	−74.24	-	8.58	2.84

		Absorptivity Solar: 0.3 vs. 0.9						Absorptivity Solar: 0.4 vs. 0.9					
		Gas		Electr.		Pump		Gas		Electr.		Pump	
		Savings	PayB	Savings	PayB	Savings	PayB	Savings	PayB	Savings	PayB	Savings	PayB
	$i = 0\%$	−6.52	-	−45.54	-	4.64	2.85	−4.69	-	−34.912	-	3.95	3.001
d=0.3%	$i = 3\%$	−6.73	-	−58.62	-	8.11	2.78	−4.32	-	−44.506	-	7.17	2.91
	$i = 6\%$	−7.02	-	−77.39	-	13.12	2.71	−3.77	-	−58.27	-	11.82	2.83
	$i = 0\%$	−6.4	-	−41.06	-	3.52	2.92	−4.76	-	−31.61	-	2.91	3.07
d=1.5%	$i = 3\%$	−6.58	-	−52.13	-	6.46	2.84	−4.46	-	−39.73	-	5.63	2.98
	$i = 6\%$	−6.82	-	−67.91	-	10.66	2.76	−4.01	-	−51.31	-	9.539	2.9
	$i = 0\%$	−6.25	-	−36.41	-	2.37	3.01	−4.83	-	−28.18	-	1.85	3.17
d=3%	$i = 3\%$	−6.41	-	−45.46	-	4.76	2.92	−4.58	-	−34.82	-	4.075	3.075
	$i = 6\%$	−6.61	-	−58.25	-	8.17	2.84	4.21	-	−44.2	-	7.22	2.98

Table A13. LCA savings for non-aged cool roofs vs. the reference absorptivity = 0.8.

		Absorptivity Solar: 0.1 vs. 0.8						Absorptivity Solar: 0.2 vs. 0.8					
		Gas		Electr.		Pump		Gas		Electr.		Pump	
		Savings	PayB	Savings	PayB	Savings	PayB	Savings	PayB	Savings	PayB	Savings	PayB
	$i = 0\%$	−29.65	-	−116.8	-	−4.7	-	−22.21	-	−92.06	-	−2.22	-
$d=0.3\%$	$i = 3\%$	−36.51	-	−153.25	-	−3.09	-	−26.55	-	−120.11	-	0.22	19.62
	$i = 6\%$	−46.39	-	−205.78	-	−0.77	-	−32.8	-	−160.533	-	3.76	16.17
	$i = 0\%$	−27.34	-	−104.53	-	−5.24	-	−20.75	-	−82.62	-	−3.05	-
$d=1.5\%$	$i = 3\%$	−33.14	-	−135.34	-	−3.88	-	−24.42	-	−106.33	-	−0.97	-
	$i = 6\%$	−41.44	-	−179.44	-	−1.93	-	−29.66	-	−140.26	-	1.99	17.54
	$i = 0\%$	−24.95	-	−91.84	-	−5.8	-	−19.25	-	−72.85	-	−3.9	-
$d=3\%$	$i = 3\%$	−29.68	-	−116.99	-	−4.69	-	−22.24	-	−92.2	-	−2.21	-
	$i = 6\%$	−36.4	-	−152.65	-	−3.12	-	−26.48	-	−119.65	-	0.18	19.69
		Absorptivity Solar: 0.3 vs. 0.8						Absorptivity Solar: 0.4 vs. 0.8					
		Gas		Electr.		Pump		Gas		Electr.		Pump	
		Savings	PayB	Savings	PayB	Savings	PayB	Savings	PayB	Savings	PayB	Savings	PayB
	$i = 0\%$	−16.577	-	−70.78	-	−1.05	-	−12.42	-	−52.63	-	−0.91	-
$d=0.3\%$	$i = 3\%$	−18.99	-	−91.61	-	1.79	17.44	−13.43	-	−67.29	-	1.97	17.2
	$i = 6\%$	−22.48	-	−121.63	-	5.89	14.61	−14.89	-	−88.42	-	6.15	14.45
	$i = 0\%$	−15.763	-	−63.77	-	−2.01	-	−12.08	-	−47.69	-	−1.89	-
$d=1.5\%$	$i = 3\%$	−17.8091	-	−81.38	-	0.39	19.3	−12.94	-	−60.09	-	0.55	19.02
	$i = 6\%$	−20.73	-	−106.57	-	3.837	15.76	−14.16	-	−77.82	-	4.06	15.57
	$i = 0\%$	−14.92	-	−56.52	-	−3	-	−11.73	-	−42.59	-	−2.9	-
$d=3\%$	$i = 3\%$	−16.59	-	−70.89	-	−1.04	-	−12.43	-	−52.7	-	−0.9	-
	$i = 6\%$	−18.95	-	−91.27	-	1.74	17.54	−13.42	-	−67.05	-	1.93	17.31

Table A14. LCA savings for non-aged cool roofs vs. the reference absorptivity = 0.9.

		Absorptivity Solar: 0.1 vs. 0.9						Absorptivity Solar: 0.2 vs. 0.9					
		Gas		Electr.		Pump		Gas		Electr.		Pump	
		Savings	PayB	Savings	PayB	Savings	PayB	Savings	PayB	Savings	PayB	Savings	PayB
	$i = 0\%$	−29.14	-	−124.53	-	−1.84	-	−21.71	-	−99.79	-	0.63	18.69
$d=0.3\%$	$i = 3\%$	−35.83	-	−163.61	-	0.74	18.85	−25.87	-	−130.47	-	4.06	15.01
	$i = 6\%$	−45.47	-	−219.92	-	4.46	15.62	−31.87	-	−174.67	-	9	12.85
	$i = 0\%$	26.89	-	−111.38	-	−2.71	-	−20.31	-	−89.46	-	−0.51	- -
$d=1.5\%$	$i = 3\%$	−32.55	-	−144.41	-	−0.52	-	−23.83	-	−115.73	-	2.38	16.41
	$i = 6\%$	−40.64	-	−191.69	-	2.59	15.62	−28.86	-	−152.83	-	6.52	13.76
	$i = 0\%$	−24.56	-	−97.78	-	−3.61	-	−18.86	-	−78.78	-	−1.7	-
$d=3\%$	$i = 3\%$	−29.18	-	−124.73	-	−1.82	-	−21.73	-	−99.94	-	0.65	18.7
	$i = 6\%$	−35.72	-	−162.96	-	0.7	18.93	−25.8	-	−129.96	-	4.01	15.14
		Absorptivity Solar: 0.3 vs. 0.9						Absorptivity Solar: 0.4 vs. 0.9					
		Gas		Electr.		Pump		Gas		Electr.		Pump	
		Savings	PayB	Savings	PayB	Savings	PayB	Savings	PayB	Savings	PayB	Savings	PayB
	$i = 0\%$	−16.0722	-	−78.5192	-	1.8	16.71	−11.92	-	−60.362	-	1.94	16.5
$d=0.3\%$	$i = 3\%$	−18.32	-	−101.97	-	5.62	13.71	−12.76	-	−77.65	-	5.81	13.56
	$i = 6\%$	−21.56	-	−135.76	-	11.13	11.87	−13.97	-	−102.56	-	11.39	11.76
	$i = 0\%$	−15.31	-	−70.6	-	0.52	18.8	−11.63	-	−54.54	-	0.64	18.53
$d=1.5\%$	$i = 3\%$	−17.21	-	−90.45	-	3.74	14.88	−12.34	-	−69.15	-	3.91	14.72
	$i = 6\%$	−19.93	-	−118.81	-	8.37	12.66	−13.363	-	−90.06	-	8.59	12.54
	$i = 0\%$	−14.53	-	−62.46	-	−0.81	-	−11.34	-	−48.5	-	−0.7	-
$d=3\%$	$i = 3\%$	−16.39	-	−81.92	-	2.35	16.76	−11.92	-	−60.45	-	1.96	16.55
	$i = 6\%$	−18.28	-	−101.58	-	5.56	13.84	−12.74	-	−77.36	-	5.75	13.7

Table A15. LCA marginal savings for non-aged cool roofs vs. the reference absorptivity = 0.8.

Absorptivity Solar: 0.1 vs. 0.8

		Gas		Electr.		Pump	
		Savings	PayB	Savings	PayB	Savings	PayB
d=0.3%	*i* = 0%	−21.64	-	−108.79	-	3.3	5.94
	i = 3%	−28.5	-	−145.24	-	4.916	5.54
	i = 6%	−38.38	-	−197.77	-	7.23	5.22
d=1.5%	*i* = 0%	−19.33	-	−96.52	-	2.76	6.2
	i = 3%	−25.13	-	−127.33	-	4.12	5.76
	i = 6%	−33.431	-	−171.4	-	6.07	5.4
d=3%	*i* = 0%	−16.94	-	−83.83	-	2.2	6.57
	i = 3%	−21.60	-	−108.98	-	3.31	6.05
	i = 6%	−28.30	-	−144.64	-	4.88	5.65

Absorptivity Solar: 0.2 vs. 0.8

		Gas		Electr.		Pump	
		Savings	PayB	Savings	PayB	Savings	PayB
d=0.3%	*i* = 0%	14.2	-	−84.05	-	5.784	3.89
	i = 3%	−18.54	-	−112.10	-	8.23	3.73
	i = 6%	−24.79	-	−152.52	-	11.77	3.59
d=1.5%	*i* = 0%	−12.74	-	−74.61	-	4.959	4.01
	i = 3%	−16.414	-	−98.32	-	7.03	3.839
	i = 6%	−21.65	-	−132.25	-	10	3.69
d=3%	*i* = 0%	−11.24	-	−64.84	-	4.10	4.16
	i = 3%	−14.23	-	−84.19	-	5.79	3.97
	i = 6%	−18.47	-	−111.64	-	8.19	3.81

Absorptivity Solar: 0.3 vs. 0.8

		Gas		Electr.		Pump	
		Savings	PayB	Savings	PayB	Savings	PayB
d=0.3%	*i* = 0%	−8.56	-	−62.7788	-	6.95	3.34
	i = 3%	−10.98	-	−83.6	-	9.80	3.23
	i = 6%	−14.47	-	−113.62	-	13.9	3.13
d=1.5%	*i* = 0%	−7.75	-	−55.769	-	5.99	3.43
	i = 3%	−9.79	-	−73.37	-	8.4	3.31
	i = 6%	−12.72	-	−98.56	-	11.84	3.21
d=3%	*i* = 0%	−6.52	-	−54.45	-	7.19	2.61
	i = 3%	−8.07	-	−70.62	-	9.8	2.55
	i = 6%	−10.27	-	−93.57	-	13.57	2.49

Absorptivity Solar: 0.4 vs. 0.8

		Gas		Electr.		Pump	
		Savings	PayB	Savings	PayB	Savings	PayB
d=0.3%	*i* = 0%	−4.41	-	−44.62	-	7.09	3.29
	i = 3%	−5.42	-	−59.28	-	9.98	3.18
	i = 6%	−6.88	-	−80.41	-	14.16	3.09
d=1.5%	*i* = 0%	−4.07	-	−39.687	-	6.1171	3.38
	i = 3%	−4.93	-	−52.08	-	8.56	3.26
	i = 6%	−6.15	-	−69.81	-	12.07	3.16
d=3%	*i* = 0%	−3.72	-	−34.58	-	5.1	3.49
	i = 3%	−4.42	-	−44.69	-	7.1	3.36
	i = 6%	−5.41	-	−59.04	-	9.94	3.25

Table A16. LCA marginal savings for non-aged cool roofs vs. the reference absorptivity = 0.9.

Absorptivity Solar: 0.1 vs. 0.9

		Gas		Electr.		Pump	
		Savings	PayB	Savings	PayB	Savings	PayB
d=0.3%	*i* = 0%	−21.13	-	−116.52	-	6.16	3.69
	i = 3%	−27.82	-	−155.60	-	8.7	3.55
	i = 6%	−37.46	-	−211.91	-	12.47	3.42
d=1.5%	*i* = 0%	−18.88	-	−103.37	-	5.29	3.8
	i = 3%	−24.54	-	−136.4	-	7.482	3.64
	i = 6%	−32.63	-	−183.67	-	10.6	3.51
d=3%	*i* = 0%	−16.55	-	−89.77	-	4.39	3.94
	i = 3%	−21.17	-	−116.72	-	6.18	3.77
	i = 6%	−27.71	-	−154.95	-	8.709	3.63

Absorptivity Solar: 0.2 vs. 0.9

		Gas		Electr.		Pump	
		Savings	PayB	Savings	PayB	Savings	PayB
d=0.3%	*i* = 0%	−13.7	-	−91.78	-	8.64	2.78
	i = 3%	−17.86	-	122.46	-	12.07	2.7
	i = 6%	−23.86	-	−166.66	-	17.01	2.64
d=1.5%	*i* = 0%	−12.3	-	−81.45	-	7.49	2.84
	i = 3%	−15.82	-	−107.38	-	10.39	2.77
	i = 6%	−20.85	-	−144.49	-	14.53	2.7
d=3%	*i* = 0%	−10.85	-	−70.77	-	6.302	2.93
	i = 3%	−13.7	-	−91.93	-	8.66	2.84
	i = 6%	−17.79	-	−121.95	-	12.01	2.77

Absorptivity Solar: 0.3 vs. 0.9

		Gas		Electr.		Pump	
		Savings	PayB	Savings	PayB	Savings	PayB
d=0.3%	*i* = 0%	−8.06	-	−70.5	-	9.81	2.49
	i = 3%	−10.31	-	−93.96	-	13.63	2.43
	i = 6%	−13.55	-	−127.75	-	19.14	2.38
d=1.5%	*i* = 0%	−7.3	-	−62.61	-	8.52	2.54
	i = 3%	−9.2	-	−82.44	-	11.75	2.48
	i = 6%	−11.92	-	−110.8	-	16.38	2.43
d=3%	*i* = 0%	−6.52	-	−54.45	-	7.19	2.61
	i = 3%	−8.07	-	−70.62	-	9.8	2.55
	i = 6%	−10.27	-	−93.57	-	13.57	2.49

Absorptivity Solar: 0.4 vs. 0.9

		Gas		Electr.		Pump	
		Savings	PayB	Savings	PayB	Savings	PayB
d=0.3%	*i* = 0%	−3.91	-	−52.35	-	9.95	2.46
	i = 3%	−4.75	-	−69.64	-	13.82	2.41
	i = 6%	5.96	-	−94.55	-	19.4	2.35
d=1.5%	*i* = 0%	−3.62	-	−46.53	-	8.65	2.51
	i = 3%	−4.33	-	−61.14	-	11.92	2.45
	i = 6%	−5.35	-	−82.05	-	16.6	2.4
d=3%	*i* = 0%	−3.33	-	−40.51	-	7.3	2.58
	i = 3%	−3.91	-	−52.44	-	9.97	2.52
	i = 6%	−4.73	-	−69.35	-	13.76	2.46

References

1. Financing the Energy Renovation of Buildings with Cohesion Policy Funding. ISBN 978-92-79-35999-6. Available online: https://op.europa.eu/en/publication-detail/publication/4ef2fc24-67d0-45c2-a6d1-c677574e6d26/language-en (accessed on 14 January 2020).
2. European Commission. DG Energy. Available online: https://ec.europa.eu/clima/policies/strategies/2030-en (accessed on 14 January 2020).
3. Spanish Royal Decrete 2429/1979. *Approving the Basic Building on Thermal Conditions in Buildings NBE-CT-79*; Gobierno de Espaã: Madrid, Spain, 1979; pp. 24524–24550.

4. Synnefa, A. *Cool Roofs Standars and the ECRC Product Rating Program*; European Cool Roofs Council: Brussels, Belgium, 2016.

5. Hernández-Pérez, I.; Alvarez, G.; Xamán, J.; Zavala-Guillén, I.; Arce, J.; Simá, E. Thermal performance of reflective materials applied to exterior building components—A review. *Energy Build.* **2014**, *80*, 81–105. [CrossRef]

6. Lu, X.; Xu, P.; Wang, H.; Yang, T.; Hou, J. Cooling potential and applications prospects of passive radiative cooling in buildings: The current state-of-the-art. *Renew. Sustain. Energy Rev.* **2016**, *65*, 1079–1097. [CrossRef]

7. ES-ENERGY STAR. Roof Products Specification. 2007. Available online: https://www.energystar.gov/products/building-products/roof-products (accessed on 24 January 2020).

8. APEC Energy Working Group. Cool Roofs in APEC Economies: Review of Experience, Best Practices and Potential Benefits. 2011. Available online: https://www.coolrooftoolkit.org/wp-content/uploads/2012/05/APEC-Cool-Roofs.pdf (accessed on 24 January 2020).

9. Akbari, H.; Levinson, R.; Rainer, L. Monitoring the energy-use effects of cool roofs on California commercial buildings. *Energy Build.* **2005**, *37*, 1007–1016. [CrossRef]

10. Romeo, C.; Zinzi, M. Impact of a cool roof application on the energy and comfort performance in an existing non-residential building. A Sicilian case study. *Energy Build.* **2013**, *67*, 647–657. [CrossRef]

11. Akbari, H.; Bretz, S.; Kurn, D.M.; Hanford, J. Peak power and cooling energy savings of high-albedo roofs. *Energy Build.* **1997**, *25*, 117–126. [CrossRef]

12. Akbari, H.; Levinson, R. Evolution of cool-Roof standards in the US. *Adv. Build. Energy Res.* **2008**, *2*, 1–32. [CrossRef]

13. Akbari, H.; Konopacki, S. Calculating energy-saving potentials of heat-island reduction strategies. *Energy Policy* **2005**, *33*, 721–756. [CrossRef]

14. Akbari, H.; Matthews, H.D. Global cooling updates: Reflective roofs and pavements. *Energy Build.* **2012**, *55*, 2–6. [CrossRef]

15. Xu, T.; Sathaye, J.; Akbari, H.; Garg, V.; Tetali, S. Quantifying the direct benefits of cool roofs in an urban setting: Reduced cooling energy use and lowered greenhouse gas emissions. *Build. Environ.* **2012**, *48*, 1–6. [CrossRef]

16. Bozonnet, E.; Doya, M.; Allard, F. Cool roofs impact on building thermal response: A French case study. *Energy Build.* **2011**, *43*, 3006–3012. [CrossRef]

17. Santamouris, M. Cooling the cities—A review of reflective and green roof mitigation technologies to fight heat island and improve comfort in urban environments. *Sol. Energy* **2014**, *103*, 682–703. [CrossRef]

18. Seifhashem, M.; Capra, B.R.; Milller, W.; Bell, J. The potential for cool roofs to improve the energy efficiency of single storey warehouse-type retail buildings in Australia: A simulation case study. *Energy Build.* **2018**, *158*, 1393–1403. [CrossRef]

19. Boixo, S.; Diaz-Vicente, M.; Colmenar, A.; Castro, M.A. Potential energy savings from cool roofs in Spain and Andalusia. *Energy* **2012**, *38*, 425–438. [CrossRef]

20. Pisello, A.L.; Santamouris, M.; Cotana, F. Active cool roof effect: Impact of cool roofs on cooling system efficiency. *Adv. Build. Energy Res.* **2013**, *7*, 209–221. [CrossRef]

21. Kolokotroni, M.; Gowreesunker, B.L.; Giridharan, R. Cool roof technology in London: An experimental and modelling study. *Energy Build.* **2013**, *67*, 658–667. [CrossRef]

22. De Masi, R.F.; Ruggiero, S.; Vanoli, G.P. Acrylic white paint of industrial sector for cool roofing application: Experimental investigation of summer behavior and aging problem under Mediterranean climate. *Sol. Energy* **2018**, *169*, 468–487. [CrossRef]

23. Levinson, R.; Akbari, H. Potential benefits of cool roofs on commercial buildings: Conserving energy, saving money, and reducing emission ofgreenhouse gases and air pollutants. *Energy Effic.* **2010**, *3*, 53–109. [CrossRef]

24. Hernández-Pérez, I.; Xamán, J.; Macías-Melo, E.V.; Aguilar-Castro, K.M. Reflective materials for cost-effective energy-efficient retrofitting of roofs. In *Cost-Effective Energy-Efficient Building Retrofitting*; Woodhead Publishing: Mexico City, Mexico, 2017; Chapter 4, pp. 119–139.

25. Zhang, Z.; Tong, S.; Yu, H. Life Cycle Analysis of Cool Roof in Tropical Areas. *Procedia Eng.* **2016**, *169*, 392–399. [CrossRef]

26. Jo, J.H.; Carlson, J.D.; Golden, J.S.; Bryan, H. An integrated empirical and modeling methodology for analyzing solar reflective roof technologies on commercial buildings. *Build. Environ.* **2010**, *45*, 453–460. [CrossRef]

27. Yuan, J.; Farnham, C.; Emura, K.; Alam, M.A. Proposal for optimum combination of reflectivity and insulation thickness of building exterior walls for annual thermal load in Japan. *Build. Environ.* **2016**, *103*, 228–237. [CrossRef]

28. Saafi, K.; Daouas, N. A life cycle cost analysis for an optimum combination of cool coating and thermal insulation of residential building roofs in Tunisia. *Energy* **2018**, *152*, 925–938. [CrossRef]

29. Favi, C.; Di Giuseppe, E.; D'Orazio, M.; Rossi, M.; Germani, M. Building Retrofit Measures and Design: A Probabilistic Approach for LCA. *Sustainability* **2018**, *10*, 3655. [CrossRef]

30. Trovato, M.R.; Nocera, F.; Giuffrida, S. Life-Cycle Assessment and Monetary Measurements for the Carbon Footprint Reduction of Public Buildings. *Sustainability* **2020**, *12*, 3460. [CrossRef]

31. Falcone, P.M.; González García, S.; Imbert, E.; Lijó, L.; Moreira, M.T.; Tani, A.; Tartiu, V.E.; Morone, P. Transitioning towards the bio-economy: Assessing the social dimension through a stakeholder lens. *Corp. Soc. Responsib. Environ. Manag.* **2019**, *26*, 1135–1153. [CrossRef]

32. Walton, G.N. *Thermal Analysis Research Program Reference Manual*; NBSSIR 83-2655; National Bureau of Standards: Washington, DC, USA; National Engineering Lab., Department of Energy: Washington, DC, USA, 1983.

33. Clark, G.; Allen, C. The Estimation of Atmospheric Radiation for Clear and Cloudy Skies. In Proceedings of the 2nd National Passive Solar Conference (AS/ISES), Philadelphia , PA, USA, 16–18 March 1978; pp. 675–678.

34. Mirsadeghi, M.; Costola, D.; Blocken, B.; Hensen, J.L.M. Review of external covective heat transfer coefficient models in building energy simulation programs: Implementation and uncertainty. *Appl. Therm. Eng.* **2013**, *56*, 134–151. [CrossRef]

35. Hagishima, A.; Tanimoto, J. Field measurements for estimating the convective heat transfer coefficient at building surfaces. *Build. Environ.* **2003**, *38*, 873–881. [CrossRef]

36. American Society of Heating, Refrigerating and Air Conditioning Engineers Inc. *ASHRAE Fundamentals Handbook (S.I.)*; American Society of Heating, Refrigerating and Air Conditioning Engineers Inc.: Atlanta, GA, USA, 1997.

37. Sanjuan, C.; Suárez, M.J.; Blanco, E.; Heras, M.D.R. Development and experimental validation of a simulation model for open joint ventilated facades. *Energy Build.* **2011**, *43*, 3446–3456. [CrossRef]

38. Mastrapostoli, E.; Santamouris, M.; Kolokotsa, D.; Vassilis, P.; Venieri, D.; Gompakis, K. On the ageing of cool roofs: Measure of the optical degradation, chemical and biological analysis and assessment of the energy impact. *Energy Build.* **2016**, *114*, 191–199. [CrossRef]

39. Xue, X.; Yang, J.; Zhang, W.; Jiang, L.; Qua, J.; Xu, L.; Zhang, H.; Song, J.; Zhang, R.; Li, Y.; et al. The study of an energy efficient cool white roof coating based on styrene acrylate copolymer and cement for waterproofing purpose. Part I: Optical properties, estimated cooling effect and relevant properties after dirt and accelerated exposures. *Constr. Build. Mat.* **2015**, *98*, 176–184. [CrossRef]

40. Durability, Pausing for Reflection on Aging of the Cool Roof. Journal of Architect Coatings. June/July 2008. Available online: https://www.paintsquare.com/library/articles/Durability_Pausing_for_reflection_on_aging_of_the_cool_roof.pdf (accessed on 12 January 2020).

41. Bretz, S.E.; Akbari, H. Long-term performance of high-albedo roof coatings. *Energy Build.* **1997**, *25*, 159–167. [CrossRef]

42. Eilert, P. *High Albedo (Cool) Roofs: Codes and Standards Enhancement (CASE) Study*; Pacific Gas and Electric Company: San Francisco, CA, USA, 2000.

43. Prontuario de Soluciones Constructivas del CTE. Instituto de Ciencias de la Construcción Eduardo Torroja. Available online: http://cte-web.iccl.es/materiales.php (accessed on 12 March 2020).

44. Díaz, P.F. Centrales Térmicas. Biblioteca Sobre Ingeniería Energética. Universidad de Cantabria. Available online: https://pfernandezdiez.es/es/libro?id=15 (accessed on 15 March 2020).

45. Agencia Estatal de Meteorología de España. Guía Resumida del Clima en España (1981–2010). 2010. Available online: http://www.aemet.es/es/conocermas/recursos_en_linea/publicaciones_y_estudios/publicaciones/detalles/guia_resumida_2010 (accessed on 15 January 2020).

46. Auliac, S.; Le Hyaric, A.; Morice, J.; Hecht, F.; Ohtsuka, K.; Pironneau, O. FreeFem++. Third Edition, Version 3.31-2. Available online: http:www.freefem.orgr (accessed on 19 January 2020).

47. Energy Plus Software. U.S. Department of Energy (DOE). Building Technologies Office (BTO). Available online: https://energyplus.net/downloads (accessed on 19 January 2020).
48. Çengel, Y. *Heat and Mass Transfer. A Practical Approach*, 3rd ed.; McGraw Hill: New York, NY, USA, 2007.
49. Gentle, A.R.; Aguilar, J.L.C.; Smith, G.B. Optimized cool roofs: Integrating albedo and thermal emittance with R-value. *Sol. Energy Mater. Sol. Cells* **2011**, *95*, 3207–3215. [CrossRef]
50. Rossi, M.; Rocco, V.M. External walls design: The role of periodic thermal transmittance and internal areal heat capacity. *Energy Build.* **2014**, *68*, 732–740. [CrossRef]
51. Al-Sanea, S.A.; Zedan, M.F.; Al-Ajlan, S.A. Effect of electricity tariff on the optimum insulation-thickness in building walls as determined by a dynamic heat-transfer model. *Appl. Energy* **2005**, *82*, 313–330. [CrossRef]
52. Ozel, M. Effect of wall orientation on the optimum insulation thickness by using a dynamic method. *Appl. Energy* **2011**, *88*, 2429–2435. [CrossRef]
53. Daouas, N. Impact of external longwave radiation on optimum insulation thickness in Tunisian building roofs based on a dynamic analytical model. *Appl. Energy* **2016**, *177*, 136–148. [CrossRef]
54. Arumugam, R.S.; Garg, V.; Ram, V.V.; Bhatia, A. Optimizing roof insulation for roofs with high albedo coating and radiant barriers in India. *J. Build. Eng.* **2015**, *2*, 52–58. [CrossRef]
55. Synnefa, A.; Saliari, M.; Santamouris, M. Experimental and numerical assessment of the impact of increased roof reflectance on a school building in Athens. *Energy Build.* **2012**, *55*, 7–15. [CrossRef]
56. Reglamento de Instalaciones Térmicas en los Edificios. Ministerio para la Transición Ecológica y el Reto Demográfico. Available online: https://energia.gob.es/desarrollo/EficienciaEnergetica/RITE/Paginas/InstalacionesTermicas.aspx (accessed on 27 January 2020).
57. Duffie, J.A.; Beckman, W. *Solar Engineering of Thermal Processes*; John Wiley and Sons Inc.: Hoboken, NJ, USA, 2006.
58. CYPE Engineers. CYPE Ingenieros Construction Price Generator. 2017. Available online: http://generadorprecios.cype.es/(accessed on 28 March 2020).
59. Vincelas, F.F.C.; Ghislain, T.; Robert, T. Influence of the types of fuel and building material on energy savings into building in tropical region of Cameroon. *Appl. Therm. Eng.* **2017**, *122*, 806–819. [CrossRef]
60. Available online: https://ec.europa.eu/eurostat/statistics-explained/index.php/Natural-gas-price-statistics (accessed on 22 March 2020).
61. Nydahl, H.; Andersson, S.; Astrand, A.P.; Olofsson, T. Energy and Buildings Including future climate induced cost when assessing building refurbishment performance. *Energy Build.* **2019**, *203*, 109428. [CrossRef]
62. Commission Delegated Regulation (EU) No 626/2011 of 4 May 2011 Supplementing Directive 2010/30/EU of the European Parliament and of the Council with Regard to Energy Labelling of Air Conditioners. Available online: https://ec.europa.eu/info/energy-climate-change-environment/standards-tools-and-labels/products-labelling-rules-and-requirements/energy-label-and-ecodesign/energy-efficient-products/air-conditioners-and-comfort-fans_en (accessed on 8 March 2020).
63. Kloepffer, W. Life cycle sustainability assessment of products. *Int. J. Life Cycle Assess.* **2008**, *13*, 89–95. [CrossRef]
64. Falcone, P.M.; Imbert, E. Social Life Cycle Approach as a Tool for Promoting the Market Uptake of Bio-Based Products from a Consumer Perspective. *Sustainability* **2018**, *10*, 1031. [CrossRef]

 *sustainability*

Article

Reversibility and Durability as Potential Indicators for Circular Building Technologies

Ernesto Antonini [1], Andrea Boeri [1], Massimo Lauria [2] and Francesca Giglio [2,*

[1] Department of Architecture, University of Bologna, 40136 Bologna, Italy; ernesto.antonini@unibo.it (E.A.); andrea.boeri@unibo.it (A.B.)

[2] Department of Architecture and Territory dArTe, Mediterranea University of Reggio Calabria, 89124 Reggio Calabria, Italy; mlauria@unirc.it

* Correspondence: francesca.giglio@unirc.it; Tel.: +39-0965-1697131

Received: 29 July 2020; Accepted: 13 September 2020; Published: 16 September 2020

Abstract: According to the Circularity Gap Report 2020, a mere 8.6% of the global economy was circular in 2019. The Global Status Report 2018 declares that building construction and operations accounted for 36% of global final energy use and 39% of energy–related carbon dioxide (CO_2) emissions. The Paris Agreement demands that the building and construction sector decarbonizes globally by 2050. This requires strategies that minimize the environmental impact of buildings and practices extending the lifecycle of their constituents within a circular resource flow. To ensure that effective measures are applied, a suitable method is needed to assess compliance in materials, processes, and design strategies within circular economy principles. The study's assumption is that synthetic and reliable indicators for that purpose could be based on reversibility and durability features. The paper provides an overview of building design issues within the circular economy perspective, highlighting the difficulty in finding circular technologies which are suitable to enhance buildings' service life while closing material loops. The results identify reversibility and durability as potential indicators for assessing circular building technologies. The next research stage aims to further develop the rating of circularity requirements for both building technologies and entire buildings.

Keywords: Circular Design; circular technologies; reversibility; durability; circular potential; indicators; service life; closed material loops

---

1. Introduction

Global demand for materials increased over the past century, driven by the steady economic growth of the Organisation for Economic Co-operation and Development Countries (OECD), the industrialization of emerging economies, and the expansion of the world population [1].

According to the Circularity Gap Report 2020, only 8.6% of the world economy is circular. This means that only 8.6% of the minerals, fossil fuels, metals, and biomass which enter it every year are subsequently reused. This share had been estimated at 9.1% in the 2018 Report; a fall of 0.5% in two years [2].

Since the global use of raw materials has increased by almost double the population growth rate, the OECD finds that efficient use of resources and furthering the transition to a circular economy can help not only material supply security but also improve environmental and economic outcomes as well. There is no single accepted definition of the circular economy, but it is often defined as the emerging opposition to the so-called linear economy (take-make-dispose model). The most recognized definition comes from the Ellen MacArthur Foundation: it "is restorative and regenerative by design and aims to keep products, components, and materials at their highest utility and value at all times," aiming to decouple economic growth from resource consumption [3]. The aim of design for a circular economy is to maintain product integrity over multiple use cycles (for instance, through repair, refurbishment, and remanufacturing) and to focus on closing loops (through recycling), while at the same time building economically viable product–service systems [4].

The UN Sustainable Development Goals and the 2015 Paris Climate Agreement target to increase the circular share of the economy by addressing several key actions on energy and material efficiency improvement.

In particular, and as far as the building sector is concerned, the Paris Agreement demands global decarbonization by 2050, with the aim of avoiding the impact of a +2 degree rise in temperature. With this objective in mind, the Global Status Reports published annually by the UN Environment, the International Energy Agency (IEA), and the Global Alliance for Buildings and Construction (GABC) focus on the state of the building and construction sector and tracking the progress of this sector. In particular, in 2018, the report highlighted that "building construction and operations [account for] 36% of global final energy use and 39% of energy–related carbon dioxide (CO2) [emissions]" [5].

Around 374 million tons of construction and demolition waste (C&DW) were generated in 2016, making it the largest waste stream in the EU by weight of which the mineral content formed the largest fraction [6].

Data demonstrate that the construction industry has traditionally followed a linear economy process.

This has led to a number of challenges requiring solutions through those enabling factors which can maximize value from the circular economy by applying Circular Business Models (CBMs) in the current value chain.

By adopting circular economy business models, the focus will shift to sustainable sourcing, maintaining material productivity over the lifecycle of developments, and reducing loss in non-renewable materials. This will produce financial, social, and environmental benefits [7].

The European Commission emphasizes the role of design in the EU action plan for the circular economy: "better design can make products more durable or easier to repair, upgrade or remanufacture" (European Parliament Research Service [8]).

To implement these strategies, criteria and indicators are needed to assess the circular potential of those technical options which can be adopted and to assess their efficiency.

The concepts of reversibility and durability, as highlighted in the current sector literature, are already widely covered in all Circular Design strategies, including those addressing the construction sector in particular.

Reversibility is the property of a process, a system, or a device to return it to the original state. This means that the transformations performed to make that object suitable for a specific purpose can be reversed, restoring previous conditions to the transformation implemented. Meanwhile durability is the property of a process, a system, or a device to maintain its capability to provide the features it is designed for overtime. Both notions are thus closely time-related but by an antithetical relation, as the first implies transience, and the second permanence.

Nevertheless, no clear definitions are yet available to link concepts related to Circular Design strategies to indicators which are suitable to measure the circular economy and, specifically, circular building technologies.

This creates a barrier which obstructs the need for extending the circularity assessment of construction technologies to the whole functional complexity of an entire building (circular building) from properly considering any further performance involved.

For this reason, the study focuses on the notion of circular building technologies that embody the need to redefine how materials, processes, and products are designed and used within circular economy principles [9].

This leads to a focus on reversibility and durability as possible, suitable indicators for integrating a definition of circular building which complies with circular economy principles.

This paper addresses those topics through a methodology structured into three parts. The first part provides an overview of building design issues from the circular economy perspective. The second part highlights, from the literature overview conducted, the difficulties in finding circular technologies suitable for enhancing a building's service life and closing material loops. Reversibility and durability are then proposed in the third part (results) as possible indicators for assessing the circular potential of building technologies, while the next research stage is aimed at further developing the rating of both construction materials and whole buildings through the application of Circular Design strategies.

The study's assumption was that reversibility and durability can represent reliable indicators for this purpose as they integrate both quantitative aspects related to circular technologies and qualitative aspects referring to the service life and closed loop.

The study aimed to identify reversibility and durability as possible, synthetic, and reliable indicators to assess the compliance of building technologies with circular economy principles.

The expected research contribution was to integrate the advancement and dissemination of knowledge of tools for the application of Circular Design to the building sector.

The result may be useful for a further, undeveloped future phase of the study in which the indicators could enhance the principal sustainability rating systems referring to the circularity requirements of both building technologies and entire buildings.

2. Designing Buildings in the Context of the Circular Economy: An Overview

2.1. From Building Design Principles Applying the Circular Economy to Circular Design

Through the European Green Deal [10], the European Commission has prompted the application of the principles of the circular economy and resource efficiency in the construction sector aimed at reducing the use of primary natural resources in the future. As summarized in the document "Circular Economy—Principles for Buildings Design" (2020) [11], within the Construction 2020 strategy [8] and in line with the upcoming new circular economy action, the European Commission focuses on a set of principles for sustainable building design, designed to prevent and reduce construction and demolition waste while facilitating the reuse and recycling of building materials. This is expected to mitigate the environmental impact of building and reduce life cycle costs.

The measures address all the construction industry actors contributing to the building value chain, including building users; investors, developers, and insurance providers; design teams (engineering and architecture of buildings); manufacturers of construction products; contractors and builders; facility managers and owners; deconstruction and demolition teams; governmental and local regulation authorities.

The document refers to "Level(s)"—the Common European Framework of Sustainability Indicators—and the EC Communication [12] "A new Circular Economy Action Plan for a cleaner and more competitive Europe", which establishes a new comprehensive strategy for the sustainability of the built environment, material efficiency, and climate impact reduction. In particular, the document contributes to Level(s) Macro-objective 2: "Resource efficient and circular material life cycles" which targets waste reduction, material use optimization, and the reduction of environmental impact throughout the life cycle through suitable design choices. To reach this macro-objective, specific measures are envisaged called "life cycle tools: scenarios for building lifespan, adaptability and deconstruction". For building design, the general principles suggested are:

1. Durability: focus the building and planning of basic living services on a medium-to-long lifespan of the major building elements, while also considering their maintenance and replacement cycles;
2. Adaptability: extend the service life of whole buildings, either by facilitating the continuation of their original use or possible future changes performed by replacement and refurbishment;
3. Reduce waste and facilitate high-quality waste management: facilitate the future circular use of dismissed building elements, components, and parts, paying special attention to both reduced waste production and exploiting the potential for reuse and high-quality recycling of major building elements exiting deconstruction [10]. As for adaptability, i.e., the disposition of a building to transformation; the design usually targets specific client requirements, while future changes

in users and context needs are often neglected. As a result, buildings must be heavily modified and sometimes extensively demolished over their lifetime, often before the removed elements have reached their expected service life. This leads to unnecessary social, financial, and environmental costs, while the service life of the whole building and their elements could be optimized by taking future adaptations in the design stage into account.

In addition, a building can adapt more easily and efficiently to changing needs if versatile spaces are designed and demountable building elements are implemented. In 2012, through "Design for Change: Development of an Assessment and Transitional Framework", OVAM—the living lab on circular construction - developed a design and construction strategy based on the principle that our requirements and aspirations concerning built environment will always change, thus introducing the crucial conception that buildings are able to stand change more efficiently.

Considering the enormous materials and waste flows generated when constructing, maintaining, and refurbishing buildings, Design for Change (also known as dynamic building) can play a key role in reducing the environmental impact of the construction industry. Table 1 provides an overview of the Design for Change guidelines, in which the actions are observed both by object (interfaces, sub-components, composition) and by scale (element, building, neighborhood), according to the holistic approach adopted. This provides a comprehensive and clear framework on which a qualitative assessment of both the building design and construction can be based [13].

Table 1. An overview of the Design for Change guidelines.

	Interfaces	Sub-Components	Composition
Element	Reversibility Simplicity Speed	Durability Reused Compatibility	Pace-layered Independence Prefabrication
Building	Accessibility	Demountability Reusability Extensibility	Versatility
Neighborhood	Clear Adaptable	Retrofitted Dimensioned Removable	Unified Multipurpose Diverse Densificable

Note: Galle W., Vandenbroucke M. (2015).

The key role of rethinking design and the need at all process and product levels thus emerges as an effect driven by both regenerative circular economy principles and the EU's focus on these issues. Tangible goods are obviously the most involved in the changing economy, but services, business models, exchange relationships, and markets should also be redefined and redesigned, as should many further elements. The global economy is showing many fragilities and disruptive technologies which challenge conventional business models. The Ellen MacArthur Foundation defined this revolutionary transformation in the design scope as "Circular design, i.e., improvements in materials selection and product design (standardisation/modularisation of components, purer materials flows, and design for easier disassembly), lie at the heart of a circular economy" [14]. Circular Design requires a wide shift from the current product centered on issues concerning material flows, production processes and conditions, use, and reuse. This need adopts a wide and systemic view, as well as a deep understanding of ecological principles. Compared to other Design for Circularity (DfC) strategies, circular building design is specific to buildings, thus easily compatible with their peculiarities [14]. Circular design can indeed be described as an approach leading to "a building that is designed, planned, built, operated, maintained, and deconstructed consistently with Circular Economy principles" [15]. This includes optimizing the buildings' useful lifetime and integrating the end-of-life phase in the design [16]. Circular design challenges the development of products and materials that minimize the use of primary raw materials. As the name implies, the focus of the circular design is on curtailing the value loss of these embedded products by maintaining their circulation in closed loops. These loops, such as reuse, repair, remanufacture, refurbishment, or recycling, extend the product's life cycle and improve resource productivity. As happens in nature, the product, its parts, or constituent materials at the end of their life will so become a resource feeding new cycles of use, within or even outside of the original application scope [17]. A number of studies on Circular Design applied to the construction sector have been carried out in recent years, developing strategies, approaches, and visions that are pushing the transition from sustainable to circular buildings. This is triggering disruptive innovations acting as enablers for the circular economy. In this context, aimed at disseminating knowledge among stakeholders, the Vrije Universiteit Brussel (VUB) project "Le Bâti Bruxellois: Source de nouveaux Matériaux" (BBSM) has collected 16 Circular Design qualities for circular building which give practical support to the designers and all stakeholders in making coherent technical choices. The BBSM research was conducted by VUB (Vrije Universiteit Brussel) Architectural Engineering and was financed by the European Regional Development Fund (ERDF) and the Brussels-Capital Region. Project partners: UCLouvain, Rotor, and Belgian Building Research Institute.

Nine of these features and their related practices are especially helpful in shaping the concept of circular building technologies and in defining added value in Circular Design strategies [18].

The circular design qualities are the following:

- Reused: using building parts and components already present on-site or re-claimed elsewhere;
- Recycled: looking for building components made of low-value by-products or waste materials;
- Renewed: using materials that are replenished continuously by responsible agriculture and forestry;
- Compostable: choosing materials that can be biologically degraded into natural substances;
- Durable: using components that resist the wear and tear of use and reuse;
- Pure: favoring components that consist of a single material instead of a blend;
- Simple: going for low-tech, legible solutions rather than complicated ones;
- Reversible: making it possible to undo connections without damaging joined components;
- Location and site: recognizing and responsibly developing the qualities of a place.

Although less relevant for our purposes, the additional Circular Design qualities of the Belgian study identified are: safe and healthy, manageable, accessible, independent, compatible, multi-purpose, varied.

Circular Design qualities enable more effective reuse, recycling, or renewal of buildings and building components; a proven framework generating added value from the start of the design process onwards, regardless of whether that process adopts a conventional programming scheme, scenario planning, or a co-creation approach [18].

2.2. Circular Design Strategies: A Growing Concept

Extending building service life and closing material loops by consistent design choices are the key instruments in transitioning towards a circular construction economy. This requires both an increase in the whole building lifespan and the assurance that all its components can also circulate in endless loops, thus preserving the value as a whole of the resources involved. If properly recovered at the end of its technical, functional, or economic service life, each element can so re-enter a use cycle, being reused for the same or another purpose.

This study refers in particular to three main Circular Design approaches [18,19]:

- Design for Longevity (or Durability): New constructions can be avoided by adopting a circular design and construction practice, which instead focusses on review and up-valuing, upgrading, and refurbishing what already exists. Several architectural features can maintain a building's value over time, facilitating maintenance and repair, while enabling the extension of its service life. Among those features, some are related to the estate strategies, such as building location, but many others concern design choices, such as a multi-purpose spatial layout.

- Design for Disassembly and Deconstruction: By adopting this principle, the building elements can be disassembled and each of their components can be easily removed without damage to others. This leads to closing material flows, since components and materials can be recovered using quick and cheap processes, then reclaimed, preserving their value and minimizing waste. Various technical design features, therefore, act as key factors in applying that strategy, such as component durability, assembly methods, and the reversibility of their connections. Since it relies on design purposes such as reversibility, independence, and simplicity of connections and components, Design for Disassembly, often referred to as Reversible Building Design, allows for resource-efficient repairs, maintenance, replacements, as well as the reuse of construction materials, products, and components.
- Design for Reuse: The reclaimed building components and materials can be used again, repaired, remanufactured, or recycled in order to reduce the consumption of virgin, non-renewable resources. To ensure suitability, the elements must be safe and healthy to reuse and made up of a single substance which is easily recycled.

Alternative business models can provide these design approaches with favorable conditions, as shown by the BAMB project, a research action funded by the EU-HORIZON 2020 program which developed the emerging concept of "Buildings as Material Banks". Project BAMB—Buildings as Material Banks—is a consortium of 16 partners from eight European countries (2015/2019). Coordination: Institut Bruxellois pour la gestion de l'environnement—Brussels Instituut Voor Milieubeheer.

Referring to both Design for Disassembly and Design for Reuse strategies applied to the building sector, this new comprehensive approach considers buildings as "material banks", i.e., repositories or stockpiles of valuable, high-quality materials that can be easily taken apart and recovered, having been designed for disassembly.

Since preserving the material value is the key to their circular use, the ways to harvest this value are the circular economy drivers. Instead of disposal, buildings must therefore become banks of valuable materials—slowing down resource withdrawal to a rate compliant with planet restoring capacity. The BAMB project has developed and integrated tools—such as Material Passports and Reversible Building Design schemes—that will enable the shift from static buildings to circular buildings [20]. In order to sprawl the circular building concepts into practice, BAMB also investigated how new circular business models can bring new business opportunities, and how policies, such as data management tools (BIM, Building Information Modelling) and decision-making models can support the implementation of circular building concepts.

2.3. Circular Technologies for Circular Buildings: Extending Service Life and Closing Material Loops

For circular technologies there is, at the moment, no recognized definition but, with reference to the Building Design principles applied to Circular Design, the notion of circular building technologies summarises and embodies the need for redefining how materials, processes, and products are designed and used within the circular economy principles [9].

The concept of circular building was introduced by Arup Associates with the prototype designed for the London Design Festival 2016. This was the first building in the UK to comply with circular economy principles while creating a comfortable and aesthetic environment for the user. In order to achieve these goals, designers and engineers worked together to refine the application of a prefabricated construction based on a low-waste, self-supporting, and structurally demountable integrated panel (SIP) wall system, connected to the recycled steel structural frame by reusable clamps. The heat-treated timber for the cladding and decking was sustainably sourced too (Figure 1).

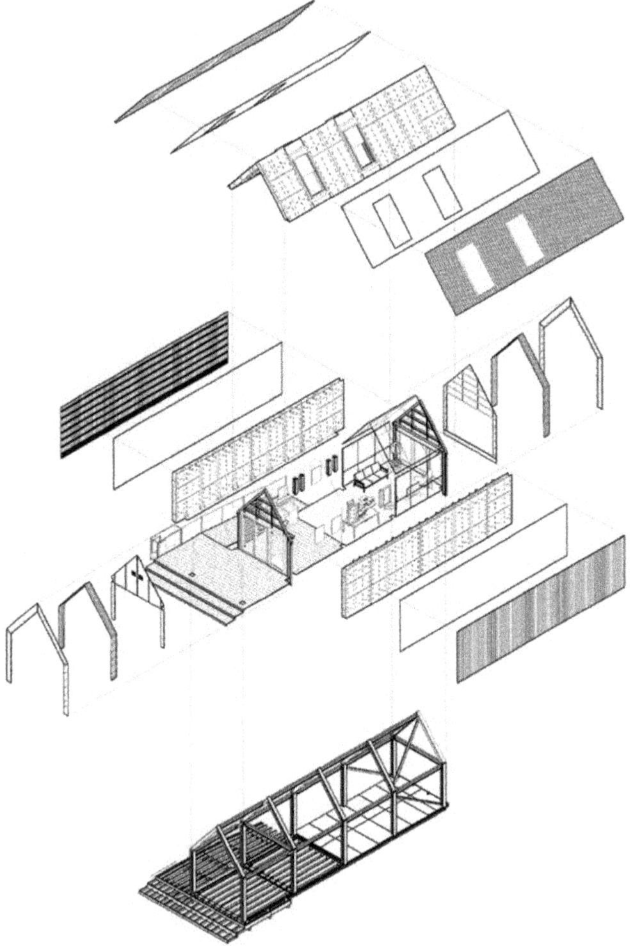

Figure 1. Exploded axonometric of the circular building prototype. Source: @ Arup Associates.

The simple "house-shaped" architecture (which refers to the 6S framework by Stewart Brand) promotes a familiar archetypical geometry at a scale which makes it immediate to perceive (Figure 2). Some definitions of circular building can be found in literature, such as those provided by the Dutch "Agenda for the transition to a circular built environment" or by the Report "A Framework for Circular Buildings: Indicators for possible inclusion in BREEAM" (2018).

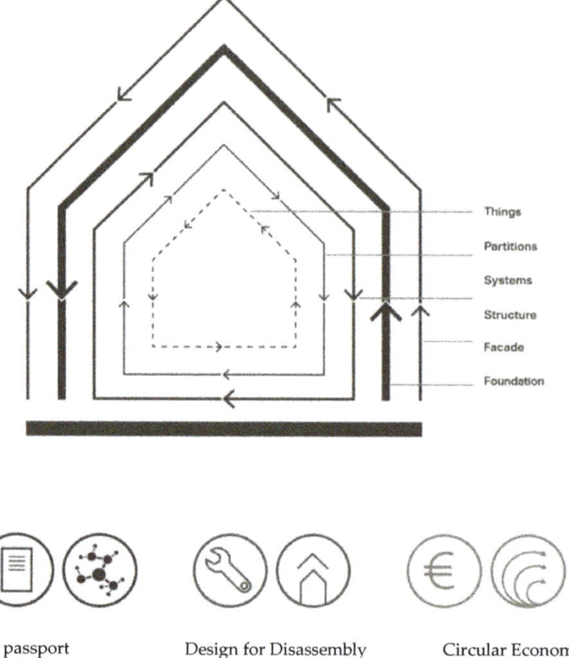

Figure 2. Reworking of Stewart Brand's Six S's diagram and Circular Design Strategies. Source: @ 3XN_GXN Architects.

A shared description states: "A building that is developed, used and reused without unnecessary resource depletion, environmental pollution and ecosystem degradation. It is constructed in an economically responsible way and contributes to the wellbeing of people and the biosphere. Here and there, now and later. Technical elements are demountable and reusable, and biological elements can also be brought back into the biological cycle" [21].

This is in line with the Ellen MacArthur Foundation's definition of circularity, where the preservation of the building's value and its components is assured by optimizing use and reuse cycles with minimal use of virgin resources. According to the 6S framework by Stewart Brand, different layers can be distinguished within a circular building, namely site, structure, skin, space plan, services, and stuff (furnishing and fittings), that are all part of a circular system of products, components, and materials [21]. Unlike circular building, the definition of circular technologies is still not well established but is deduced from the technical and functional specifications of the building in which those technologies are applied, aimed at enhancing the life cycle of each component and implementing circular potential.

According to Kanters [22], Circular buildings could significantly reduce environmental impact and the pressure on natural resources that buildings currently produce, as they are not designed according to circular economy principles. When considering the metrics by which this enhancement may be assessed, most literature focused on the lifecycle analysis of buildings and building materials, but not much is known about the design process of circular buildings [22].

Since design choices strongly jeopardize the subsequent life cycle stages of a building and its components, the role of the design must be questioned within a system that is no longer linear. Although extending service lives is necessary, recycling materials alone does not close all the loops, while more life cycle stages have to be reconnected, leaving virgin material consumption as the last option and avoiding waste production.

The circular technologies and design choices meeting these requirements are those making it easier to close the material loops. Their main features are summarized below:

- Durability: Since the materials can be only recovered if they remain intact during their use and disassembly, they have to withstand the typical wear and tear of repeated recovery and reuse cycles. Moreover, durable and hardwearing materials often age beautifully, thus maintaining or even increasing their value over time. Choosing the right material at the initial design stage is therefore crucial.
- Compatibility: Building components are compatible if they can be easily reused in different assembling stages, reconfigured, and recombined time and again. The simple geometry of the connection devices, versatile fixing systems, and standardized dimensions are the features maximizing component compatibility to their reuse potential.
- Reversibility: Even durable elements can only be recovered when they can be removed from their location by technically and economically feasible operations. Reversibility describes this requirement; the attitude of a component to be easily disconnected from the building elements that surround them. Dry, mechanic connections such as bolts and screws are thus more effective than wet and chemical joints such as mortar or glue. The high reversibility of the fixings not only allows the recovery of the components at the end of their useful life but also facilitates their maintenance and repair within the operational stage [23].

The notion of service life is in the background of both reversibility and durability features, as it is the time reference which makes the two requirements measurable, also allowing better perception of their qualitative connotations. Due to such complex implications, the service life cannot be established abstractly as just a certain number of expected years of life but must be quantified when planning the intervention by considering the multiple variables affecting it (economic, cultural, etc.), with reference to the process under design. Both scientific literature and regulatory measures often combine terms such as "cost", "efficiency", and "sustainability", which identify the main challenges the building sector is facing today. They are also the main pillars of the ISO 15686 Standard on Buildings and constructed assets, which is included within the service life planning series and provides a comprehensive guide to both service life planning and assessment of building components and systems. The series defines and systematizes methodologies and tools to ensure the most efficient combination of investment, maintenance, and management costs during the life cycle of the building. In particular, it introduces the estimation of service life in the design phase (ISO 15686, Part 2), costs, LCC (ISO15686, Part 5), and environmental impacts in the life cycle (ISO 15686, Part 6) [24].

This is particular for a circular building whose useful life does not end upon its dismantling but will continue through the programmed removal of its constituents feeding the circular process.

By this logic, the convergence of the above-mentioned areas into a single regulatory instrument allows for new ways of operating and critical analysis. It is important to underline, however, that although the main economic and environmental costs are paid during the building management phase, according to the dominant literature, these costs derive from the repeated inability to plan, from the project phase, correct life cycle management.

This highlights the key role that the design phase plays in shaping circular building by implementing effective circular technologies which cause positive effects across the various use cycles of all the involved.

3. Lack of Innovative Circular Technologies and Indicators for Their Evaluation

From the analysis of the current literature, it is possible to summarize two main aspects that are particularly relevant: the steady presence of reversibility and durability as a reference point in the current literature concerning Circular Design strategies; and the key role of the Design for extending the service life of buildings and closing material loops.

From these considerations we wish to highlight:

- The difficulty in finding circular building technologies suitable for enhancing the building's service life and closing material loops;
- The lack of indicators suitable to assess the circular potential of building technologies.

To support the transition to the circular economy, governance, regulations, and business models will play a crucial role. According to the research report Circular Business Models for the Built Environment maximizing value from a circular economy requires several enabling factors. Generally, these fall into three categories—design, information, and collaboration. As a result, additional value will be created through the operation, with benefits for asset value and waste production.

The report explores ways Circular Business Models (CBMs) [7] would provide added benefits throughout the value chain in construction, focusing on three principal models:

- Circular Design Model (development and planning phase of a built asset);
- Circular Use Model (operational phase of an asset);
- Circular Recovery Model (end of the product's service life).

At each stage of a development life cycle, there are opportunities and challenges which need to be addressed. Successful implementation of these business models will require action from designers, suppliers, service providers, contractors, and end-of-life companies by sharing materials, systems, energy, as well as information and services.

The aim of the paper, with particular reference to the Circular Design Model—as opportunities—products, and systems, is that the entire built structure should be designed to last longer with a higher residual value. Therefore, they should be easier to maintain, repair, upgrade, refurbish, remanufacture, or recycle with respect to traditional ones. Additionally, new materials can be developed and sourced, particularly bio-based, that are less resource-intensive or fully recyclable [7]. This highlights that design choices, extending the service life of buildings, and fostering closed material loops can be a key instrument for the transition toward a circular economy [25].

However, suitable evaluation criteria are still lacking, given the limited innovative circular building technologies identified—with reference to the concept previously given as a set of materials, products, and processes [9]—despite the urgent need for their implementation into the circular economy. According to Jawahir and Bradley [26], the technologies that are needed to feed sustainable value in the circular economy have not yet been given due consideration. As a result, expected circular economy benefits are not being achieved. Therefore, a large gap needs to be filled between scarce availability and the large demand for innovative circular technologies, whose expansion is expected in all industries worldwide [27].

Ensuring the successful transition to a circular economy requires, however, measuring circularity through suitable criteria, indicators, and assessment tools [28,29], whose development is still at an early level of definition [30,31]. Ghisellini et al. [32] showed that only 10 out of 155 studies on circular economy strategies include a focus on assessment indicators.

As the circular economy operates on three systemic levels (macro, meso, micro), the tools and indicators for measuring the ircular economy differ depending on the level of application (Su et al., 2013). Among the three levels or scopes at which the circular economy can be observed, some indicators have been developed for the macro-level (global, national, regional, city) and meso-level (industrial symbiosis, eco-industrial parks), while the micro-level (single firm, product) is still largely under-investigated. Although the development of indicators has progressed in recent years, there is currently no standardized way to measure the circularity at a micro-level [33]. According to Kristensen and Mosgaard [34], a detailed understanding of how to measure and document progress toward a circular economy is lacking, especially on a micro-level. This is a barrier for both producers who want to provide circular products and services and for the consumers who want to know how to compare products. The authors reviewed 30 circular economy indicators at the micro-level. The 30 indicators found in literature and practice were categorized according to the main focal point(s) of each indicator. This revealed the core circular economy principles considered by the indicators, which enabled analysis of the current scope of circular economy indicators at a micro-level. Nine circular economy categories emerged from this data-driven coding: recycling, remanufacturing, reuse, disassembly, lifetime extension, resource-efficiency, waste management, end-of-life management, and, lastly, multidimensional indicators, which are indicators that do not fall into the other categories as they cover a broader circular economy perspective. The majority of micro-level indicators are developed to assess individual products and materials, and thus function to support decision-making processes in companies.

The conclusion of their study is that there is no shared way of measuring circularity at the micro-level, and this lack is a barrier to further absorption and implementation. The development of standardized criteria suitable for micro-level circularity assessment would be useful, which should be explored either as industry-specific or circular strategy-specific [31].

In line with Kristensen and Mosgaard's study, the paper's focus on micro-level indicators, compared to other levels, is due to the compatibility of the field of application of the nine categories of indicators identified. The field of application of building technologies and their requirements are useful for measuring their circular potential.

4. Results

The result confirms the identification and delimitation of indicators that can contribute to the advancement of the state of knowledge which is useful for assessing the circular potential of construction solutions for circular buildings.

This refers to construction solutions that follow the principles of Circular Design, therefore they concern that set of materials, systems, and products, within the principles of the circular economy, summarized as circular building technologies. In particular, they identify and delimit reversibility and durability as possible indicators for assessing circular building technologies suitable for enhancing the building's service life and closing material loops.

The study summarized by this paper aimed to discuss whether reversibility and durability may represent suitable indicators for potential circular building technologies.

A definition is firstly provided of the two terms and their meanings:

- Durability should be understood as the ability of the building system and its materials not to exhibit significant deterioration that implies the loss of functionality for which they were designed. The durability of a building must meet, at least, the working life design, and during this period the necessary maintenance procedures that should be specified for the system must be met [35]. This feature depends on the reliability of the object and its constituents, the environment in which it is placed, and some conditions related to the socio-economic context, such as user behavior. Together, these conditions determine how long the building can properly remain in use.

- Reversibility is a feature or a transforming or dismantling process of a building or parts of it, preserving the maximum integrity of the removed elements and assuring minimum damage to those kept in place [36]. Reversibility, thus, refers mainly to the attitude of easily disassembled systems: this means going back step by step through the building assembly process, recovering most of the elements that had entered the primary process previously. To comply with this requirement, the dismantled element via disassembly must not be disposed of but re-enter into a cycle of use.

Once the requirements are defined, some suitable indicators must be identified to measure the compliance of a technical building configuration. By including these elements within the sustainability protocols, or tools, currently in use, it should be possible to avoid the proliferation of further procedures and better integrate the assessment of circular features within comprehensive and already tested building rating systems.

This will enhance the effectiveness of those tools in assessing the environmental impact and energy performance of buildings directly from the design phase and whose adoption has been suggested since the early 2000s [37] as part of a necessary integrated approach to accompany technical policies. Since these systems are typically based on a multi-criteria analysis (Criteria Decision Analysis Fines, MCDA), they are set to manage numerous and contrasting variables, which are considered in parallel then fused into an overall score which is very useful to quickly and effectively compare different alternatives. This should allow the addition of some new indicators which specifically refer to circularity requirements, preserving both the structure and the scoring scheme of each rating system. Among the most adopted in current practice, the rating systems based on these principles are LEED (Leadership in Energy and Environmental Design), BREEAM (Building Research Establishment Environmental Assessment Method for Buildings), and SBMethod.

The integration of circularity indicators within the rating system is the third step of this work. It is currently a work in progress whose outcome cannot be made available yet.

5. Discussion and Conclusions

The analyzed literature shows great potential for the advancement of knowledge in the field of Circular Design strategies applied to the construction sector, opening several areas of research which are yet to be fully explored with reference to circular building technologies. With regard to this issue, in order to advance knowledge, it is necessary to make a contribution, both to overcome the difficulty in finding circular building technologies suitable for improving the useful life of buildings and to close materials loops; the lack of indicators to assess the circular potential of building technologies. Reversibility and durability, as highlighted in the analyzed studies, are already widely included in all Circular Design strategies, including those addressing the construction sector in particular. Nevertheless, no clear definitions are yet available to link these concepts to indicators suitable for measuring circular economy and, specifically, circular building technologies. The notion of circular building technologies, however, embodies the need to redefine how materials, processes, and products are designed and used within circular economy principles. This absence is a barrier which prevents the extension of circularity assessment of construction technologies to the whole functional complexity of the entire building (circular building). Reversibility and durability can represent reliable indicators for this purpose, as they integrate both quantitative aspects related to circular technologies and qualitative aspects referring to service life and closed loops.

Most literature provides evidence that the building management phase causes most of the economic and environmental costs of the building process due to the recurrent inability to plan an ineffective life cycle management from the design stage. The imperatives of service life and closed loops are key issues which are fundamental for reaching the circular economy targets. They need to be endowed with reliable indicators to allow their assessment and the strengthening of their mutual relationship. Providing micro-level indicators for circular building technologies contributes to the advancement of knowledge on the measurability of circular economy processes, products, and business models applied to the construction sector.

Integrating the sustainable building rating systems currently in use with circular economy indicators, with particular emphasis on reversibility and durability at the micro-level, can provide useful assessment tools to designers to assist their choices early on in the project stages.

Whilst promising, these integrations need further study to develop assessment schemes fit for the operational level, as a wide application of these schemes would accelerate the market penetration of circular building technologies, thus acting as a catalyst for the success of the circular economy.

Author Contributions: Conceptualization, investigation, and data curation, F.G. and M.L.; methodology and formal analysis, E.A.; validation and supervision, A.B.; writing—original draft preparation: for reversibility, E.A.; for durability and service life, M.L.; for circular economy principles, A.B.; for circular building and technologies, F.G.; writing—review and editing, F.G. All authors have read and agreed to the published version of the manuscript.

Funding: This study did not receive external funding.

Acknowledgments: The authors are grateful to the following contributors for providing copyright permissions for the table and images: Waldo Galle, researcher and policy advisor on sustainability transitions, VUB Architectural Engineering; Arup Associates Europe press office; Tom Castle, 3XN_GXN Architects Business Development Coordinator.

Conflicts of Interest: The authors declare no conflict of interest.

References

1. OECD. *Improving Resource Efficiency and the Circularity of Economies for a Greener World*; OECD Environment Policy Papers, No. 20; OECD Publishing: Paris, France, 2020. [CrossRef]
2. Circle Economy. *The Circularity Gap Report*; Ruparo: Amsterdam, The Netherlands, 2020.
3. Ellen MacArthur Foundation. *Towards the Circular Economy: Economic and Business Rationale for an Accelerated Transition Volume 1*; Ellen MacArthur Foundation: Cowes, UK, 2013.
4. Den Hollander, M.C. *Design for Managing Obsolescence: A Design Methodology for Preserving Product Integrity in a Circular Economy*; Delft University of Technology: Delft, The Netherlands, 2018.
5. IEA, UNEP. Global Status Report—Towards a Zero-Emission, Efficient and Resilient Buildings and Construction Sector. 2018. Available online: https://www.unenvironment.org/resources/publication/2019-global-status-report-buildings-and-construction-sector (accessed on 4 July 2020).
6. Eurostat, Generation of Waste by Waste Category, Hazardousness and NACE Rev. 2 Activity [env_wasgen]. Available online: https://appsso.eurostat.ec.europa.eu/nui/show.do?dataset=env_wasgen&lang=en,2019a (accessed on 20 August 2020).
7. Carra, G.; Magdani, N. *Circular Business Models for the Built Environment*; BAM CE100; Arup: Bristol, UK, 2017.
8. European Parliament Research Service (EPRS). The Ecodesign Directive: European Implementation Assessment. In *Ex-Post Evaluation Unit of the Directorate for Impact Assessment and European Added Value*; European Parliament Research Service (EPRS): Brussels, Belgium, 2017.
9. Abbasova, Z. Adopting Circular Innovative Technologies in the Construction Supply Chain of the MRA. Master's Thesis, Tu Delft, Delft, The Netherlands, 2018.
10. European Commission Communication. The European Green Deal, Brussels, 11.12.2019 COM 640 Final. Available online: https://ec.europa.eu/info/sites/info/files/european-green-deal-communication_en.pdf (accessed on 4 July 2020).
11. European Commission. Circular Economy principles for Building Design, 21/02/2020. Available online: https://ec.europa.eu/docsroom/documents/39984 (accessed on 4 July 2020).
12. European Commission Communication. A new Circular Economy Action Plan for a Cleaner and More Competitive Europe, Brussels, 11.03.2020 COM 98 final. 2020. Available online: https://eur-lex.europa.eu/legal-content/EN/TXT/DOC/?uri=CELEX:52020DC0098&from=en (accessed on 9 July 2020).
13. Galle, W.; Vandenbroucke, M. *Design for Change: Development of a Policy and Transitional Framework*; Summary; OVAM: Mechelen, Belgium, 2015; pp. 3–6.
14. Jelmer, A. *A Guidance Tool for Circulr Building Design*; Delft University of Technology: Delft, The Netherlands, 2019.
15. Pomponi, F.; Moncaster, A. Circular Economy for the built environment: A research framework. *J. Clean. Prod.* **2017**, *143*, 710–718.
16. Leising, E.; Quist, J.; Bocken, N. Circular Economy in the building sector: Three cases and a collaboration tool. *J. Clean. Prod.* **2017**, *176*, 976–989. [CrossRef]
17. Fifield, B.; Medkova, K. Circular Design—Design for Circular Economy. In *Lahti CleanTech Annual Review*; Lahti University of Applied Sciences: Lahti, Finland, 2016.
18. Cambier, C.; Elsen, S.; Galle, W.; Lanckriet, W.; Poppe, J.; Tavernier, I.; Vandervaeren, C. *Building a Circular Economy*; Vrije Universiteit Brussel VUB Architectural Engineering: Brussel, Belgium, 2019.

19. Galle, W.; De Temmerman, N.; Vandervaeren, C.; Cambier, C.; Tavernier, I.; Poppe, J.; Elsen, S.; Lanckriet, W.; Verswijver, K. Building a Circular Economy: Design Qualities to Guide and Inspire Building Designers and Clients; VUB—Faculty of Engineering. 2019. Available online: https://www.vub.be/arch/files/circular_design_qualities/VUB%20Architectural%20Engineering%20-%20Circular%20Design%20Qualities%20(2019.12).pdf (accessed on 18 June 2020).
20. Durmisevic, E. *Reversible Building Design Guideline*; BAMB Wp3document; ebook; University of Twente: Enschede, The Netherlands, 2018.
21. DGBC, Metabolic, SGS Search, Circle Economy & Redevco Foundation. *A Framework for Circular Buildings: Indicators for Possible Inclusion in BREEAM*; Dutch Green Building Council: The Hague, The Netherlands, 2018.
22. Kanters, J. Circular Building Design: An Analysis of Barriers and Drivers for a Circular Building Sector. *Buildings* **2020**, *10*, 77. [CrossRef]
23. Galle, W. *Design for Change, towards a Circular Economy (Presentation Handout)*; Vrije Universiteit Brussel: Brussels, Belgium, 2017.
24. ISO 15686-5:2017. Buildings and Constructed Assets—Service Life Planning Life-Cycle Costing. Available online: https://www.iso.org/obp/ui/#iso:std:iso:15686:-5:ed-2:v1:en (accessed on 19 June 2020).
25. Galle, W.; Cambier, C.; Denis, F.; Vandervaeren, C.; De Temmerman, N. *Dynamic Design Matrix, A Major Review of the Flemish design Principles for Closed MateriSal Loops*; Presented at the WASCON Conference, RIL, Tampere; Vrije Universiteit Brussel: Brussels, Belgium, 2018; 14p.
26. Jawahir, I.S.; Bradley, R. Technological Elements of Circular Economy and the Principles of 6R-Based Closed-loop Material Flow in Sustainable Manufacturing. *Sci. Procedia CIRP* **2016**, *40*, 103–108. [CrossRef]
27. Baporikar, N. Advances in Business Strategy and Competitive Advantage. In *Handbook of Research on Entrepreneurship Development and Opportunities in Circular Economy*; IGI Global, Namibia & University of Pune: Pune, India, 2020; p. 353.
28. Cayzer, S.; Griffiths, P.; Beghetto, V. Design of indicators for measuring product performance in the circular economy. *Int. J. Sustain. Eng.* **2017**, *10*, 289–298. [CrossRef]
29. Saidani, M.; Yannou, B.; Leroy, Y.; Cluzel, F.; Kendall, A. A taxonomy of circular economy indicators. *J. Clean. Prod.* **2019**, *207*, 542–559. [CrossRef]
30. Giurco, D.; Littleboy, A.; Boyle, T.; Fyfe, J.; White, S. Circular economy: Questions for responsible minerals, additive manufacturing and recycling of metals. *Resources* **2014**, *3*, 432–453. [CrossRef]
31. Mesa, J.; Esparragoza, I.; Maury, H. Developing a set of sustainability indicators for product families based on the circular economy model. *J. Clean. Prod.* **2018**, *196*, 1429–1442. [CrossRef]
32. Ghisellini, P.; Cialani, C.; Ulgiati, S. A review on circular economy: The expected transition to a balanced interplay of environmental and economic systems. *J. Clean. Prod.* **2016**, *114*, 11–32. [CrossRef]
33. Elia, V.; Gnoni, M.G.; Tornese, F. Measuring circular economy strategies through index methods: A critical analysis. *J. Clean. Prod.* **2017**, *142*, 2741–2751. [CrossRef]
34. Kristensen, H.S.; Mosgaard, M.A. A review of micro level indicators for a circular economy e moving away from the three dimensions of sustainability? *J. Clean. Prod.* **2020**, *243*, 10–17. [CrossRef]
35. Parsekian, G.; Ramos Roman, H.; Silca, C.O.; Santos Faria, M. *Concrete Block in Long-Term Performance and Durability of Masonry Structures*; Bahman, G., Lourenço, P.B., Eds.; Woodhead Publishing: Cambridge, UK, 2019; pp. 21–57. [CrossRef]
36. Durmisevic, E. Transformable Building Structures: Design for Disassembly as a Way to Introduce Sustainable Engineering to Building Design & Construction. Ph.D. Thesis, TU Delft University of Technology, Delft, The Netherlands, 2006.
37. European Commission Communication Towards a Thematic Strategy on the Urban Environment, Brussels, 11.02.2004 COM 60 Final. 2004. Available online: https://www.europarl.europa.eu/meetdocs/committees/rett/20040316/com_com(2004)0060en.pdf (accessed on 22 July 2020).

 *sustainability*

Article

Assessment Model of End-of-Life Costs and Waste Quantification in Selective Demolitions: Case Studies of Nearly Zero-Energy Buildings

Eduardo Vázquez-López [1,*], Federico Garzia [2], Roberta Pernetti [2], Jaime Solís-Guzmán [1] and Madelyn Marrero [1]

[1] ArDiTec Research Group, Departament of Architectural Constructions II, Higher Technichal School of Building Engineering, Universidad de Sevilla, 41012 Seville, Spain; jaimesolis@us.es (J.S.-G.); madelyn@us.es (M.M.)

[2] Institute of Renewable Energy, EURAC Research, 39100 Bolzano, Italy; federico.garzia@eurac.edu (F.G.); roberta.pernetti@eurac.edu (R.P.)

* Correspondence: evazquez6@us.es; Tel.: +34-686-253-770

Received: 7 July 2020; Accepted: 31 July 2020; Published: 3 August 2020

Abstract: Innovative designs, such as those taking place in nearly zero-energy buildings, need to tackle Life Cycle Cost, because reducing the impact of use can carry other collateral and unexpected costs. For example, it is interesting to include the evaluation of end-of-life costs by introducing future activities of selective dismantling and waste management, to also improve the environmental performance of the demotion project. For this purpose, it is necessary to develop methods that relate the process of selective demolition to the waste quantification and the costs derived from its management. In addition, a sensitivity analysis of end-of-life parameters allows different construction types, waste treatment options, and waste management costs to be compared. The assessment of end-of-life costs in the present work is developed by a case-based reasoning. Cost data are obtained from three actual studies which are part of the H2020 CRAVEzero project (Cost Reduction and Market Acceleration for Viable Nearly Zero-Energy Buildings). Results show that end-of-life costs are similar to traditional building typologies. The most influential materials are part of the substructure and structure of the building, such as concrete and steel products.

Keywords: selective demolition; waste quantification; nearly zero-energy building; End-of-Life Cost

---

1. Introduction

The study of the life cycle of buildings has generally focused first on the use phase. Studies about hotels' performance consider that this phase can represent between 70%–80% of the total life cycle energy consumption [1] and 80–90% of CO_2 emissions [2]; in the case of residential typology, building operation phase consume between 81–94% of total energy of life cycle and 75–90% of CO_2 emissions [3]. Once the use phase analysis is delimited, it is then focused on the construction phase. In this stage, studies have been centered on the manufacture of construction products and the building construction itself, which last a relatively short period of time (1–2 years) but cause a very high environmental impact. This is largely due to the intensive use of concrete and steel for the construction of building structures, which represent a very high percentage of the CO_2 emissions generated [4,5]. For instance, Du et al. computed embodied energy to be equal 4.9% of the total life cycle energy of building. This value was due to covering only the initial embodied energy of the building [6].

However, it is also important to evaluate the whole life cycle of buildings, including end-of-life, mainly related to demolition and waste management. Rosselló-Batle et al. [1] calculated that around 40–50% of the whole energy of this stage is used to demolish the structure of the building, and to transport construction and demolition waste (CDW), which is between 16–32%. The authors also asserted that this result is highly dependent on the distance travelled to the disposal or recycling plant. In the European Union, Michailidou et al. [7] highlighted the data unavailability to perform end-of-life studies. Marrero et al. [8] analyzed urbanization and demolition phases and concluded that both generate 90% of CDW of whole life cycle; the former is due to earthworks and the latter is due to the elimination of all buildings materials.

A building's end-of-life takes place when it is considered out of service [9], implies a total demolition, selective or not, and includes all its processes and waste management [10,11]. Most building materials are considered waste in this phase [12]. As defined in the EN-16627 standard, End-of-Life Cost (EOLC) includes deconstruction/demolition, transport, waste processing for reuse or recycle, and waste disposal [11].

Several studies related to EOLC can be found in the literature. Islam et al. [13] did a complete review of Life Cycle Assessment (LCA) and Life Cycle Cost (LCC) on residential buildings. Practical models and systems of LCA and LCC also were analyzed. The main limitation of the case studies was that EOLC was not studied in detail. Dwaikat and Ali [14] carried out LCC analysis for a sustainable building and how to identify and use LCC variables to perform the assessment, considering a useful life of 60 years. In that case, a deeper study about EOLC was developed. Pelzeter [15] tried to demonstrate that the optimization of the results obtained with the LCC analysis largely depended on the assumptions that were made during the calculation process of LCC, but in this study, EOLC were not determined.

EOLC are important for the construction sector because they can incorporate the concept of life cycle from the design, in order to promote good practices such as reuse and recycling in selective demolition works or on-site separation processes for the proper treatment of CDW. Recent European regulations related to the management of CDW include the concept of circular economy, which focused on the waste hierarchy: Reduce, reuse, and recycle [16]. Dismantling or selective demolition are therefore promoted [17], for example, through the Green Public Procurement Program developed by the European Union.

Furthermore, the European legislation about public procurement has considered since 2014 that the LCC evaluation is necessary to find the most advantageous economical tender [18]. In the most important standards, LCC is normally divided into four stages, that is, construction, operation, maintenance, and end-of-life [10], thus allowing, in Net Present Value (NPV) terms, different options of building design to be compared.

To determine the EOLC of new building designs, the definition of a quantification method of the materials making up the building is required. Furthermore, its corresponding costs should be known. Other important aspects to incorporate in the assessment of end-of-life cost are the identification of technological and administrative barriers to adequately manage CDW, which is widely studied by the environmental community [19–21]. To analyze the waste generation processes in construction, quantification procedures should be defined. A recent piece of research divides these procedures into six categories, namely site visit, generation rate calculation, lifetime analysis, classification system accumulation (CSA), variables modelling, and others particular methods [22].

The cumulative systems are the most used in the literature [23–25]. In these methods, a construction works classification system is essential; this system is the structure on which the calculations are developed to estimate the costs and quantities of each construction material. These systems offer a rigorous way to quantify waste streams, allowing decision-making to define the most appropriate treatment strategies for each waste stream. They generally combine computer tools with calculation tables. Akinade et al. [26] studied the environmental evaluation of building design, and described four quantification tools, which include the demolition stage in the analysis, as well as its corresponding cost estimation and material quantification. The first tool was Building for Environmental and Economic Sustainability (BEES) [27], which allows building products to be individually evaluated. The second tool was One-click LCA, a software that assesses capital costs, LCC, and environmental impacts of 3D models of construction materials and systems [28]. The third tool was Demolition and Renovation Waste Estimation (DRWE), which is focused on the evaluation of material quantification and waste management options such as reuse, recycle, and disposal [29]. The fourth one was Integrate Material Profile and Costing Tool (IMPACT), which measures the environmental impact of construction phase and the LCC analysis of buildings [30].

In this paper, EOLC will be calculated using the ISO 15686-5:2008 standard [10], which is the main source for the building LCC calculation. According to this standard, the LCC of a building is the NPV, which is the sum of the discounted costs and revenue streams during the phases of the selected period of the life cycle. The phases of the life cycle that are included in the calculation are the initial investment cost (design and construction), the cost for operation and maintenance, and the EOLC [31].

An assessment model is proposed to study the end-of-life phase, including quantification and management of demolition waste. These assessments are needed in economic and environmental analyses. This study will be applied to Nearly Zero-Energy Buildings (NZEB), as part of the H2020 CRAVEzero research project, of which the objective is to identify the additional economic costs of NZEB from the point of view of the complete life cycle, in order to be able to make proposals that minimize these costs. As part of the project, Pernetti et al. have developed a tool that implements a global and structured methodology [32]. This tool analyzes a set of NZEB case studies, representing the most technologically advanced buildings across Europe. The projects data have been provided by the companies Bouygues, Skanska, Köhler & Meinzer, ATP sustain, and Moretti, which participated as designers, general contractors, or technology providers in the construction processes of buildings. NZEB have the challenge of reducing energy consumption in the stage of use of the life cycle of buildings, the longest and most impacting phase [33,34].

European NZEB are defined as "Very high energy performant buildings with a very low amount of energy required covered to a very significant extent by energy from on-site or nearby renewable sources" [35]. In the European Union, each country can established his own energy consumption and emissions requirement, e.g., new single-family house in Madrid (Spain) should not surpass 44.6 kWh/m^2 of primary energy consumption or 10.1 kgCO$_2$/m^2 of emissions per year [36]. Furthermore, each level of energy performance will be joined to cost-optimality solutions to evaluate the building energy performance [33,37]. A NZEB typology will be mandatory in new buildings in the European Union States (EU) by 2021 [33,38]. These improvements achieve an energy demand reduction from non-renewable energy sources [34,38], e.g., in Spain, a NZEB could reduce approximately 75% of energy consumption and emissions [36].

In the early stages of the design, the analysis of the EOLC should allow all costs to be estimated. For this purpose, construction materials, dismantling processes, and waste management should be quantified. In this work, an EOLC assessment is proposed for the CRAVEzero project, which is based on the standard classification system of construction work of the Andalusian Construction Cost Database (ACCD) [39]. The methodology developed by the ARDITEC research group [8,12,25] for the quantification and environmental assessment of CDW is also implemented.

However, we can say that there are insufficient studies or information regarding the EOLC. Therefore, one of the objectives of this article is to propose a methodology to estimate the costs and the quantification of waste generated by this phase because the impact from the environmental point of view is crucial; consequently, its analysis is pertinent. In addition, a sensitivity analysis of end-of-life parameters allows for the comparison of different construction types, waste treatment options, and waste management costs. The estimation model is developed by a case-based reasoning and cost data are obtained from three actual studies that are part of the CRAVEzero project.

2. Methodology

The proposed model calculates the EOLC through the quantification of demolished materials by using an indirect estimation method per Kim classification [40]. This method predicts the waste generated by construction work units and accumulation units. The quantification during the project design stage is based on a forecast of the CDW production and is helpful for defining the potential recycling benefit and disposal costs.

In Spain, a waste quantification method, included in the category of CSA methods, allows the volumes of waste generated by a building demolition to be known [25,41]. In addition, if this method is combined with a cost classification method of work units, a waste cost estimation is likely to be obtained [42]. The ACCD is employed in this work for the work unit definition [39] as it has been successfully used in previous works on waste quantification [43]. EOLC includes costs of demolition (TDC), load (TLC), transport (TTC), and waste (TWC) [9]

$$EOLC = TDC + TLC + TTC + TWC \tag{1}$$

The structure of the EOLC calculation is divided into three groups (Figure 1). The first group is demolition costs (DC), the second group is made up of load (LC), transport (TC), and waste management costs (WC), and the third one is EOLC in terms of NPV. NPV represents the actual value of a future cost, so the project time analysis and economic factors, such as discount rate and inflation, should be considered [44]. These costs categories are compatible with the indications of the ISO 15686-5 and EN 15643-4 standards [10,45].

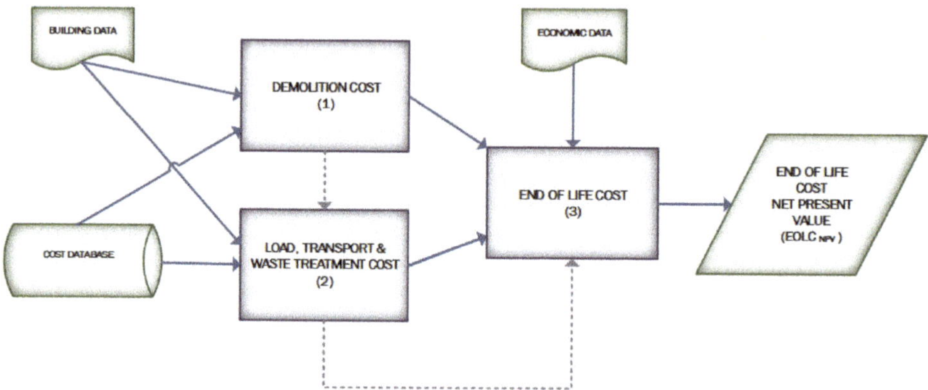

Figure 1. Groups and data flow for the end-of-life cost calculation.

2.1. Demolition Costs

The demolition activity is mainly conducted in two ways, massive demolition or dismantling and selective demolition. The former is not considered in this work because it does not meet European policies [35,46] about CDW management due to the mixed waste that it produces. In economic terms, there is no consensus about which demolition (massive or selective) is the best economic option; on the one hand, Dantata et al. indicate that selective demolition can be 17–25% greater than massive demolition costs [47], and on the other hand, some studies consider selective demolition more profitable than massive demolition [12,48,49]. When the demolition method is massive, costs can be calculated by means of the volumetric or area data of the building. However, if a selective demolition takes place, detailed information of the building design should be available.

ACCD [39] is employed in this paper for the work unit definition. Its most widespread use is for estimating costs in the construction sector and is mandatory in public works in Andalusia (Spain) [50]. It uses a hierarchical organization for work units, where the highest level is the construction site, followed by categories called chapters, each representing a construction process (for example, demolitions, foundations, installations, waste management, etc.), which then in turn are divided into subchapters. The base of this structure is formed by the Basic Costs, which correspond to elementary resources (materials, machinery, and labor), which are added to form Auxiliary Costs, generally mixtures of materials such as cement mortar. The aggregation of several basic and auxiliary costs give rise to Simple Costs, which represent the various construction activities or work units. Putting together Simple Costs generates Complex Costs, which represent more complicated work units.

In this research, demolition cost is calculated (Figure 2) by assigning Demolition Complex Costs to each building element (BE). The last are parts in which the building's project is divided, and they were used in CRAVEzero project to calculate LCC. Some examples are foundation, facilities, and roofs. Each of them is expressed in a certain unit of measurement: m^2, kg, m, etc. Thirty-nine new Demolition Complex Costs and thirty-one Simple Costs were defined using ACCD. These costs have been defined to adapt the demolition process to the work units in the case studies, such as wooden structures and renewable energy systems.

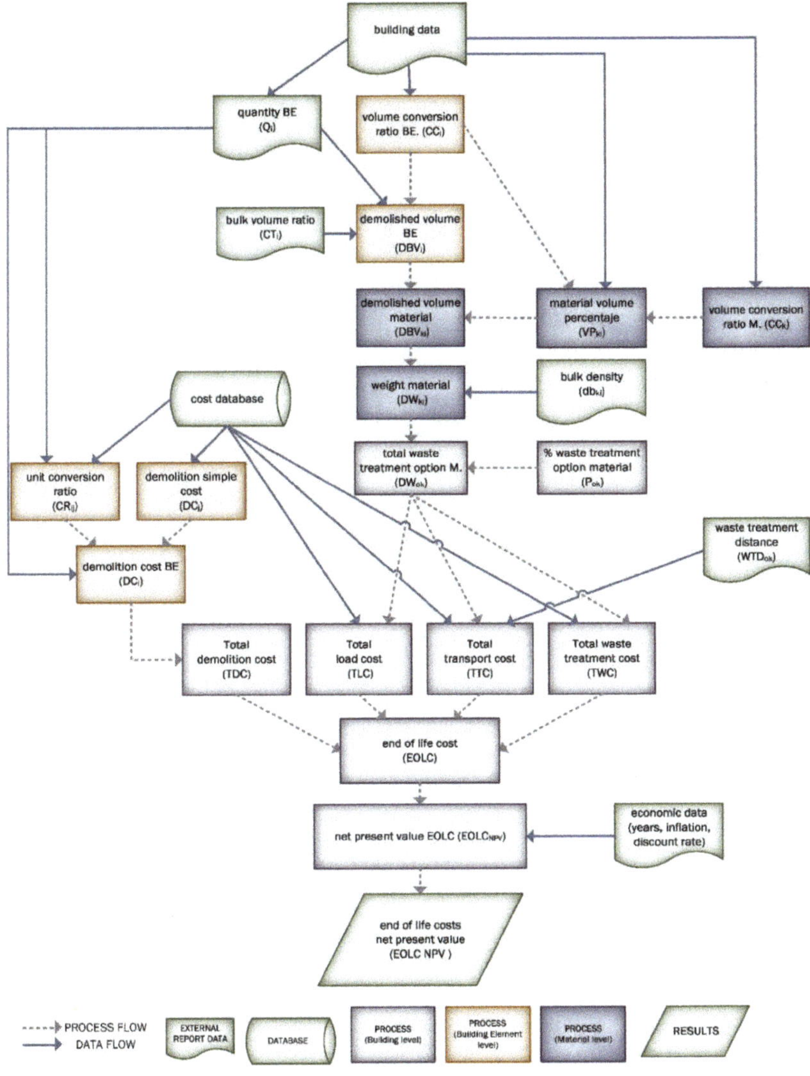

Figure 2. Groups and data flow for the end-of-life cost calculation.

In many cases, the unit of measurement of BE could be different to the one of the corresponding demolition work unit, thereby a unit conversion ratio (CR) is required. The demolition cost of each BE (DC_i) is calculated according Equation (2),

$$DC_i = Q_i \times \sum (CR_{i,j} \times DC_j) \tag{2}$$

where DC_i = Demolition Cost of BE_i, and Q_i = Quantity of BEi, expressed in its unit of measurement. CR_{ij} = Unit Conversion Ratio. DC_j = Demolition Simple Cost j, expressed in €/unit of measurement. The sum of the DC_i that are part of the BE, leads to the total demolition costs (TDC) of the building according to Equation (3),

$$TDC = \sum DC_i \tag{3}$$

Table 1 shows the BE, roof, and, the Demolition Complex Cost generated from the Demolition Simple Costs, and how the use of CR is necessary to be able to relate the units of both elements (BE and Demolition Costs). Finally, the cost of the roof with cold attic (DC) is 33.67 €/m2.

Table 1. Example of Demolition Complex Cost of building elements (BEs).

Description		Q_i	umBE
Roof with cold attic		1541	m^2
Components		Height (cm)	
Gypsum board		1.30	
Wood/air		2.80	
Plastic foil		0.02	
Isolation + Wood frames		60.00	
Wood		2.20	
Bituminous sheet		3.25	
TOTAL height		69.57	

Breakdown cost calculation					
$umDC_j$	Concept	CR_{ij}	€/$umDC_j$	€/umBE
m^2	Selective demolition wood roof	1.00 m^2/m^2	7.46 €/m^2	7.46 €/m^2
m^2	Selective demolition wood frames	1.00 m^2/m^2	14.72 €/m^2	14.72 €/m^2
m^2	Selective demolition suspended ceiling	1.00 m^2/m^2	2.24 €/m^2	2.24 €/m^2
m^3	Selective demolition isolation panels	0.60 m^3/m^2	15.42 €/m^2	9.20 €/m^2
			TOTAL	33.67 €/m^2

2.2. Load, Transport, and Waste Management Costs

Load, transport, and management activities represent the conversion of demolished BEs into waste and their treatment cost. Previously to cost calculation, waste should be quantified, and waste management options should be defined. The first part of the calculation process (waste quantification) is performed with the support of the ARDITEC waste volume calculation [25].

For each BE, a volume conversion coefficient (CC_i) is applied to change the unit of measurement of the BE to cubic meter. This ratio indicates how much apparent volume (m^3_a) is contained in a unit of the BE before being demolished. In example (Table 1), a wooden roof has 0.6957 m^3_a/m^2. However, gaps are discounted by measuring the project to obtain a value of $0.6957 - 0.021 = 0.6747$ m^3_a/m^2. This last figure (0.6747) is what it is defined as CC_i.

After BE is demolished, its volume is changed and is transported as bulk volume. This conversion is made by a bulk volume ratio (CT_i). Representative CT_i values are in parenthesis, for reinforced concrete (1.30), walls-partitions (1.30), tiling (1.20), ceiling (1.35), wood doors (1.15), heaters (1.00), and glass (1.10) [8,25]. The demolished bulk volume (DBV_i) of the BE is defined as the demolished volume generated by each BE of the project according to Equation (4).

$$DBV_i = Q_i \times CC_i \times CT_i \qquad (4)$$

In the example of Table 1, $DBV_{\text{wooden roof}} = 1541 \times 0.6747 \times 1.30 = 1351.98$ m$^3{}_b$; CT_i in this case is 1.30. An average value of CT_i is stablished for each BE.

It is necessary to calculate the total waste of each material that can be generated by each Demolition Complex Cost. Therefore, the next step is to determine the weight of different material components by nature in the BE. The waste nature is defined by using the classification from the European Waste List [51], included in Chapter 17 Construction and demolition wastes. The codes are 17 01 01 concrete, 17 01 07 ceramic, 17 02 01 wood, 17 02 02 glass, 17 02 03 plastic, 17 03 02 bituminous, 17 04 07 metal, 17 04 11 cable electric, 17 05 04 soil, 17 06 04 isolation, 17 08 02 gypsum-based, 17 09 04 mixed inert, 17 09 04 mixed non-inert, and hazardous materials.

Firstly, the percentage of each material in the BE, the material volume percentage (VP_{ki}), should be determined according to Equation (5),

$$VP_{ki} = q_{ki} \times (CC_k/CC_i) \qquad (5)$$

where q_{ki} is the quantity of the material included in the BE per unit of BE. It is obtained from the bill of quantities of the project. CC_k is a conversion ratio that relates the unit of measurement of the material included in the BE to volume unit (m^3). CC_i comes from Equation (4).

For example (Table 1), the percentage of wood in BE (wooden roof) will be calculated. CC_i is calculated before (0.6747 m$^3{}_a$/m^2); $CC_k = 1$, since wood is measured in m^3; and q_{ki} is obtained from the bill of quantities of wooden roof, in this case 0.039 m^3/m^2, $VP_{\text{wood, wooden roof}} = 0.039 \times 1/0.6747 = 5.78\%$. A similar analysis is done with the other materials that are part of the BE. Thus, it is possible to determine the quantity of demolished bulk volume for each material (DBV_{ki}), according to Equation (6),

$$DBV_{ki} = DBV_i \times VP_{ki} \qquad (6)$$

To determine the demolished weight material (DW_{ki}), the bulk density (db_{ki}) is employed according to Equation (7),

$$db_{ki} = da_k / CT_i \qquad (7)$$

where da_k is the apparent density. Finally, DW_{ki} was calculated according to Equation (8),

$$DW_{ki} = DBV_{ki} \times db_{ki} \qquad (8)$$

In the example (Table 1), the bulk density of wood = 0.5 t/m^3 and $DW_{\text{wood, wooden roof}} = 0.0578 \times 1351.98 \times 0.5 = 39.07$ t. This process is done with all the materials that make up the BE. In this example, the materials are wood, bituminous, plastic, insulation, and gypsum. This process would be replicated for all BEs in the project.

After determining all weights, the treatment option percentage can be indicated for each material. The waste treatment options are chosen by considering the waste treatment hierarchy, reuse (PU_k), recycle (PY_k), and disposal (PD_k) [16], thus allowing different percentage combinations to be compared. Quantities are obtained (in weight) of reused (DWU_k), recycled (DWY_k), and disposed (DWD_k) waste with the Equations (9), (10), and (11) respectively,

$$DWU_k = PU_k \times \sum DW_{ki} \qquad (9)$$

$$DWY_k = PY_k \times \sum DW_{ki} \qquad (10)$$

$$DWD_k = PD_k \times \sum DW_{ki} \tag{11}$$

After determining all material weights, the costs of load, transport, and waste treatment can be established. The total load cost (TLC) includes in situ transport of waste and being loaded on a truck, according to Equation (12),

$$TLC = LC \times \sum (DWU_k + DWY_k + DWD_k) \tag{12}$$

The calculation is made by multiplying the unitary load cost (LC) per tonne to the waste weight, after adding the weight of different waste treatment options. The total transport cost (TTC) includes all the costs incurred from the construction to the treatment site. The calculation is made with the Equation (13),

$$TTC = \sum (TC \times WTDok \times DWok) \tag{13}$$

where TC is the unitary transport cost (TC), WTD_{ok} is the distance between the worksite and the waste treatment site, and DW_{ok} is the total amount of waste and waste treatment option (the options are recycling or disposal). The total waste treatment cost (TWC) includes all the cost related to the waste treatment site, fees, and special taxes included according to Equation (14),

$$TWC = \sum (WC_{ok} \times DW_{ok}) \tag{14}$$

which is the summation of each unitary waste treatment cost (WC_{ok}) per tonne times the total amount of wasted materials for each treatment option DW_{ok} (recycling or disposal).

2.3. EOLC

The EOLC is calculated through the aggregation of demolition, load, transport, and waste treatment costs (Equation (1)). This cost will be therefore affected by economic factors over time and indicates the investor's time value of money [44]. The NPV of EOLC is obtained from Equation (15),

$$EOLC_{NPV} = EOLC \times ((1 + i)^y/(1 + dr)^y) \tag{15}$$

where $EOLC_{NPV}$ represents the present worth of an investment, which will take place in the future, so the period of time (y), the value of discount rate (dr), and the inflation rate (i) that the investor considered in the LCC assessment should be known. This equation is used for the LCC assessment in ASTM 917, EN-16627, and ISO 15686-5 [10,11,44]. Additionally, if the data cost is not recent, an updated cost factor is also required [10].

2.4. Cost Normalization

To compare the results of EOLC of the case studies and to add into previous costs calculated of NZEBs, a cost normalization should be performed [32], because these case studies come from different countries in Europe. For this purpose, the European construction cost index is employed [52] to establish an equivalence value of the EOLC obtained from the ACCD and the CRAVEzero project. It is defined as $EOLC_{NPVn}$ and obtained in Equation (16),

$$(EOLC_{NPV})_n = EOLC_{NPV}/icd \tag{16}$$

$EOLC_{NPV}$ represents the end-of-life cost calculated with ACCD in this research, and icd is the cost index of Spain where the ACCD is defined.

2.5. Sensitivity Analysis

The main weak point of an LCC analysis are the assumptions that should be made for the input parameters. The difficult access to data, especially in the case of economic boundary conditions, leads to uncertainty in LCC calculation, thus limiting its application. To determine the most relevant input parameters and to tackle in this way the uncertainty issue, a sensitivity analysis is performed, so decision makers could concentrate on the analysis of the most critical parameters [53].

First, the set of input parameters and their variation range must be selected. On the one hand, a fixed range can be adopted from the technical literature, norms, and the data collection of case studies, i.e., interest rates. On the other hand, if a fixed range is not available, e.g., the price of building features, the baseline value is arbitrary varied, ±10%.

One of the simplest screening techniques, the differential sensitivity analysis, is adopted. This method belongs to the class of the one factor at a time methods [54–56]. The impact on the LCC of one parameter at a time is studied, keeping the other parameters set equal to their baseline value. In spite of the influence of inflation rate and discount rate, it is very intense in terms of present worth. This affirmation is widely supported in existing literature [13,44,56,57]. Thus, we focus the sensitivity analysis on evaluation of parameters related directly with EOL: Recycle percentage (PY_k), Recycle distance (DY_k), Disposal distance (DD_k), Unitary recycle cost (CY_k), and Unitary disposal cost (CD_k).

2.6. LCC CRAVEzero Calculation

For the LCC calculation of the other life cycle phases (design, construction, and operation) CRAVEzero methodology [32] was adopted, which is resumed in this section. The first step for the calculation of the LCC is establishing the time period. Following the framework of the ISO 15686-5:2008 standard, the largest period possible is 100 years [10] but shorter periods lead to more reliable assessments as time-uncertainties have a smaller impact. For the phases up to the operation phase, a period of 40 years was selected. A common discount rate value for all case studies was adopted to updating future costs over the 40-year life. The selected value was taken from the FRED Economic Database [58], which provided an average discount rate of 1.51% for the time period going from 2009 (the year of construction of the oldest case study) to 2017 (time period during case studies were built). Average values from 2009–2016 (year of construction of the case studies) of interest rate were used. Interest rate is taken from CRAVEzero project, which is necessary to maintain for further comparison and aggregation. Inflation rate was not considered in this calculation since it affects all results in the same way.

In a second step, energy costs related to building operation were obtained from CRAVEzero project. As the official energy bills were not available in most cases, the evaluation was based on the energy demand calculated. In particular, the energy consumption and production was analyzed using the PHPP evaluation tool [59]. Energy prices derived from Eurostat [60], considering the average values from 2010 to 2017.

The analyzed buildings were built between 2009 and 2016, and there are not enough data for maintenance cost yet. Therefore, the analysis within CRAVEzero of maintenance and use costs was based on the standard values from the literature. In particular, the EN 15459:2017 standard [32,61] provided annual maintenance costs for each item, including operation, repair, and service, as a percentage of the initial construction cost. For passive construction elements, an annual average value, representing 1.5% of the construction cost, was taken for the evaluation, and it was verified with average values derived from the experience of industry partners in the CRAVEzero project.

3. Case Studies Description, Scenario

3.1. Case Studies Description

Three buildings with different construction characteristics were chosen for this study (Table 2). Their main difference was their structural material, ranging from concrete, wood, and mixed. Also, they have a high degree of thermal insulation in their envelope, use low-consumption installations, and have their own system of photovoltaic power production.

Table 2. Nearly Zero-Energy Buildings (NZEB) case studies description and equipment. Adapted from Pernetti et al. [32].

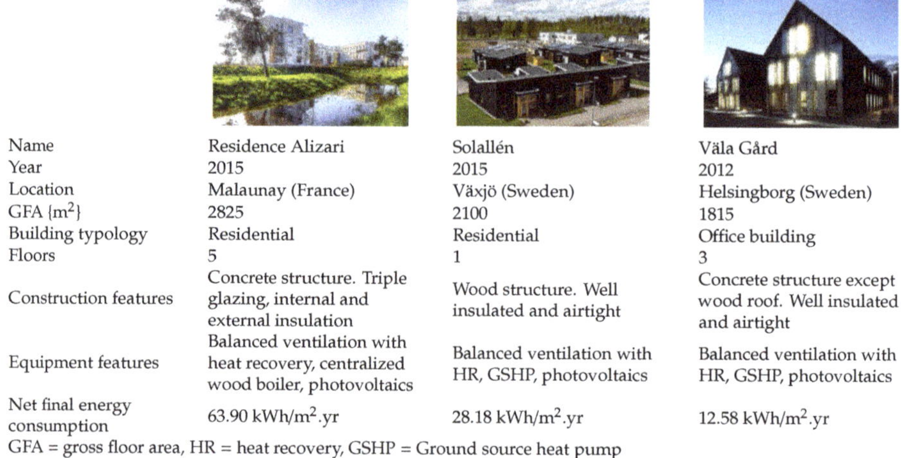

Name	Residence Alizari	Solallén	Väla Gård
Year	2015	2015	2012
Location	Malaunay (France)	Växjö (Sweden)	Helsingborg (Sweden)
GFA {m^2}	2825	2100	1815
Building typology	Residential	Residential	Office building
Floors	5	1	3
Construction features	Concrete structure. Triple glazing, internal and external insulation	Wood structure. Well insulated and airtight	Concrete structure except wood roof. Well insulated and airtight
Equipment features	Balanced ventilation with heat recovery, centralized wood boiler, photovoltaics	Balanced ventilation with HR, GSHP, photovoltaics	Balanced ventilation with HR, GSHP, photovoltaics
Net final energy consumption	63.90 kWh/m^2.yr	28.18 kWh/m^2.yr	12.58 kWh/m^2.yr

GFA = gross floor area, HR = heat recovery, GSHP = Ground source heat pump

3.2. Scenario

To compare different buildings correctly, economic and geographical estimates were made, including the normalization of the economic values of the LCC (Section 2). Furthermore, the idealization of the location of waste treatment plants included the unification of the distances between the construction site and the plant. In addition, the costs of each type of waste management, such as reuse, recycle, and/or disposal, for each waste stream were defined by employing current fees in Spanish municipalities and ACCD [39,62,63]. Reuse did not take place due to the complex factors making it difficult to define its corresponding costs because of both the opacity in the market of construction companies and the lack of public data. Recycling and disposal rates at European level were obtained from the project called Resource Efficient Use of Mixed Wastes Improving Management of Construction and Demolition Waste [64] as well as from the Institute of Civil Engineers [65]. Main data and its sources are summarized in Table 3; no estimation has been made in the present work besides the cost normalization as explained in Section 2.4.

Table 3. Data scenario and assumptions.

Parameter	Value		Source
Life cycle cost building	Various		[32]
Time period	40 years		[32]
Inflation rate	0		-
Discount rate	1.51%		[58]
Cost normalization index Spain	70.52%		[52]
Load Cost (LC)	0.65 €/t		[39]
Transport Cost (TC)	0.69 €/t km		[39]
Demolition Cost (DC)	Various		[39]
Waste Treatment Cost (WC) Material(k)	Recycle (WCY$_k$)	Disposal (WCD$_k$)	[39,62,63]
Mix non-inert	-	70.00 €/t	
Mix inert	9.50 €/t	30.00 €/t	
Concrete	4.00 €/t	30.00 €/t	
Ceramics	6.00 €/t	30.00 €/t	
Wood	25.00 €/t	70.00 €/t	
Glass	30.00 €/t	70.00 €/t	
Bituminous	3.50 €/t	70.00 €/t	
Metal	−80.00 €/t	30.00 €/t	
Cable	−900.00 €/t	70.00 €/t	

Table 3. *Cont.*

Parameter	Value		Source
Soil	3.00 €/t	30.00 €/t	
Isolation	60.00 €/t	80.00 €/t	
Gypsum based	60.00 €/t	80.00 €/t	
Paper	3.50 €/t	70,00 €/t	
Hazardous	-	80,00 €/t	
Waste Treatment Distance (WTD)	15 km		-
Recycle waste percentage			
Mixed inert, concrete, ceramic	75%		[65]
Gypsum-based	10%		[64]
Wood	57%		[65]
Metal	80%		[64]
Soil, glass, paper	50%		[64]
Plastic, bituminous, isolation	25%		[64]

4. Results and Discussions

4.1. EOLC Results

The importance of the EOLC is established by comparing it with the rest of the LCC previously calculated in the CRAVEzero project [32]. The indicator chosen for comparison was the cost per Gross Floor Area (GFA).

EOLC is divided into demolition, load, transport, and waste management costs. In Figure 3, EOLC in the three cases is approximately 100 €/m2 and 5% of the LCC.

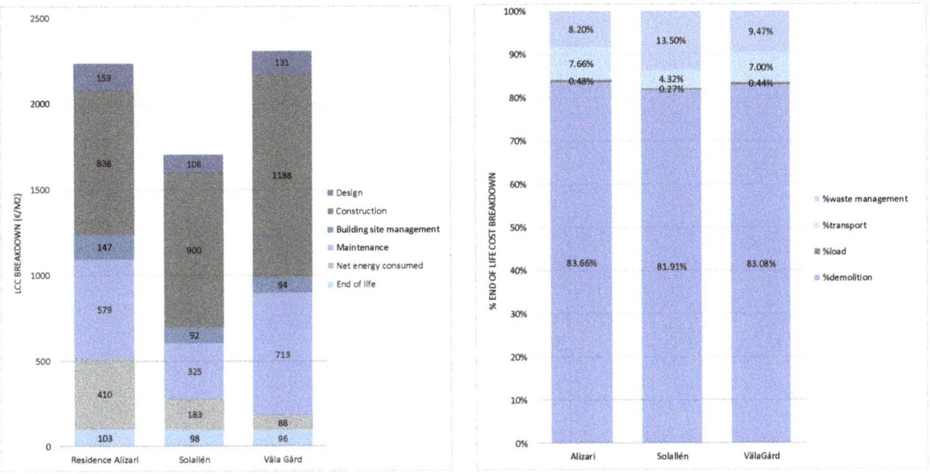

Figure 3. Life Cycle Cost (LCC) breakdown and End-of-Life Cost (EOLC) breakdown percentage.

Comparing these results with other research, Banias et al. determined the transport and management cost is in the range between 31–42 €/m^2 [66], but cost normalization or NPV were not applied. In terms of the importance of the EOLC with respect to the LCC, Islam et al. established those in 2.69% (a discount rate of 6% and a study period of 50 years) [13], Pelzeter in 1% (a discount rate of 5% and a study period of 90 years) [15], and Dwaikat in 1% (a discount rate of 0% and a study period of 60 years) [14]. The difference was mainly due to different discount rates and the number of years employed.

The cost distribution (Figure 3) of EOLC was as follows: Demolition cost from 82 to 84%, waste management from 8 to 14%; transport cost from 4 to 7%; and load cost, 1%.

4.2. Waste Streams and EOLC

Table 4 shows the total amount of waste streams obtained in this work and in others [23]. The insulation material was that showing significant differences, as expected in NZEB.

Table 4. Weight of waste/Gross Floor Area (GFA) (kg/m^2).

Waste	Alizari	Residential. Reinforced. Concrete *	Solallén	Residential. Wood *	Väla Gård	Non-Residential. Reinforced Concrete *
Concrete	810.76	492–840	278.95	137–300	683.50	401–768
Wood	3.08	12–58	92.81	70–275	61.93	20–159
Metal	45.49	9.8–28.4	9.84	4.8–22.5	34.69	28.4–53
Isolation	9.69	0.1–2.2	47.32	0.1–2.2	19.58	0.1–2.2
Gypsum	27.57	10.8–64.3	40.61	10.9–105.4	12.67	10.8–75.7

* [23].

Another work analyzed the construction waste generated during the building life cycle of social housing projects [8] named P1, P2, and P3; the main streams of P2 and P3 (concrete, plastic, and metallic waste) were similar to those included in this work, the Residence Alizari project, but P1 was significantly different due to its urbanization characteristics.

In the three case studies, a detailed analysis of the influence of the materials within the EOLC, in terms of the percentage per weight and costs of transport and management, is shown in Figure 4. The main materials were concrete, metallic, insulation, wooden, and gypsum-based materials, representing between 88 and 95% per weight. Moreover, NZEB buildings increased the amounts of insulation waste, thus making the recycling of these materials important.

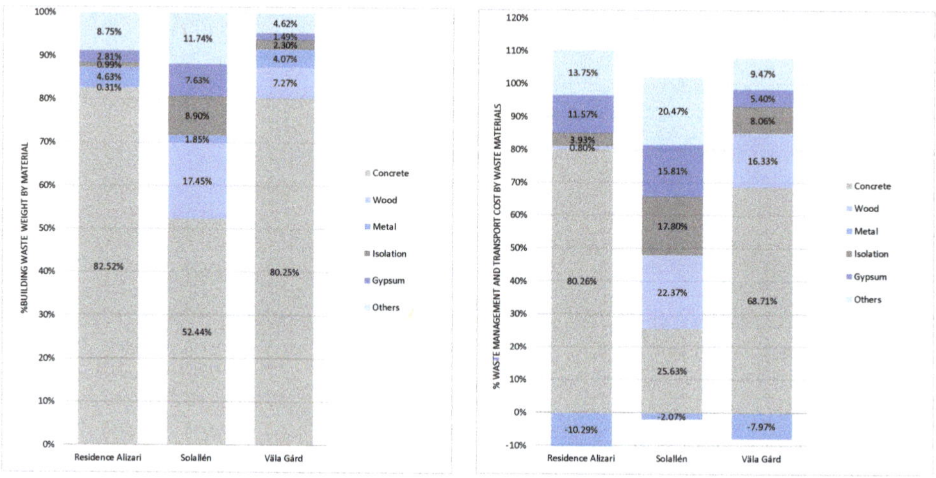

Figure 4. Waste fraction per weight and its influence on transport and waste management cost.

In the percentage comparison of the material weight, the importance of concrete was stressed as it represented between 80 and 82% in buildings with a reinforced concrete structure and façade, and 52% when it was employed only in the foundation.

The cost of transport and waste management represented between 79 and 90% of the EOLC. In turn, the metallic materials which were sold were between −3 and −10%.

4.3. Sensitivity Analysis

The sensitivity analysis compared the percentage of variation of the EOLC with respect to waste parameters, when the parameter studied changes 1%. In terms of the EOLC and the building weight, the five materials most influencing were as follows (see Figure 5): Concrete, metallic, insulation, wooden, and gypsum-based materials. Moreover, the three case studies were compared.

The first parameter evaluated was the waste recycling percentage. The results showed that concrete was the most influential, which was between 4 and 12%, followed by metallic materials which were between 1 and 3%, and wooden materials which were between 0 and 2%. The second parameter assessed was the distance to the recycling plant. The waste most influencing the EOLC was also concrete, which was between 2 and 5%; the remaining waste did not significantly affect the EOLC.

The third parameter analyzed was the distance to the disposal plant. The residue most affecting the EOLC was also concrete, which was between 1 and 2%; the remaining waste did not significantly affect the EOLC. The fourth parameter evaluated was the unit cost of waste recycling. The results showed that the cost of recycling metallic waste varied up to 2%. Then, concrete influenced between 1 and 2%, and finally, the wooden waste reached up to 1%.

The fifth parameter was the unit cost of disposal. Concrete waste was again the most influential, between 2 and 5%. Wooden, insulation, and gypsum-based materials were up to 2% of EOLC. The results showed, in general, that the EOLC was mainly influenced by the materials most presented in the building, and in particular those which were part of the structure and substructure.

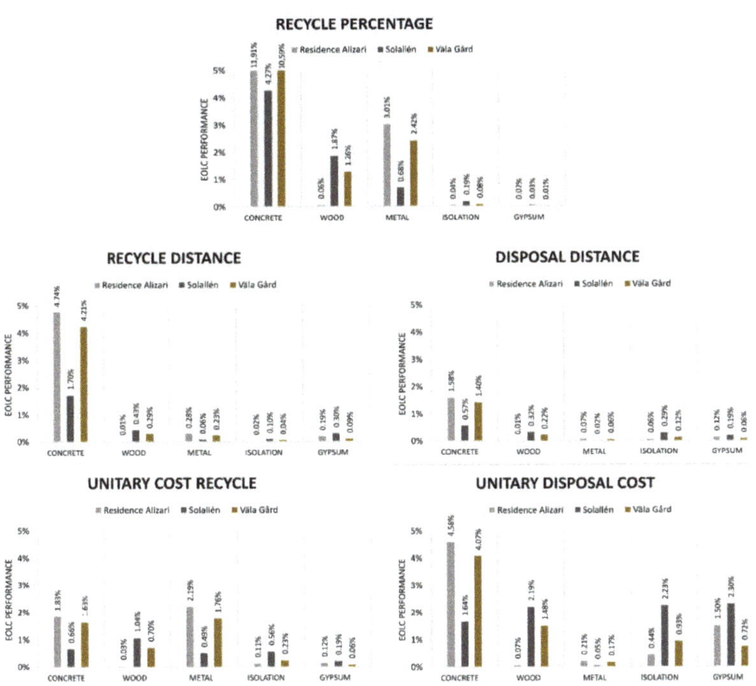

Figure 5. Sensitivity analysis. EOLC performance.

Also, it is possible to analyze the differences between concrete and wooden buildings, verifying that in the latter (Solallén), the influence of wood on the sensitivity analysis is much greater than in the other two buildings (Alizari and Väla Gård). The uncertainties in the main material volume calculations (i.e., concrete) were low because this material was an essential part of the building structure and was always precisely defined in construction projects due to safety issues.

5. Conclusions

This work defines an assessment model for the evaluation of the EOLC of buildings, based on a cost model that implements selective demolition. It is aimed at designing stage decision making, by knowing in advance what type of waste is to be generated at the end-of-life and its economic impact. These aspects could be considered during the materials selection in the building design stage. In addition, the influence of the distance to the recycling plants is also assessed in economic terms, as well as the influence of waste management fees.

Firstly, the results showed that the EOLC represented approximately 5% of the total of the LCC. That percentage significantly changed with modifications in time span and discount rate, thus making the comparison to other results difficult. Furthermore, among the elements describing the EOLC (i.e., demolition, load, transport, and waste management costs), demolition costs were those most influencing the EOLC as they constituted between 82 and 84% of the total. In addition, these percentages were not affected by the typology of the structure or by the number of floors.

Secondly, the results showed that the waste generated was mainly from the foundation and structure. In particular, concrete, wood, metals, insulation, and plaster-based materials. Also, the indicator of weight per floor area obtained in the three case studies was in line with those calculated with other methodologies in the scientific literature. Moreover, NZEB buildings increased the amounts of insulation waste, thus making the recycling of these materials more important than in conventional projects. The presence of waste that generates income, such as metallic waste, showed the importance of the recycling market.

Thirdly, the sensitivity analysis indicated that the parameters most influencing the EOLC were the percentages of recycled waste and disposal unit cost. In this regard, it was clearly shown that if the difference between disposal costs with respect to recycling costs increased, then recycling would be an economic advantageous. So, the sensitivity analysis confirmed, as previously mentioned, that those materials constituting the building structure and foundation were the most influencing, and consequently, recycling policies would be focused on these materials.

In future works, the complete LCC of the building will be calculated for a bigger sample of building typologies.

Author Contributions: Conceptualization, E.V.-L., R.P., and J.S.-G.; data curation, E.V.-L.; formal analysis, J.S.-G. and M.M.; investigation, E.V.-L.; methodology, E.V.-L., F.G., and J.S.-G.; project administration, R.P. and M.M.; resources, F.G. and R.P.; software, E.V.-L and F.G.; supervision, R.P., J.S.-G., and M.M.; validation, R.P., J.S.-G., and M.M.; visualization, E.V.-L.; writing—original draft, E.V.-L and F.G.; writing—review and editing, J.S.-G. and M.M. All authors have read and agreed to the published version of the manuscript.

Funding: This paper and the costs for its publication in open access have been funded by the CIRCULAR-BIM project (code 2019-1-ES01-KA203-065962), an ERASMUS+ project co-funded by the European Union and within the framework of an initiative of 2019 (KA2, Strategic partnerships in the field of higher education), with the support of the Servicio Español Para la Internacionalización de la Educación (SEPIE, Spain). This research was funded by the ERASMUS+ Programme and the VI Own Research and Transfer Plan of the University of Seville (VI PPIT-US).

Acknowledgments: The present work has been carried within the project CRAVEzero—Grant Agreement No. 741223 (www.cravezero.eu), financed by the Horizon 2020 Framework Programme of the European Union. The authors would like to acknowledge the support of the following contributors that collaborated in the collection of the data: David Venus, Anna Maria Fulterer, Klara Meier, Jens Gloeggler, Bjorn Berggren; Gerold Koehler; Thomas Stoecker; Christian Denacquard, Marine Thouvenot, Cristina Foletti6, Gianluca Gualco, Mirco Balachia (AEE-INTEC, ATP Sustain, Skanska, Koeler&Meinzer, Bouygues Construction, Moretti 73i engineering.

Abbreviations

Equation Parameter List

CCi	Volume conversion ratio of building element $\{m^3_a/umBEi\}$
CCk	Volume conversion ratio of material $\{m3_a/umMk\}$
CRij	Unit conversion ratio. $\{umDC_j /umBE_i\}$
CTi	Bulk volume ratio $\{m^3_b/ m^3_a\}$
DCi	Demolition cost of building element $\{€\}$
TDC	Total demolition cost $\{€\}$
da_k	Apparent density of material $\{t/m^3_a\}$
db_{ki}	Bulk density of material $\{t/m^3_b\}$
dr	Discount rate $\{dimensionless\}$
DBV_i	Demolished bulk volume of building element $\{m3_b\}$.
DBV_{ki}	Quantity of demolished bulk volume for each material $\{m3_b\}$.
DC_j	unit demolition cost $\{€/umDCj\}$
DW_{ki}	Quantity of demolished weight for each material $\{t\}$
DWU_k	Demolition waste reuse $\{t\}$
DWY_k	Demolition waste recycle $\{t\}$
DWD_k	Demolition waste disposal $\{t\}$
EOLC	End-of-life cost $\{€\}$
$EOLC_{NPV}$	End-of-life cost net present value $\{€\}$
$(EOLC_{NPV})n$	End-of-life costs net present value and normalized. $\{€\}$
i	Inflation rate $\{dimensionless\}$
icd	Cost index $\{dimensionless\}$
P_{ok}	Treatment option percentage $\{dimensionless\}$
PU_k	Reuse percentage $\{dimensionless\}$
PY_k	Recycle percentage $\{dimensionless\}$
PD_k	Disposal percentage $\{dimensionless\}$
Q_i	Quantity of building element $\{umBEi\}$.
q_{ki}	Quantity of the material included in the BE $\{umM_k/umBE_i\}$.
TLC	Total Load costs $\{€\}$
TTC	Total Transport cost $\{€\}$
LC	Unitary load cost $\{€/t\}$
TC	Unitary transport cost $\{€/t·km\}$
TWC	Total Waste treatment cost $\{€\}$
V_{ki}	Volume percentage of material k in building element i. $\{dimensionless\}$
WC_{ok}	Unitary waste treatment cost $\{€/t\}$
WTD_{ok}	Waste treatment distance $\{km\}$
y	Period of time $\{years\}$

Measure Unit List

m^3_a	Apparent volume.
m^3_b	Bulk volume.
$umBE_i$	Unit of measurement of building elements $\{m, m2, m3 \dots \}$
$umDC_j$	Unit of measurement of demolition work unit $\{m, m2, m3 \dots \}$
umM_k	Unit of measurement material $\{m, m2, m3, t \dots \}$
t	tonns

References

1. Rosselló-Batle, B.; Moià, A.; Cladera, A.; Martínez, V. Energy use, CO_2 emissions and waste throughout the life cycle of a sample of hotels in the Balearic Islands. *Energy Build.* **2010**, *42*, 547–558. [CrossRef]
2. Radhi, H.; Sharples, S. Global warming implications of facade parameters: A life cycle assessment of residential buildings in Bahrain. *Environ. Impact Assess. Rev.* **2013**, *38*, 99–108. [CrossRef]

3. Nematchoua, M.K.; Teller, J.; Reiter, S. Statistical life cycle assessment of residential buildings in a temperate climate of northern part of Europe. *J. Clean. Prod.* **2019**, *229*, 621–631. [CrossRef]

4. Asif, M.; Muneer, T.; Kelley, R. Life cycle assessment: A case study of a dwelling home in Scotland. *Build. Environ.* **2007**, *42*, 1391–1394. [CrossRef]

5. Dimoudi, A.; Tompa, C. Energy and environmental indicators related to construction of office buildings. *Resour. Conserv. Recycl.* **2008**, *53*, 86–95. [CrossRef]

6. Du, P.; Wood, A.; Stephens, B.; Song, X. Life-Cycle Energy Implications of Downtown High-Rise vs. Suburban Low-Rise Living: An Overview and Quantitative Case Study for Chicago. *Buildings* **2015**, *5*, 1003–1024. [CrossRef]

7. Michailidou, A.V.; Vlachokostas, C.; Moussiopoulos, N.; Maleka, D. Life Cycle Thinking used for assessing the environmental impacts of tourism activity for a Greek tourism destination. *J. Clean. Prod.* **2016**, *111*, 499–510. [CrossRef]

8. Marrero, M.; Rivero-Camacho, C.; Alba-Rodríguez, M.D. What are we discarding during the life cycle of a building? Case studies of social housing in Andalusia, Spain. *Waste Manag.* **2020**, *102*, 391–403. [CrossRef]

9. EN-15643-1. *Sustainability of Construction Works—Sustainability Assessment of Buildings—Part 1: General Framework*; Comite Europeen de Normalisation: Geneva, Switzerland, 2010.

10. ISO 15686-5. *Buildings and Constructed Assets—Service Life Planning—Part 5: Life-Cycle Costing*; International Organization for Standardization: Geneva, Switzerland, 2017.

11. EN-16627. *Sustainability of Construction Works—Assessment of Economic Performance of Buildings—Calculation Methods*; Comite Europeen de Normalisation: Geneva, Switzerland, 2015.

12. Alba-Rodríguez, M.D.; Martínez-Rocamora, A.; González-Vallejo, P.; Ferreira-Sánchez, A.; Marrero, M. Building rehabilitation versus demolition and new construction: Economic and environmental assessment. *Environ. Impact Assess. Rev.* **2017**, *66*, 115–126. [CrossRef]

13. Islam, H.; Jollands, M.; Setunge, S. Life cycle assessment and life cycle cost implication of residential buildings—A review. *Renew. Sustain. Energy Rev.* **2015**, *42*, 129–140. [CrossRef]

14. Dwaikat, L.N.; Ali, K.N. Green buildings life cycle cost analysis and life cycle budget development: Practical applications. *J. Build. Eng.* **2018**, *18*, 303–311. [CrossRef]

15. Pelzeter, A. Building optimisation with life cycle costs—The influence of calculation methods. *J. Facil. Manag.* **2007**, *5*, 115–128. [CrossRef]

16. European Commission. *Report from the Commission to the European Parliament on the Implementation of the Circular Economy Action Plan COM/2019/190 Final*; European Union: Brussels, Belgium, 2019.

17. European Commission. *EU Construction & Demolition Waste Management Protocol; Dublin, 22 June 2017*; European Union: Brussels, Belgium, 2016; Volume 62, ISBN 1866505041.

18. European Parliament. *Directive 2014/24/EU of The European Parliament and of The Council of 26 February 2014 on Public Procurement and Repealing Directive 2004/18/EC (Text with EEA Relevance)*; European Parliament: Brussels, Belgium, 2014; Volume 94.

19. Kofoworola, O.F.; Gheewala, S.H. Estimation of construction waste generation and management in Thailand. *Waste Manag.* **2009**, *29*, 731–738. [CrossRef] [PubMed]

20. Wang, J.; Yuan, H.; Kang, X.; Lu, W. Critical success factors for on-site sorting of construction waste: A china study. *Resour. Conserv. Recycl.* **2010**, *54*, 931–936. [CrossRef]

21. Yuan, H.; Shen, L. Trend of the research on construction and demolition waste management. *Waste Manag.* **2011**, *31*, 670–679. [CrossRef]

22. Wu, Z.; Yu, A.T.W.; Shen, L.; Liu, G. Quantifying construction and demolition waste: An analytical review. *Waste Manag.* **2014**, *34*, 1683–1692. [CrossRef]

23. Mália, M.; De Brito, J.; Pinheiro, M.D.; Bravo, M. Construction and demolition waste indicators. *Waste Manag. Res.* **2013**, *31*, 241–255. [CrossRef]

24. Coelho, A.; De Brito, J. Generation of construction and demolition waste in Portugal. *Waste Manag. Res.* **2011**, *29*, 739–750. [CrossRef]

25. Solís-Guzmán, J.; Marrero, M.; Montes-Delgado, M.V.; Ramírez-de-Arellano, A. A Spanish model for quantification and management of construction waste. *Waste Manag.* **2009**, *29*, 2542–2548. [CrossRef]

26. Akinade, O.O.; Oyedele, L.O.; Omoteso, K.; Ajayi, S.O.; Bilal, M.; Owolabi, H.A.; Alaka, H.A.; Ayris, L.; Henry Looney, J. BIM-based deconstruction tool: Towards essential functionalities. *Int. J. Sustain. Built Environ.* **2017**, *6*, 260–271. [CrossRef]

27. NIST BEES Online. Available online: https://ws680.nist.gov/bees/(A(YRWmdd3E1Q\protect\kern+.1667em\relaxEkAAAAZWE1NmE3MzUtMWM4Zi00OTA1LWI0MTctN2RkODkzOWYyNGMzu9uo8zgO971\protect\kern+.1667em\relaxrP3Ett3WWtXZwV6A1))/AnalysisParametersBuildingProds.aspx (accessed on 7 July 2020).

28. BIONOVA Ltd. One Click LCA. Available online: https://www.oneclicklca.com/ (accessed on 7 July 2020).

29. Cheng, J.C.P.; Ma, L.Y.H. A BIM-based system for demolition and renovation waste estimation and planning. *Waste Manag.* **2013**. [CrossRef] [PubMed]

30. BRE Group. IMPACT. Available online: https://www.bregroup.com/impact/ (accessed on 7 July 2020).

31. Garzia, F.; Pernetti, R.; Weiss, T.; Venus, D.; Knotzer, A.; Thouvenot, M.; Berggren, B. *D2.1: Report on EU Implementation of nZEBs*; EURAC research: Bolzano, Italy, 2018.

32. Pernetti, R.; Garzia, F.; Paoletti, G. *D2.2: Spreadsheet with LCCs A Database for Benchmarking Actual NZEB Life-Cycle Costs of the Case Studies*; EURAC research: Bolzano, Italy, 2018.

33. European Parliament. Directive 2010/31/EU, of european parliament and of the council of 19 May 2010 on the energy performance of building (recast). *Off. J. Eur. Union* **2010**, *153*, 13–35.

34. European Commission. Guidelines accompanying Commission Delegated Regulation (EU) No 244/2012 of 16 January supplementing Directive 2010/31/EU of the European Parliament and of the Council. *Off. J. Eur. Union* **2012**, *115*, 1–28.

35. European Commission. Commission recommendation (EU) 2016/1318, on guidelines for the promotion of nearly zero-energy buildings and best practices to ensure that, by 2020, all new buildings are nearly zero-energy buildings. *Off. J. Eur. Union* **2016**, *208*, 46–57.

36. IDAE. *Escala de calificación energética—Edificios de nueva construcción. Labelling schemes for new buildings*; IDAE—Instituto para la Diversificación y Ahorro de la Energía: Madrid, Spain, 2009.

37. D'Agostino, D.; Mazzarella, L. What is a Nearly zero energy building? Overview, implementation and comparison of definitions. *J. Build. Eng.* **2019**, *21*, 200–212. [CrossRef]

38. Karlessi, T.; Kampelis, N.; Kolokotsa, D.; Santamouris, M.; Standardi, L.; Isidori, D.; Cristalli, C. The Concept of Smart and NZEB Buildings and the Integrated Design Approach. *Proc. Eng.* **2017**, *180*, 1316–1325. [CrossRef]

39. ACCD—Andalusia Construction Cost Database. 2017. Available online: http://www.juntadeandalucia.es/organismos/fomentoyvivienda/areas/vivienda-rehabilitacion/planes-instrumentos/paginas/bcca-sept-2017.html (accessed on 7 July 2020).

40. Kim, Y.-C.; Hong, W.-H.; Park, J.-W.; Cha, G.-W. An estimation framework for building information modeling (BIM)-based demolition waste by type. *Waste Manag. Res.* **2017**, *35*, 1285–1295. [CrossRef]

41. Ramirez de Arellano Agudo, A.; Llatas-Oliver, C.; García-Torres, I. *Retirada Selectiva de Residuos: Modelo de Presupuestacion*; Coleccion Nivel 6; Fundacion Cultural del Colegio Oficial de Aparejadores y Arquitectos Tecnicos: Sevilla, Spain, 2002.

42. Marrero, M.; Ramirez-De-Arellano, A. The building cost system in andalusia: Application to construction and demolition waste management. *Constr. Manag. Econ.* **2010**, *28*, 495–507. [CrossRef]

43. González-Vallejo, P.; Muñoz-Sanguinetti, C.; Marrero, M. Environmental and economic assessment of dwelling construction in Spain and Chile. A comparative analysis of two representative case studies. *J. Clean. Prod.* **2019**, *208*, 621–635. [CrossRef]

44. ASTM E917. *Stardard Practice for Measuring Life-Cycle Costs of Buildings and Building Systems*; ASTM International: West Conshohocken, PA, USA, 2013.

45. EN-15643-4. *Sustainability of Construction Works. Assessment of Buildings. Framework for the Assessment of Economic Performance*; Comite Europeen de Normalisation: Geneva, Switzerland, 2012.

46. European Parliament. Directive 2008/98/EC, On Waste and Repealing Centain Directives. *Off. J. Eur. Union* **2008**, *312*, 3–30.

47. Dantata, N.; Touran, A.; Wang, J. An analysis of cost and duration for deconstruction and demolition of residential buildings in Massachusetts. *Resour. Conserv. Recycl.* **2005**, *44*, 1–15. [CrossRef]

48. Solís-Guzmán, J.; Marrero, M.; Guisado García, D. Model for the quantification and budgeting of the construction and demolition waste. Application to roads|Modelo de cuantificación y presupuestación en la gestión de residuos de construcción y demolición. Aplicación a viales. *Carreteras* **2014**, *4*, 6–18.

49. Pun, S.K.; Liu, C.; Langston, C. Case study of demolition costs of residential buildings. *Constr. Manag. Econ.* **2006**, *24*, 967–976. [CrossRef]

50. Freire-Guerrero, A.; Alba-Rodríguez, M.D.; Marrero, M. A budget for the ecological footprint of buildings is possible: A case study using the dwelling construction cost database of Andalusia. *Sustain. Cities Soc.* **2019**, *51*. [CrossRef]

51. European Commission. 2000/532/EC European waste catalogue. *Off. J. Eur. Communities* **2000**, *226*, 3–24.

52. European Constructions Costsd Ltd. European Cost Index. Available online: http://constructioncosts.eu/cost-index/ (accessed on 7 July 2020).

53. Langdon, D. Life Cycle Costing (LCC) as a contribution to sustainable construction: A common methodology. Literature review. *Davis Langdon Manag. Consult.* **2007**, *1*, 1–113.

54. Heiselberg, P.; Brohus, H.; Hesselholt, A.; Rasmussen, H.; Seinre, E.; Thomas, S. Application of sensitivity analysis in design of sustainable buildings. *Renew. Energy* **2009**, *34*, 2030–2036. [CrossRef]

55. Tian, W. A review of sensitivity analysis methods in building energy analysis. *Renew. Sustain. Energy Rev.* **2013**, *20*, 411–419. [CrossRef]

56. Weiss, T.; Moser, C.; Venus, D.; Garzia, F.; Pernetti, R.; Köhler, G.; Stöcker, T. *D6.1: Parametric Models for Buildings and Building Clusters: Building Features and Boundaries*; EURAC research: Bolzano, Italy, 2019.

57. Marenjak, S.; Krstić, H. Sensitivity analysis of facilities life cycle costs. *Teh. Vjesn.* **2010**, *17*, 481–487.

58. Federal Reserve Bank of St. Louis Interest Rates, Discount Rate for Euro Area (INTDSREZQ193N)|FRED|St. Louis Fed. Available online: https://fred.stlouisfed.org/series/INTDSREZQ193N (accessed on 7 July 2020).

59. Feist, W.; Pfluger, R.; Schneieders, J.; Kah, O.; Kaufman, B.; Krick, B. *Passive House Planning Package Version 7*; Passive House Institute: Darmstadt, Germany, 2012.

60. Eurostat Electricity Prices for Households in the European Union 2010–2017, Semmi-Annually. Available online: http://epp.eurostat.ec.europa.eu (accessed on 7 July 2020).

61. EN15459-1:2017. *Energy Performance of Buildings—Economic Evaluation Procedure for Energy Systems in Buildings—Part 1: Calculation Procedures, Modules M1-M14*; Comite Europeen de Normalisation: Geneva, Switzerland, 2017.

62. SGC. *Tarifa de Precios de Reciclaje de Residuos de la Construcción*; SGC: Fresno de la Ribera, Spain, 2016.

63. RESIDUALIA. Tarifas—Centro de Gestión Medioambiental Sierra Segura. Available online: http://www.residualia.net/tarifas/ (accessed on 8 November 2019).

64. BRE; Deloitte; FCT; ICEDD; RPS; VTT, D.-G. *Resource Efficient Use of Mixed Wastes Improving Management of Construction and Demolition Waste*; Final report; European Commission: Brussels, Belgium, 2017.

65. ICE. *Demolition Protocol*; Institution of Civil Engineeres: London, UK, 2008.

66. Banias, G.; Achillas, C.; Vlachokostas, C.; Moussiopoulos, N.; Papaioannou, I. A web-based Decision Support System for the optimal management of construction and demolition waste. *Waste Manag.* **2011**, *31*, 2497–2502. [CrossRef]

 *sustainability* 

Article

The Revaluation of Uninhabited Popular Patrimony under Environmental and Sustainability Parameters

Begoña Blandón [1,*]**, Luís Palmero** [2] **and Giacomo di Ruocco** [3]

[1] Architectural Construction Department 1, University of Seville, 41004 Sevilla, Spain
[2] Architectural Construction Department, University Politècnica of València, 46022 València, Spain; lpalmero@csa.upv.es
[3] Departamento de Ingeniería Civil, Universitat degli Studi di Salerno, 84084 Fisciano, Italy; gdiruocco@unisa.it
* Correspondence: bblandon@us.es

Received: 25 June 2020; Accepted: 10 July 2020; Published: 13 July 2020

Abstract: Abandoning rural areas requires promoting their repopulation. In Europe, wealth and life in these enclaves are valued. However, the current state of these houses does not meet actual needs and requires interventions to actualize current standards. Therefore, decisions in the design and execution of the works will generate a volume of construction and demolition waste (CDW), which must be managed sustainably out of respect towards its origin, the architecture, and the surrounding environment. This paper examines the prevention and management of CDW, providing control strategies and actions to monitor and plan them from the rehabilitation project itself. Some of the interventions carried out in recent years on this type of housing have been analyzed and the existing management protocols within the European Union have been reviewed, specifying their application in Mediterranean popular housing. As a result, we herein show a representative case that observes the existing reality regarding the destination of generated CDW and delves into their possibilities for use. We present these findings in order to reduce the energy cost resulting from manufacturing new materials and meeting the established sustainability and energy efficiency parameters.

Keywords: disinvestment; popular housing; rehabilitation works; sustainable construction; waste management; circular economy

---

1. Introduction

The year 2020 began by recognizing the end of an era. More specifically, this era was marked by important scientific and technological advances, such as the expansion of communications, interesting archaeological, astronomical, or biological findings that tackled challenges in medicine, biotechnology, or nanotechnology. Moreover, new discoveries supported technological progress in the information age. Likewise, this era made room for relevant decisions regarding greater social awareness on aspects as diverse as gender equality, quality of life for the elderly, or environmental deterioration and its consequences. However, the year 2020 will be remembered most for the international health and economic emergency caused by COVID-19. However, there may be an opportunity in which new habits with concern for the future of human health now take center stage. It is obvious that our social, labor, and habitat habits require reviewing (with environmental criteria), especially with regard to certain models of architecture, urban planning, environment, and city-life [1].

For years, the growth of large European cities has caused depopulation in rural areas. Although the origin of this important demographic imbalance is located in the middle of last century, when cities housed factories and needed abundant and unspecialized labor, these areas today still have the lowest population densities in the entire continent, and the rural exodus continues to grow. Likewise, in Asia, cities have grown exponentially in the face of attractive city living. Opportunities across the Asian continent are better compared to the isolated and resource-poor rural environments of Europe. As will be seen in many cases, perhaps life is not as hopeful as promised. As a novelty, during the pandemic, rural areas have been able to defend themselves during different stages of this pandemic and this favors possible future changes in population dynamics.

On the other hand, rural areas have a significant proportion of the population aged 65 years or older. The continued decline in birth rates and the increase in elderly life expectancy distorts the population pyramids. (Europe has 20% of its population above the age quoted and there is no generational replacement.) Italy, for example, has the highest median age in Europe [2]. Considering that in the next 50 years demographic forecasts will triple these figures, the level of depopulation to which these municipalities are subjected has not yet ceased and its consequences are aggravated [3].

With this concern, Spain can serve as an example of high depopulation [4]. As in other European Union countries like France or Italy, 40% of the municipalities are classified as having a severe risk of depopulation. The growth with which abandonment develops in these zones is one of the essential lines of study in order to enact possible solutions in the next decade [5]. In recent years, initiatives have emerged that seek to revitalize rural environments at the economic, demographic, and social level [6]. In 2017, Spain confronted the demographic challenge as state policy. Thus, in 2018, the 2018/2021 state housing plan [7] launched aid to rehabilitate housing for general interventions and promoted energy efficiency in rural areas, bringing its offer closer to young people and those aged over 65. In 2019, general guidelines were established as the 2030 Agenda and National Strategy, which constructed a plan to fight against depopulation in rural areas [8].

At present, administrations are implementing improvement policies in infrastructures and transport systems, betting on new lines and models of production and trade, defending sustainable agriculture and livestock, or rural tourism. Thus, technological advances are being considered that could make field work more efficient, present work alternatives in the case of an economic crisis, improve telecommunications (which have made telework possible during the period of coronavirus confinement), or, simply, the approach to a more natural life [9]. In the incoming decade, a new repopulation phenomenon is to be expected that, fueled by the wealth of these enclaves, demands the rural world as a habitat option. Now, the rural era will take center stage. The year 2020 is a decisive time to promote actions that slow down the isolation and review living conditions in these locations.

Principles of Action and Objectives

Residence rehousing options are divided between new work (i.e., visual alteration of the environment that surrounds them) and the occupation of existing and currently unoccupied homes (i.e., as a consequence of the abandonment process). However, the buildings available in these places do not respond to the needs of a new population, who will be forced to adapt buildings to the regulatory requirements and European guidelines on comfort, energy efficiency, and environmental sustainability. From an architectural point of view, the requirements that are demanded, at a typological, constructive, or technological level, are not considered in the original home [10]. Thus, any proposal will be accompanied by essential adaptation works (punctual or comprehensive) to achieve these objectives. In this regard, the environmental impact that the construction and demolition waste (CDW) generated deserves a study on the control of its production, the promotion of prevention, or any other recovery formula according to the current legislation.

In 2017, the construction industry consumed 36% of the world's energy, which represents almost 40% of gas emissions and contributes to the current greenhouse effect and climate change. According to the Green Building Council (UKGBC) [11], more than 400 million tons of materials are used each year, 15% of fresh water and 25% of the world's wood. Annual global demand for sand and gravel is between 40 and 50 billion tons per year. On the other hand, construction produces a quarter of global waste with materials such as cement, metals, glass, or asphalt that end up in landfills or incinerators. This exorbitant expense is far from what is dealt with in international protocols. The term sustainable development (Brutland Report of the year 1978) is a concept that implies a very important change in the idea of guaranteeing an ecological attitude over time, within a framework that also emphasizes economic and social development. "Our Common Future" is the World Commission on Environment and Development's report that addresses such concerns (United Nations Headquarters, August 1987). In the years after the report was published, there were other world-renowned actions such as the famous Kyoto protocol, (Japan, December 11, 1997), whose March 2005 implementation led to an emission reduction of at least 5% between 2008–2012, compared to emissions reported in 1990. This action and other subsequent ones highlight the global situation regarding gas emissions due to the release of polluting emissions into the atmosphere. The effects have manifested themselves through a progressive phenomenon of global warming, in addition to representing uncertainty about the residual availability of non-renewable resources, with all of the consequent potential environmental risks [12].

If this disproportionate level of consumption continues, the planet is headed for an energy and pollution crisis of irreparable consequences. The construction industry should be closely examined to prevent such consequences, given its material manufacturing waste propagation. The European Directive on Energy Efficiency [13] supports the revitalization of the market for the renovation of existing buildings: Member States must renovate at least 3% of their building stock through rehabilitation and, if necessary, demolition and reconstruction. Likewise, to reach the goal of zero emissions before the end of the 21st century [14,15], all buildings, regardless of the country in which they are built, must produce not only the energy they consume but also the energy needed to build, maintain, and dismantle them via recycling their components.

Because existing rural homes lack attractive technical and comfortable conditions, we decided to study building intervention and observe the construction and demolition waste (CDW) impact on planned rehabilitation works. Thus, we propose the generated use of CDW, betting on recycling and/or reuse to incorporate materials and components back into buildings.

2. Materials and Methods

There is a clear antecedent to our work: to raise awareness for partly abandoned and depopulated heritages. Different seminars and congresses, such as the "Small Towns Conference" (STC-19) held in Salerno, Italy, emphasized the habitat situation that addresses sustainable strategies for the valorization of construction and the landscape/cultural/architectural heritage in inland areas [16]. Currently, the University of Seville's Higher Technical School of Architecture TEP 954-IN FACT (Investigation Factory) research group in coordination with professors from the Polytechnic University of Valencia's (Spain) Higher Technical School of Building Engineering and the Faculty of Engineering of the University of Salerno's, (Italy), are taking the opportunity to share and promote interventions that defend and protect the visual beauty of the rural environment.

In this work, we reviewed the European Union's existing management protocols, specifying their application in the example of popular architecture in the Mediterranean area. Our final task was to create actionable guidelines that guaranteed the sustainable quality of these interventions and to develop a monitoring guide that benefited the management and treatment of the indicated waste. This objective encompassed different stages differentiated into research phases according to their purpose and temporal state (Figure 1).

1. Phase 1: Studying risk locations. The scope of the intervention was specified by selecting locations of interest and going deeper into their architecture and recovery possibilities.
2. Phase 2: Producing CDW and studying the nature of CDW and its typology.
3. Phase 3: Managing CDW. Protocol for action in the rehabilitation of housing in small towns. After the data was collected, we analyzed the final destination of CDW in the cases studied.
4. Phase 4: Possibilities of taking advantage of the generated CDW. Reuse and study new recycled construction products. Currently, this work is still in the process of development. We continue to set new goals in this last phase.

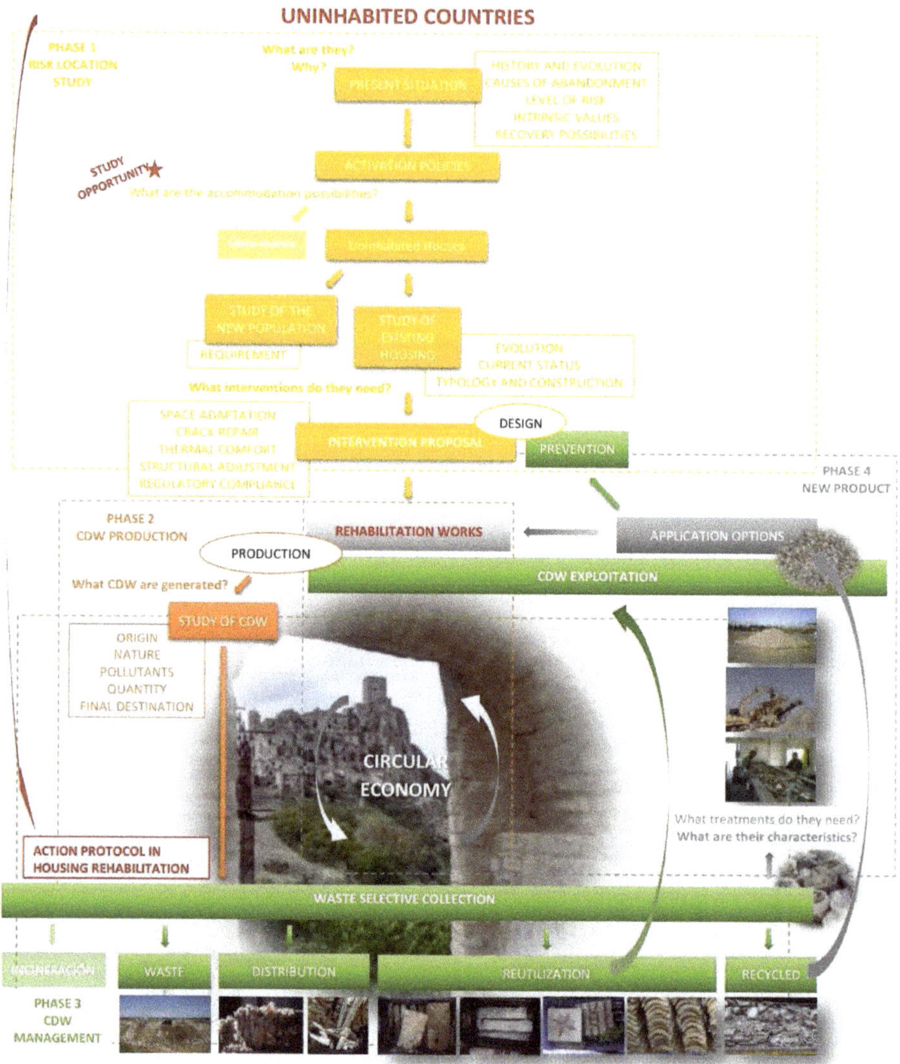

Figure 1. Outline of contents and defined stages in the research process.

2.1. Study of the Municipalities and Identification of Affected Locations

The history and evolution of rural area populations throughout the 20th century facilitates the understanding of what factors have led communities to their present situation. In the cases of warfare and agricultural production crises (i.e., labor force mitigation caused by the mechanization of working mediums), massive migrations can occur, which give rise to abandonment in many locations [17]. Young people flock to cities for job opportunities. Cities grew and benefited from the post-war baby boom, leaving rural areas forgotten. In the years 1960 to 1970, population growth in rural towns paralyzed, streets were empty, and inhabitants began aging and dying. In the 1990s, the abandonment dynamics to which many municipalities in the Mediterranean area are subjected to, led to alarming statistical data.

To specify the scope of the intervention, different levels of risk were identified to locate the affected municipalities according to the number of inhabitants, causes of abandonment, context, and opportunity for recovery. The study was enriched with intrinsic values (i.e., landscape, culture, gastronomy, and architecture) and other parameters, such as an area's geography, economy, and current activation policies [18].

The scope of this study covers the following municipalities: Andalusia, Levante, and Catalonia (Spain); Occitania and the region of Provence-Alpes-Côte d'Azur (France); and Emilia-Romagna, Piedmont, Sicily, Tuscany, and Campania (Italy). All of these regions had less than 1000 inhabitants and highlighted the need for a recovery process. Currently, they have improvements in accesses, infrastructure, and services, so the need to provide a solution to the possible rehousing is immediate and requires a detailed study and an empty housing stock.

2.2. Study of the Existing Housing

In an area as extensive as the one proposed, architecture and construction vary as a consequence of geographical location, topography, climatology, and the various impacts incurred to these locations over centuries. This section identifies existing homes within the canons of vernacular architecture. We can deepen our knowledge of traditional architecture and thus learn to reactivate these areas by understanding construction valorization, landscapes, and cultural heritage [19].

Similar models have been observed in the Mediterranean area. Although the origin of some populations date back to prehistory and Roman times, the buildings examined in this study are largely from the 18th and 19th centuries. However, the poor evolution of the rural house verifies that there are no outstanding innovations and thus proves that surviving constructions steeped in history deserve consideration and respect [20]. In any of the analyzed cases, housing reflects rural life, as it is not far removed from the idea of comfort many people in the developed world have today [21]. Tradition and environment tend to justify the predominance of these functions in existing housing prototypes.

As a starting point, we examined each houses spaces and constructive solutions to determine their evolution and deficiencies. For data collection in the different locations, we developed a visit plan, which was not always successful. These visit plans have limited our inspection of the surrounding environment and architectural enclosures. To complete the study, the sample of dwellings was expanded to include homes with similar typological and construction characteristics (belonging to the same, or neighboring municipality). Although some of these were not abandoned, collecting data was more accessible.

In this study, three types of properties were observed according to their constructive typology, their walls, and their roof coverages and claddings, allowing for a first classification summarized in Figure 2.

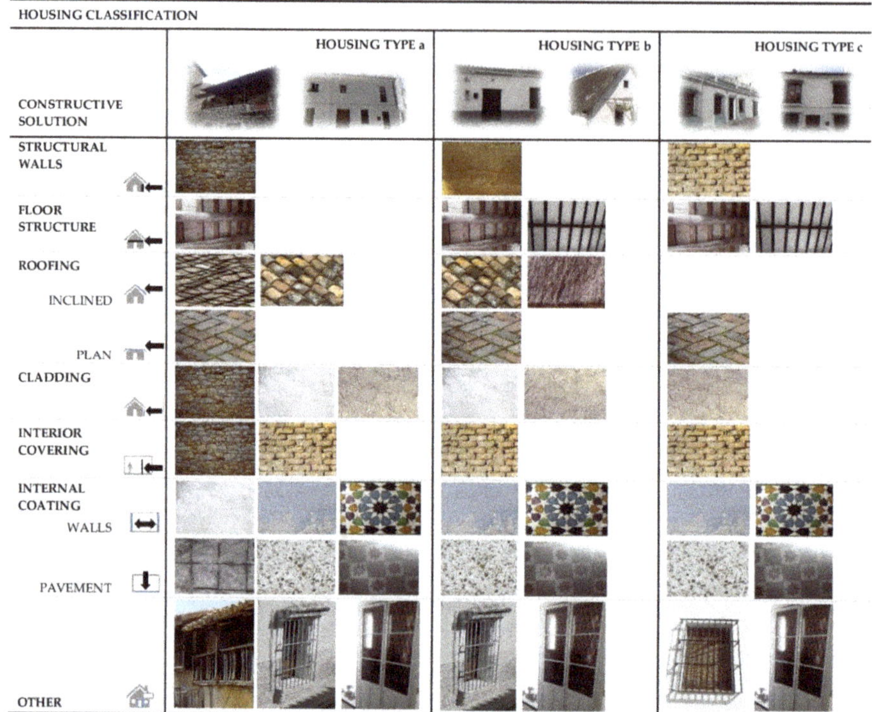

Figure 2. Typology of existing houses. Classification made according to the constructive solution and family of representative materials in each case.

2.3. Proposed Interventions on Existing Housing

The study of the original housing, its state of conservation, and the needs of new users, have confirmed the necessity of intervention. Since we did not use normative criterion (perhaps because these homes did not exist in their origins), the need for a formal study can hence be observed [22]. During the time period investigate in this study, the paradigm of environmental sustainability was hardly taken into account, which was an inevitable requirement when intervening in this type of construction. Knowledge of traditional architecture guided the search for sustainable strategies to guarantee respect for future interventions. We herein combined its technical adaptation and the value of the existing buildings.

From a construction point of view, the study of materials, products, techniques used, and plastic solutions allowed us to understand the evolution and responses over the years in comparison with their current state and nature of the CDW. Likewise, progressive location abandonment and their consequent lack of maintenance accelerated deterioration, causing defects and injuries that affected stability, habitability, and decoration. As can be extracted from Figure 3, the criteria followed in the Catalog of Constructive Rehabilitation Solutions [23] made it possible to decide which intervention was necessary. In this respect, the vertical supporting elements were scarce; thermal deficiencies detected in the surrounding elements of the property should be improved in 56% of the cases analyzed. However, solutions should be transferred and the original element was hardly affected). In 70% of houses, it was necessary to review the stability of the horizontal structure and roof. Moreover, it was necessary to guarantee the tightness of the roof in 68% of the cases. The intervention proposals included pavement revision and of interior patio conditioning in 76% of cases. When renovating the vertical cladding (interiors and exteriors), we needed to replace the carpentry and replace the majority of the facilities. The layout of the interior spaces and routes were affected by 50%. In this regard, the intervention measures were grouped into three levels according to economic, technical, and social character and criteria [24]: basic level (i.e., elementary actions that affect installations and vertical finishes), medium level (i.e., redistribution of interior spaces, pavements, and thermal comfort), and intense level (i.e., action planned on structural elements, their renewal or reinforcement). In the required interventions, cleaning and sanitation of walls and leveling of floors were included.

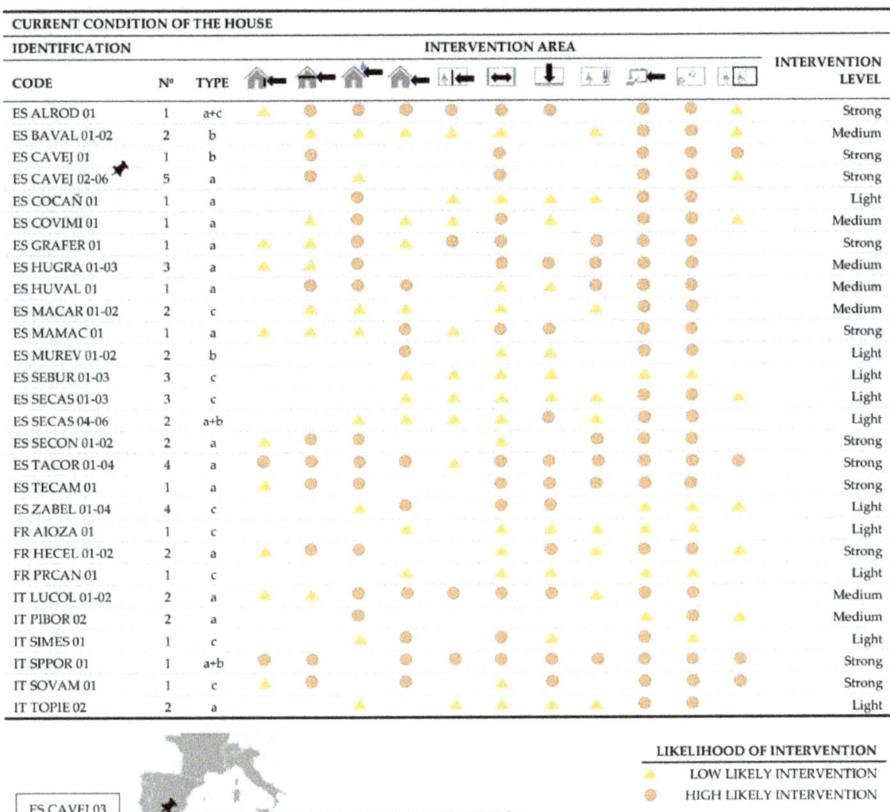

CODE	N°	TYPE	INTERVENTION LEVEL
ES ALROD 01	1	a+c	Strong
ES BAVAL 01-02	2	b	Medium
ES CAVEJ 01	1	b	Strong
ES CAVEJ 02-06	5	a	Strong
ES COCAÑ 01	1	a	Light
ES COVIMI 01	1	a	Medium
ES GRAFER 01	1	a	Strong
ES HUGRA 01-03	3	a	Medium
ES HUVAL 01	1	a	Medium
ES MACAR 01-02	2	c	Medium
ES MAMAC 01	1	a	Strong
ES MUREV 01-02	2	b	Light
ES SEBUR 01-03	3	c	Light
ES SECAS 01-03	3	c	Light
ES SECAS 04-06	2	a+b	Light
ES SECON 01-02	2	a	Strong
ES TACOR 01-04	4	a	Strong
ES TECAM 01	1	a	Strong
ES ZABEL 01-04	4	c	Light
FR AIOZA 01	1	c	Light
FR HECEL 01-02	2	a	Strong
FR PRCAN 01	1	c	Light
IT LUCOL 01-02	2	a	Medium
IT PIBOR 02	2	a	Medium
IT SIMES 01	1	c	Light
IT SPPOR 01	1	a+b	Strong
IT SOVAM 01	1	c	Strong
IT TOPIE 02	2	a	Light

Figure 3. Description of the original state of the houses and the character of the interventions. The reflected identification codes corresponded to the geographical location (country, province, or region and municipality of the locations). The type of dwelling (a, b, c) is recognized from the classification shown in Figure 2. The selected dwelling will be the protagonist of the application and results described in Point 3.

In order to adapt the house to the required technical and comfort needs, the proposed interventions contemplated sustainable strategies that improved habitability, indoor air quality, and energy efficiency [25]. Passive thermal conditioning was sought whenever possible as an energy saving functionary. As part of prevention, the choice of materials and products with low environmental impact in their management was done to minimize CDW production [26].

2.4. Construction and Demolition Waste Management Protocol CDW

The CDW management protocol was drawn up on behalf of the European Commission as part of follow-up measures and served as a starting point in this study. This document is part of the

Construction Strategy 2021 and the proposed measures will contribute to achieving the objective of the waste framework directive of recycling 70% of construction and demolition waste by 2020. These results will thus close the life cycle of products by increasing recycling and reuse, and obtaining benefits for both the environment and the economy [27,28].

By applying the CDW management protocol to different case studies, the conclusions and particularities detected allowed us to review its content and write a simplified action guide that files, orders, and classifies the CDW with the hope of rehabilitating houses in uninhabited towns. Tracking sheets were configured in a simple content format in which different sections related to the work data and description of the planned works favored the circular economy and the global competitiveness of the resulting products, i.e., CDW identification (origin and typology), CDW quantity estimation, the preventive measures to be adopted (to minimize the generation of CDW), the exploitation operations (on-site or its external management), and the final destination of the CDW surplus (specifying the place of discharge if necessary).

3. Results: Application of the Protocol on Case Studies

For the review of the CDW management protocol and its application, one of the representative cases when rehabilitating homes in southern Spain was typological; its construction characteristics made it representative in this study (see Appendix A: Figure A1). The house belonged to the municipality of Vejer de la Frontera, in the province of Cádiz (Spain). This municipality is part of the route of the White Towns in Andalusia and includes buildings defined by typical mountain locations. In this specific case, the position of the municipality gave it proximity and ease of access to the coast, which facilitated its activation in recent years. Today, it is part of the recovered villages moving away its prior status as an abandonment area. The planned interventions were developed mainly inside the house. We checked the stability of the horizontal structure, access to the rooftop, the renovation of vertical cladding and pavements (interiors), carpentry, and the facilities (see Appendix A: Figure A2). Considering waste management, the process allowed us to draw interesting conclusions about the intervention circumstances.

In compliance with RD 105/2008, the preparation of the waste management study during the drafting of the rehabilitation project [29] was carried out by a designer prior to the start of the work, which allowed us to check the veracity of the results and the greater or lesser success in the typology forecasts, quantity, and CDW destination. As part of this process, the adequacy of the prepared monitoring sheets was checked. We deduced deficiencies inherent in the data collection, such as the derived decisions and the incidents suffered during the execution of the work.

3.1. Type and Characteristics of the Waste Generated

In accordance with the European waste list [30], the study and identification of the generated waste allowed us to classify origin and place of production, stage of the process, danger, adhered contaminants, and alternatives in recovery [31]. As observed in the tracking sheets (see Appendix A: Figure A3), the planned CDW mostly came from the demolition/dismantling of existing elements in the original construction and, to a lesser extent, from the execution/replacement of the new proposed elements.

Figure 4 shows that the highest percentage of expected CDW was included in the inert group (a name that encompasses those products that do not cause environmental contamination over time). In total, 85% came from masonry, bricks, mortars, and other materials from the family of ceramics or stone. Within this percentage, ceramic waste (belonging to the elements of coverage, interior partitioning, and cladding) was found to have greater presence in the stony houses, even though it was a type of dwelling (type "a") with good supporting elements. Despite the initial classification made on the housing typology, the results obtained allowed us to unify the ceramic CDW in this type of construction. The CDW composition considered ceramic waste as a consequence of other materials and pastes with which it was in contact with. In the cases studied, the presence of lime, mortars, plasters, and paints attached to the pieces was inevitable. Although they occurred in a smaller proportion, they were considered as pollutants because they altered their original characteristics, quality, and posed a problem for new uses, i.e., plaster.

Figure 4 shows the important presence of wood as a consequence of the deterioration and rot of beams, demonstrating the need for new carpentry, glass, and metals/alloys. Unlike new construction, there were no remains from the demolition of concrete elements. Its only detection resulted from the removal of the compression layer from the original floors.

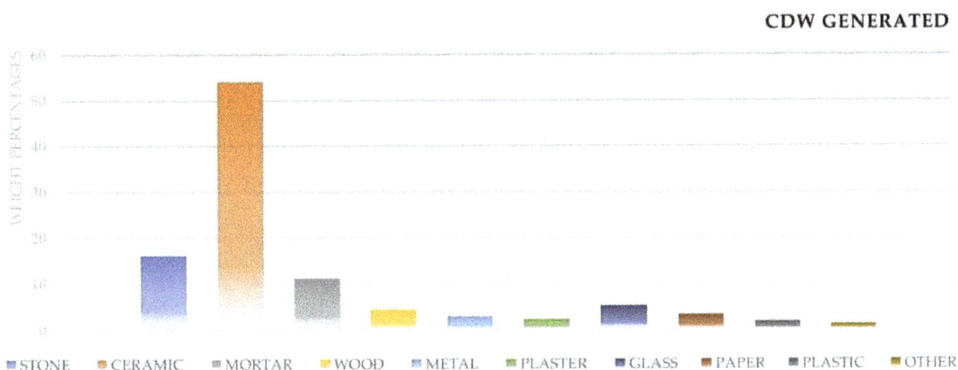

Figure 4. Proportion and nature of the generated construction and demolition waste (CDW). Coded housing application ESCAVEJ 03.

3.2. Current Management of Generated CDW

The final destination of the generated CDW in ESCAVEJ 03 (see Appendix A: Figure A4) was chosen as the only option for use. We made a first selection for waste, incorporating them almost immediately into the building. In this type of construction, due to its heritage value, the reuse of original materials/products turned out to be an inquiry demanded in many cases by the user himself, resorting to those materials and products whose state of conservation was worth incorporating. During the execution of this study, the lack of storage space made reuse impossible.

The CDW was initially considered as waste with no possibility of use. The final destination was the rubble dump in Chiclana de la Frontera (32 km). Given the volume and nature of the remaining CDW, the review and management dynamics for the generated waste was considered unsuitable for direct reuse. It was necessary to study CDW characteristics for a possible assessment that allowed for the possibility of recycling.

4. Discussion of Results and Progress on the Possibility of Recycling

The information obtained in the set of case studies provides an interesting perspective when considering the management tools relevant to decision-making. In the cases studied, i.e., the ESCAVEJ 03 application, the reuse of original construction materials and products (e.g., ceramic tiles, old bricks, glazed tiles, clay tiles, stone tiles) was considered as the only option for use. The treatment and recycling of the generated CDW was not considered in any of the studied cases. Furthermore, as in other EU countries, rural communities are exposed to illegal CDW removal, controlled discharge points, and, sometimes, reduced CDW volume, inviting bad practices on roadsides, riverbanks, easily accessible farms, or abandoned lots [32]. Trying to avoid this practice, our research reviews final destinations and regulatory positions. In this regard, it is recommended to manage the future of inert CDW within the municipality itself (specifically within the intervention itself).

This study defends the on-site use of generated CDW, as it tries to incorporate CDW back into buildings, which extends its life cycle unlike anything else. To propose a reliable recycling option, it is necessary to consider it in advance of the rehabilitation project. Therefore, it is essential to specify the classification, separation, and on-site collection (selectively) of the generated CDW [33].

The nature and possible contaminants of the main house with which it was in contact allowed us to take a first approach to the existing waste properties that facilitates final decision-making. To do this, the study and characterization of a representative sample of granular material (MG) obtained after crushing the initially discarded CDW began. We sought simple and effective treatments that would allow them to be managed in on-site facilities. The MG obtained must be valued and, always consistent with the nature of the waste, it must reduce the current lack of confidence in the quality of the recycled materials. Logically, the costs derived from its management must also be added. Based on the characteristics obtained and, at a later stage, a new construction product obtained from recycling may be proposed. Figure 1 summarizes the full content of this investigation.

Although still in process, the laboratory of the Department of Architectural Construction I of the ETSA at the University of Seville is studying the characteristics of the planned MG according to the requirements and recommendations required to form the aggregates projected on coated pastes, whether vertical, graveled, or terrazzo pieces, for flooring and baseboards [34]. Obviously, given the nature of the materials and products that are part of the original construction systems, their low reliability is deduced and their use for other purposes is ruled out (such as aggregates a part of concrete or road sub-bases) [35].

In the case studied, the MG obtained presented a high percentage of hollow brick and tiles (from partitioning and cladding elements). In its larger pieces, this origin reflected a geometry of laminar aggregates with frequent edges that limited their use in construction. The shape coefficient and its index of flagstones and needles was reduced for sizes smaller than 4 mm, which allowed the MG to be used as an for the proposed uses. In general, in other locations in the Mediterranean arch, with a greater presence of natural aggregates and/or solid brick, there was a higher percentage of rounded aggregates and the proportion of peaks and ridges was reduced, benefiting their final use for other uses. In any case, among its physical characteristics, the MG did not soften or decompose with water. Due to its ceramic nature, this value was innate to the selected material. Regarding their chemical characterization, the CDW selected for the sample contained a high percentage of clays (this result was expected during poor firing of old bricks and possible contamination of clay soils, sand, or gypsum, which, other than clays, were able to decompose and crumble during the test). Freezing was a quality offered by the studied MG for its resistance to chlorides or sulfates, which indicated its low aggressiveness towards metallic elements with which it could be in contact.

Based on the obtained results and future application of aggregate for terrazzo pavements, we can deduce that with correct selection and planning, the selected MG, after a cleaning process, would reduce the clay content and fines detected in its composition (Table 1). Likewise, the previous removal of the gypsum coatings influenced the results for certain coating solutions that were limited by the percentage of clays and light particles, which would confirm their suitability as MG.

Table 1. Recycling possibilities of the granular material (MG) obtained after crushing the planned CDW. Application and evaluation as MG. Coded housing application ESCAVEJ 03.

GRANULAR MATERIAL CHARACTERIZATION				
REGULATORY REQUIREMENTS		**GRAVEL**	**SUBFLOOR AGGREGATES**	**AGGREGATE FOR FINISHING**
PARTICLE SIZE COMPOSITION	TMA	Accepted		Accepted
GEOMETRY	FORM	Accepted		Accepted
	FRACTURE FACES, UNEVEN, SHARP EDGES, ELONGATED	Accepted with diameter limitations		Accepted
CONTENT THIN AGGREGATES		Initially rejected	Initially rejected	Initially rejected
WATER ABSORPTION			Accepted	
LIGHTWEIGHT PARTICLES		Accepted		
PLASTER CONTENT		Initially rejected	Initially rejected	
ORGANIC MATTER		Accepted	Accepted	Accepted
SOFT PARTICLES		Accepted		
CLOTS OF CLAY		Rejected		
JELIVITY				
FIRE RESISTANCE		Accepted		
CALCIUM OXIDE CONTENT			Accepted	Accepted
LOSSES THROUGH CALCINATION			Accepted	Accepted
SULPHUR COMPOUNDS		Accepted		
RATING		Accepted [1]	Accepted [1]	Rejected for its geometry

[1] Accepted by requiring excessive removal of finish and plaster.

Some of the obtained results could improve by recycling the MG obtained from CDW. Collections should be organized on-site on a selective basis, with protected storage, classification (labelling), origin, and CDW recoverability. Moreover, they must be located out of contact and access to the neighborhood to ensure the absence of organic waste from domestic rubbish. This prior selection would allow for the consideration of CDW organization, cleaning, and subsequent management. With regard to the manufacturing of terrazzo pieces, the product must be subjected to corresponding tests that guarantee its suitability for use in pedestrian traffic inside homes or patios. Moreover, it must have resistance to wear and tear, slipperiness, finish options, and textures, according to the envisaged area, insulation against impact noise, resistance to simple compression and flexotraction, resistance to certain chemical agents, etc.

It is well known that recycled products have to compete with traditional building materials. Thus, it is essential to achieve adequate levels of quality, a quality that derives directly from their origin and composition. If we consider these improvements as MG treatments, new results could be deduced to allow for betting on a new construction product that could be incorporated for building improvements, as reflected in Figure 5; Figure 6.

CDW MANAGEMENT								Section 4
STRATEGIES, SELECTIVE COLLECTION, RECOVERY AND USE								
DEPOSIT	FINAL DESTINATION			ON-SITE TREATMENTS				
ENVELOPE TYPE	RE-USE	RECYCLE	REMOVE	CLEANING and WASHING	SHREDDING	SIEVING and SIFTING	ON-SITE PRODUCTION	USE DESTINATION
EARTH and STONES	X	X		X	X	X	MG X	*Aggregate for coatings*
CERAMICS	X	X		X	X	X	MG X	*Aggregate for coatings*
WOODS	X	X		X	X	X	MG X	*Aggregate for coatings*
CONCRETE, MORTAR		X		X	X	X	MG X	*Aggregate for coatings*
IRON and STEEL	X							
GLASS		X						*Transport industry*
PLASTER		X						*Transport industry*
PAPER and BOARD		X						*Transport industry*
PLASTICS		X						*Transport industry*
DANGEROUS								
Other								

GENERAL CONSIDERATIONS:

The collection of CDW with options for reuse will be carried out independently by storing it according to its nature and final location in the building. Prior to their incorporation into the building they will be submitted to an appropriate cleaning process.

The collection of CDW with recycling options such as GM suitable for inclusion in coatings at their final destination will be carried out in a single mixed inert container over which the planned treatments will be carried out jointly.

Figure 5. Tracking sheet. Proposal for the management of CDW generated on coded housing ESCAVEJ 03 (Section 4 on strategies for the use of generated CDW).

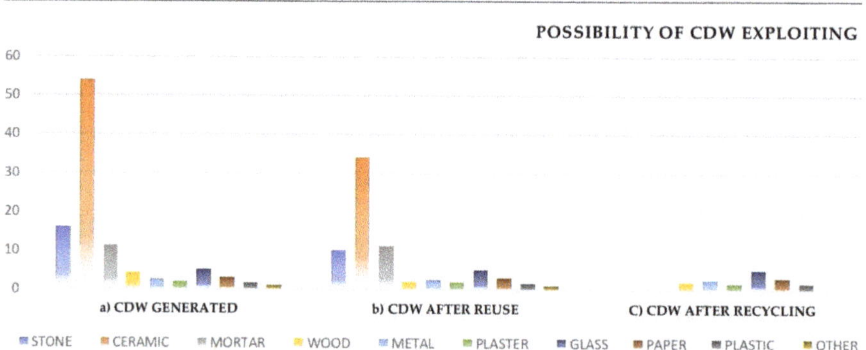

Figure 6. Results obtained after the planned use on-site for generated CDW. Comparison in the proportion and nature of the generated CDW: (**a**) surplus CDW (total); (**b**) surplus CDW after reuse; (**c**) surplus CDW after the management of reuse and recycling. Final result on CDW without the use option foreseen on-site. Coded housing application ESCAVEJ 03.

5. Conclusions

One of the great challenges of today is to better understand the vicissitudes of rural environments and building rehabilitations, specifically the prevention, reuse, and recycling of CDW in planned intervention works. The European Union's CDW protocol should be implemented in order to promote the specifications of data accumulation and decision-making surrounding housing recovery in uninhabited villages. Once we analyze our object of study, possibilities and advantages of this type of action were noted, especially for promoting a sustainable strategy beneficial to the environment (compared to new works, energy savings and environmental benefits are notable). With this type of intervention, we also pursued profitable economic and social sustainability.

Waste management provides a viable and guaranteed solution. Moreover, the historical regulatory framework represents unprecedented support. Many years have passed since the inception of Directive 91/156/CEE [36] and this process undoubtedly had a double environmental benefit. On the one hand, waste dumping should be avoided and on the other hand, the exploitation of natural resources, which in countries like Spain, has reached truly exaggerated levels, should be reduced. However, establishing an economy of rotation requires strategies specifying in recovery, recycling, and reuse, as well as timely assessments for agents in the process, whether they are of a legislative, technical, regulatory, or business nature.

Lastly, and in relation to the effective and convenient good building practice, "deconstruction" (i.e., demolition following the opposite process of construction) must be taken into account in order to demolish or selectively take down and make the most of a building's materials. In this work, we verified the process of acting on representative examples of popular architecture, which allowed us to adjust its content and structure while also guaranteeing its operation. In this respect, special emphasis was placed on identifying the origin and nature of waste, as well as the on-site measures and operations needed to minimize waste. Cumulatively, this provides sustainability benefits and a better quality of life, given the uniqueness of the housing types.

In recent years, the industry initially linked to the land has suffered a radical shift due to social changes, new materials, and new practices. Today, the capacity for pollution, contamination, energy expenditure, environmental deterioration, and resources scarcity is enormous. Therefore, it is necessary to consider new awareness programs aimed at minimizing and controlling waste. Thus, the "use" of popular architecture—which is often abandoned and in danger of becoming demolition waste—is becoming a first-rate alternative for achieving the maintenance and appreciation of architectural heritage. There is also an environmental benefit in the return of life to places with a privileged environment. This return of life offers a sense of authenticity, welfare, and positive future to its inhabitants

Author Contributions: B.B. conceived the original idea of the work; B.B. and L.P. they supervised the project and planned the applications; B.B., L.P. and G.d.R. developed the field work and interpreted the results; B.B. and L.P. wrote the manuscript. All authors have read and accepted the published version of the manuscript.

Funding: This research received no external funding.

Acknowledgments: The work developed would not have been possible without the collaboration of the Architect Jose Maria Ramos, for his knowledge and dedication to the rehabilitation of housing in the villages of Cadiz (Spain); For the data collection and study of abandoned houses, in different locations of the Mediterranean arch, we would like to thank the students from the Master in City and Sustainable Architecture MCAS, belonging to the course 2019/20 of the Superior Technical School of Architecture of the University of Seville, for their participation. The knowledge and characterization of CDW was possible thanks to the supervision of Reyes Rodriguez, who is the head of the Construction Laboratory for the School of Architecture at the University of Seville.

Conflicts of Interest: The authors declare no conflict of interest.

Appendix A

Tracking sheets. Action protocol. Coded housing application ESCAVEJ 03.

1. Figure A1: Follow-up sheet. Section 1: Identification and recognition of the original state.
2. Figure A2: Follow-up sheet. Section 2: Data collection of the intervention.
3. Figure A3: Follow-up sheet. Section 3: CDW forecast.
4. Figure A4: Follow-up sheet. Section 4: Strategies for the use of generated CDW.

Data corresponding to an example of application on a real case without considering recycling.

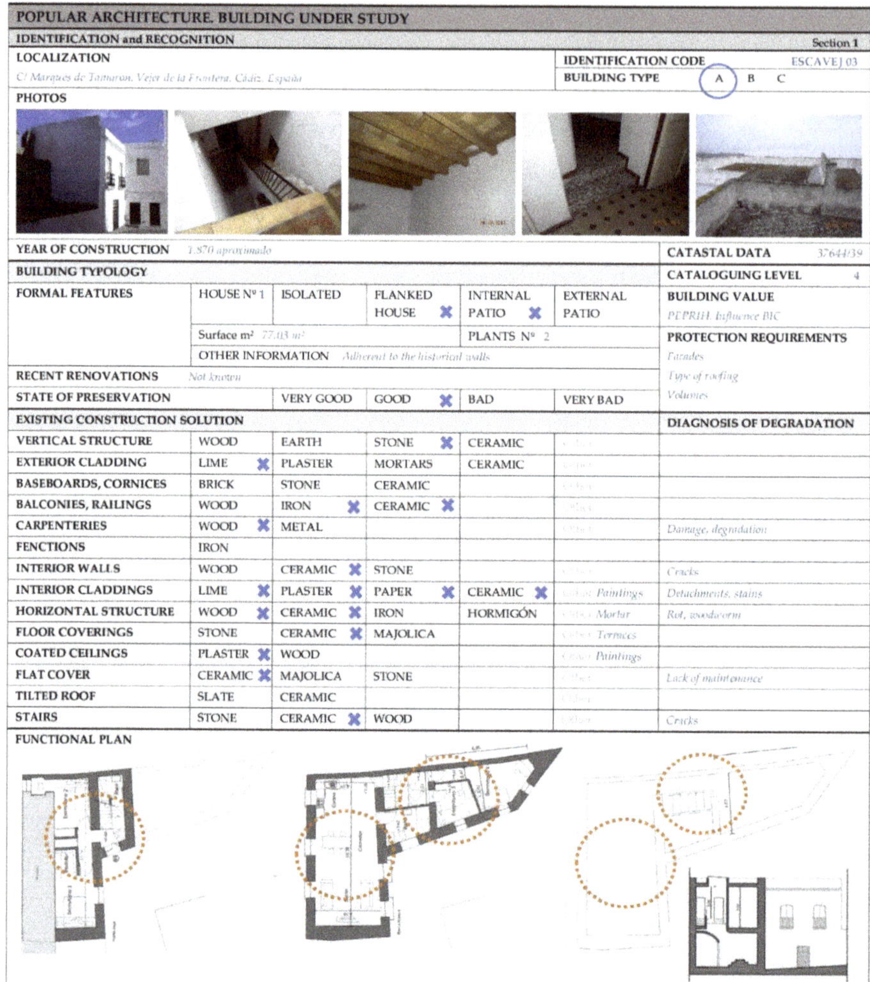

POPULAR ARCHITECTURE. BUILDING UNDER STUDY

IDENTIFICATION and RECOGNITION						Section 1
LOCALIZATION				**IDENTIFICATION CODE**		ESCAVEJ 03
C/ Marqués de Tamarón. Vejer de la Frontera. Cádiz. España				**BUILDING TYPE**		(A) B C

PHOTOS

YEAR OF CONSTRUCTION	*1.870 aproximado*				**CATASTAL DATA**	3764439
BUILDING TYPOLOGY					**CATALOGUING LEVEL**	4
FORMAL FEATURES	HOUSE Nº 1	ISOLATED	FLANKED HOUSE ✕	INTERNAL PATIO ✕	EXTERNAL PATIO	**BUILDING VALUE** *PEPRIH. Influence BIC*
	Surface m² *77.03 m²*			PLANTS Nº 2		**PROTECTION REQUIREMENTS** *Facades*
	OTHER INFORMATION *Adherent to the historical walls*					*Type of roofing*
RECENT RENOVATIONS *Not known*						*Volumes*
STATE OF PRESERVATION		VERY GOOD	GOOD ✕	BAD	VERY BAD	

EXISTING CONSTRUCTION SOLUTION					DIAGNOSIS OF DEGRADATION	
VERTICAL STRUCTURE	WOOD	EARTH	STONE ✕	CERAMIC	Other	
EXTERIOR CLADDING	LIME ✕	PLASTER	MORTARS	CERAMIC	Other	
BASEBOARDS, CORNICES	BRICK	STONE	CERAMIC		Other	
BALCONIES, RAILINGS	WOOD	IRON ✕	CERAMIC ✕		Other	
CARPENTERIES	WOOD ✕	METAL			Other	*Damage, degradation*
FENCTIONS	IRON					
INTERIOR WALLS	WOOD	CERAMIC ✕	STONE		Other	*Cracks*
INTERIOR CLADDINGS	LIME ✕	PLASTER ✕	PAPER ✕	CERAMIC ✕	Other: *Paintings*	*Detachments, stains*
HORIZONTAL STRUCTURE	WOOD ✕	CERAMIC ✕	IRON	HORMIGÓN	Other: *Mortar*	*Rot, woodworm*
FLOOR COVERINGS	STONE	CERAMIC ✕	MAJOLICA		Other: *Terraces*	
COATED CEILINGS	PLASTER ✕	WOOD			Other: *Paintings*	
FLAT COVER	CERAMIC ✕	MAJOLICA	STONE		Other	*Lack of maintenance*
TILTED ROOF	SLATE	CERAMIC			Other	
STAIRS	STONE	CERAMIC ✕	WOOD		Other	*Cracks*

FUNCTIONAL PLAN

Figure A1. Follow-up sheet. Section 1: Identification and recognition of the original state.

IMPLEMENTATION PROTOCOL FOR THE REHABILITATION OF ABANDONED COUNTRIES					
INTERVENTION DATA					Section 2
AREA OF OPERATION					
EXTERNAL FACADES ✕	ROOFING ✕	INTERNAL ROOMS ✕	PATIO ✕	URBAN CONTEXT	OTHER OUTBUILDINGS
TYPE OF INTERVENTION					
WALLS STABILITY	ENERGY SAVING ✕	FLOOR STABILITY ✕	ROOF WATERPROOFING ✕	INSTALLATIONS UPGRADING ✕	REDISTRIBUTION OF INTERNAL SPACES *On the first floor*
KITCHEN FURNISHING ✕	BATHROOM FURNISHING ✕	VERTICAL CLADDINGS ✕	HORIZONTAL COATING ✕	INCLINED ROOF	OTHER ACTIONS PLANNED *Roof access*
INTERVENTION LEVEL					**OBSERVATIONS**
DILAPIDATED BUILDING		HEAVY ✕	MEDIUM ✕	LIGHT	*Interventions at the floors are localized*
INTERVENTION PLAN					
The area of intervention is indicated on the functional plan (Section 1).					

Figure A2. Follow-up sheet. Section 2: Data collection of the intervention.

CDW MANAGEMENT										Section 3
CDW FORECAST GENERATED BY EXISTING BUILDINGS										
TYPE OF WASTE	**EWC CODE**	**CAUSE**	**ORIGINAL POSITION**	**OTHER ADHERED**	**QUANTITY M3**	**TON**	**POSSIBILITY OF USE IN THE NEW BUILDING**			
							RE-USE	**RECYCLE**	**REMOVE**	**COMMENTS**
EARTH										
VEGETATION										
ORGANIC										
STONES	17 05 03	Wall revision	Load-bearing walls	Lime		0.292				
STONE CLADDING										
BRICKS	17 01 02	Distribution of spaces	Internal partitions	Mortars Plasters Paintings		3.026	X		X	Re-use of rough bricks in pavement walls
CERAMIC and MAJOLICA CLADDING	17 01 03	Floor Renovation	Indoor flooring	Mortars		2.375	X		X	Rease in pavements
CLADDING TILES	17 01 06	Renovation of facades	Interior cladding	Mortars Plasters		0.625	X			
ROOF TILES										
WOODEN BEAMS	17 02 02	Floor replacement		Mortars Paintings		0.336	X		X	Reusing beams as decorative elements
CARPENTERIE Wood	17 02 01	Renewal doors and windows		Plasters Paintings		0.110	X		X	Reuse of doors. Removal of windows
GLASSES	17 02 02	Deterioration of window structure	Windows and doors			0.558			X	
CONCRETE and MORTAR	17 01 01	Screed revision	Compression layer			0.809			X	
TERRACE TILES	17 01 01	Soil Renovation	Interior floors			1.147	X		X	Reuse in pavements
LIME		Cladding overhaul	Adherent residuals							
CHALK and PLASTER	17 08 02	Cleaning of walls	Interior cladding	Paintings		0.223			X	They are small proportions and attached
IRON and METALS	17 04 07	Structural restoration				0.279				
Other Cables	17 04 03	Installations				0.085			X	
Other Paper	20 01 01		Interior cladding			0.334			X	
Other										
TOTAL CDW GENERATED IN THE ORIGINAL BUILDING						10.2	**OTHER POSSIBILITIES OF USE:**			
DANGER OF THE PLANNED CDW: None							*No other alternatives for external use are contemplated*			

FORECAST OF THE CDW GENERATED FROM THE NEW CONSTRUCTION										
TYPE OF WASTE	**EWC CODE**	**CAUSE**	**ORIGINAL POSITION**	**OTHER ADHERED**	**QUANTITY M3**	**TON**	**POSSIBILITY OF USE IN THE HOUSE**			
							RE-USE	**RECYCLE**	**REMOVE**	**COMMENTS**
BRICKS and other CERAMICS	17 01 07	Residues of the new construction	Partitions and coatings	Mortars		0.662			X	
PLASTER	17 08 02	Residues of the new construction	Interior cladding			0.135			X	
Other										
TOTAL CDW GENERATED IN THE ORIGINAL BUILDING						0.79	**OTHER POSSIBILITIES OF USE:**			
DANGER OF THE PLANNED CDW: None							*No other alternatives for external use are contemplated*			
OTHER RELATED WASTE										
TYPE OF WASTE	**EWC CODE**	**CAUSE**	**ORIGINAL POSITION**	**OTHER ADHERED**	**QUANTITY M3**	**TON**	**POSSIBILITY OF USE IN THE HOUSE**			
							RE-USE	**RECYCLE**	**REMOVE**	**COMMENTS**
PLASTICS	17 02 03	Packaging				0.160			X	
WOODEN PALLET										
PAPER and BOARDS	20 01 01	Packaging				0.003			X	
ALUMINIUM										
Other										
TOTAL CDW GENERATED WITH THE NEW CONSTRUCTION						0.16	**OTHER POSSIBILITIES OF USE:**			
DANGER OF THE PLANNED CDW: None							*No other alternatives for external use are contemplated*			

(Coming from the original housing and from the new intervention works)

Figure A3. Follow-up sheet. Section 3: CDW forecast.

CDW MANAGEMENT								Section 4
STRATEGIES, SELECTIVE COLLECTION, RECOVERY AND USE								
DEPOSIT	FINAL DESTINATION			ON-SITE TREATMENTS				
ENVELOPE TYPE	RE-USE	RECYCLE	REMOVE	CLEANING and WASHING	SHREDDING	SIEVING and SIFTING	ON-SITE PRODUCTION	USE DESTINATION
EARTH and STONES								
CERAMICS	✗							*External management*
WOODS	✗							*External management*
CONCRETE and MORTAR								
IRON and STEEL								
GLASS								
PLASTER								
PAPER and BOARD								
PLASTICS								
DANGEROUS								
Other Mixed *	✗		✗					*External management*
GENERAL CONSIDERATIONS:								
* Since the limit values established in RD 105/2008 are not exceeded, the CDWs are not separated on site. The waste holder (contractor) or an external agent is responsible for separation, collection and transport for subsequent treatment in the plant (no separation on site due to lack of physical space on the site).								

Figure A4. Follow-up sheet. Section 4: Strategies for the use of generated CDW.

References

1. Ren, S.Y.; Gao, R.D.; Chen, Y.L. Fear can be more harmful than the severe acute respiratory syndrome coronavirus 2 in controlling the corona virus disease 2019 epidemic. *World J. Clin. Cases* **2019**, *8*, 652–857. [CrossRef] [PubMed]

2. Eurostat. Eurostad Stadistics Explained 2020; European Commission. Available online: https://ec.europa.eu/eurostat/statistics-explained/index.php?title=Population_structure_and_ageing (accessed on 28 February 2020).

3. Circle. La Despoblación de las áreas Rurales. *Circle J. Ecoembes* **2019**, *8*. Available online: https://www.revistacircle.com/2019/02/22/la-despoblacion-de-las-zonas-rurales (accessed on 22 June 2019).

4. Atienza, J. Empty Spain Map and Southern Sparsely Populated Areas, SSPA. *ETHIC J.* **2019**, *23*. Available online: https://ethic.es/2019/04/el-mapa-de-la-espana-vacia (accessed on 12 September 2019).

5. Fiore, P.; Blandón, B.; D´Andria, E. Smart Villages for the Sustainable Regeneration of Small Municipalities. In Proceedings of the 3° International Forum on Architecture and Urbanism, IFAU 2019, Tirana, Albania, 21 November 2019; Available online: https://fau.edu.al/home-1-2/ (accessed on 4 February 2020).

6. Moore, P. The Challenges Small Towns Face. In *The Importance of Small Towns. The European Council for the Village and Small Town*; ECOVAST, 2014; Cranbrook, England: by Stationery Express UK Ltd; Stationery Express Uk Ltd: Kent, UK, 2014; Available online: http://www.ecovast.org/papers/ECOVAST%20Importance%20of%20Small%20towns.pdf (accessed on 1 June 2019).

7. Gobierno de España. Real Decreto 106/2018, de 9 de marzo, por el que se regula el Plan Estatal de Vivienda 2018–2021. Sección: I. Disposiciones generales; Publisher: Ministerio de Fomento. «BOE» núm. 61, de 10 de marzo de 2018; pp. 28868–28916. Ref. BOE-A-2018-3358. Available online: https://www.boe.es/eli/es/rd/2018/03/09/106 (accessed on 15 December 2019).

8. Gobierno de España. *Directrices Generales de la Estrategia Nacional Frente al Reto Demográfico*; Ministerio de Política Territorial y Función Pública, Comisionado del Gobierno Frente al Reto Demográfico: Madrid, Spain, 2019; Available online: https://www.mptfp.gob.es/dam/es/portal/reto_demografico/Estrategia_Nacional/directrices_generales_estrategia.pdf (accessed on 30 January 2020).

9. INE. *Informe del Instituto Nacional de Estadística y Cartografía 2019*; Consejería de Economía, Hacienda y Administraciones Públicas de la Junta de Andalucía: Andalucía, España, 2019; Available online: www.epdata.es (accessed on 20 February 2020).

10. Carrasco-Sáez, J.L.; Careaga Butter, M.; Badilla-Quintana, M.G. The New Pyramid of Needs for the Digital Citizen: A Transition towards Smart Human Cities. In Sustainable Smart Cities and Smart Villages Research. *Sustain. J.* **2018**, *9*, 2258. [CrossRef]

11. Global Alliance for Buildings and Construction (GlobalABC). *2019 Global Status Report for Buildings and Construction. Towards a Zero-Emissions, Efficient and Resilient Buildings and Construction Sector*; IEA, United Nations Environment Programme: Paris, France, 2019; ISBN 978-92-807-3768-4. Available online: https://wedocs.unep.org/bitstream/handle/20.500.11822/30950/2019GSR.pdf?sequence=1&isAllowed=y (accessed on 20 February 2020).

12. European Commission. *A Roadmap for Moving to a Competitive Low-Carbon Economy in 2050*; Communication from the Commission to The European Parliament, the Council, the European Economic and Social Committee and the Committee of the Region: Brussels, Belgium, 8 March 2011; Available online: https://eur-lex.europa.eu/LexUriServ/LexUriServ.do?uri=COM:2011:0112:FIN:EN:PDF (accessed on 3 April 2020).

13. European Parliament and the Council of the European Union. Directive 2012/27/EC of the European Parliament and of the Council of 25 October 2012 on Energy Efficiency, Amending Directives 2009/125 / EC and 2010/30 / EU, and repealing the Directives 2004/8/CE y 2006/32/CE. *Off. J. Eur. Commun.* **2012**. No. L 315/1. Available online: https://www.boe.es/doue/2012/315/L00001-00056.pdf (accessed on 15 March 2020).

14. Sicignano, E.; Di Ruocco, G.; Melella. Mitigation Strategies for Reduction of Embodied Energy and Carbon, in the Construction Systems of Contemporary Quality Architecture. *Sustain. J.* **2019**, *11*, 3806. [CrossRef]

15. Spenser, T.; Pierfederici, R.; Sartor, O.; Berghmans, N. *State of the Low-Carbon Energy Union: Assessing the EU's Progress towards its 2030 and 2050 Climate Objectives*; Study 2016, N° 08/2016; Institut du développement durable et des relations internationals IDDRI: Paris, France; Available online: https://www.actu-environnement.com/media/pdf/news-27832-rapport-iddri.pdf (accessed on 2 April 2020).

16. Fiore, P. *Small Towns...From Problem to Resource. Rehabilitation of Abandoned Villages, Proceedings of the Abstracts Small Towns Conference, Fisciano, Italy, 19–20 September 2019*; Fiore, P., D'Andria, E., Eds.; Coopetativa Universitaria Athena CUA: Salerno, Italy; ISBN 978-888944245-2-2. Available online: https://www.verderosa.it/wp-content/uploads/2019/09/STC2019_BOOK-OF-ABSTRACTS.pdf (accessed on 15 February 2020).

17. Instituto de Estadística de Andalucía (IEA), Junta de Andalucía. La transición demográfica hasta 1975. In *Un siglo de demografía en Andalucía. La población desde 1900*; Informes de la Junta de Andalucía: Andalucía, España, 2000; Available online: https://www.juntadeandalucia.es/institutodeestadisticaycartografia/sid/pub/UnSigloDeDemografiaAnd.pdf (accessed on 1 September 2019).

18. Blandón, B.; Rodriguez, R. Recovering of Abandoned Towns. A Sustanaible Strategy for Construction and Demolition Waste Management. In *I centri minori ... da problema a risorsa. Strategie sostenibili per la valorizzazione del patrimonio edilizio, paesaggistico e culturale nelle aree interne*; Nuova Serie di Architettura FrancoAngeli; Plataforma FrancoAngeli: Milano, Italy, 2019; pp. 175–185. ISBN 978-8891798428. Available online: https://iris.unibs.it/retrieve/handle/11379/527112/112010/Passamani%20I-Fasolini%20S_Pasini%20A-Ghidinelli%20N_GAVARDO_STC%202019.pdf (accessed on 10 May 2020).

19. Cillis, G.; Statuto, D.; Picuno, P. Vernacular Farm Buildings and Rural Landscape: A Geospatial Approach for Their Integrated Management. *Sustain. J.* **2020**, *12*, 4. [CrossRef]

20. Carrascosa, G.I. *Caracterización constructiva de edificaciones vernáculas de carácter doméstico en Valverde de Burguillos: Catalogación, análisis, recomendaciones de mantenimiento y reparaciones en muros de tierra cruda y mampostería*; Trabajo Fin de Grado; Departamento de Construcciones Arquitectónicas I, Universidad de Sevilla: Sevilla, Spain, 2018; Available online: https://hdl.handle.net/11441/79494 (accessed on 10 September 2019).

21. Flores López, C. Arte y arquitectura en la vivienda española. Cap. III in *La casa popular en España*; Fomento de Construcciones y Contratas, S.A. D.L.: M-40308-1996. 1996, pp. 163–321. Available online: https://dialnet.unirioja.es/servlet/articulo?codigo=6738859 (accessed on 1 October 2019).

22. Gobierno de España. Real Decreto 314/2006, de 17 de marzo por el que se aprueba el Código Técnico de Edificación (CTE) en que establece las exigencias que deben cumplir los edificios en relación con los requisitos básicos de seguridad y habitabilidad establecidos en la Ley 38/1999 de 5 de noviembre de Ordenación de la Edificación LOE; Ministerio de Vivienda «BOE» núm. 74, de 28 de marzo de 2006 Ref. BOE-A-2006-5515. Available online: https://boe.es/buscar/pdf/2006/BOE-A-2006-5515-consolidado.pdf (accessed on 10 March 2020).

23. Instituto Valenciano de la Edificación. *Catálogo de Soluciones Constructivas de Rehabilitación*; 1st ed.; DRD 07/11; Consejería de Medio Ambiente, Agua, Urbanismo y Vivienda: Valencia, Spain, 2011; ISBN 978-84-96602-72-4. Available online: http://www.anerr.es/images/stories/pdf/catalogo-soluciones.pdf (accessed on 10 December 2019).

24. Barrios Padura, A. Criterios de intervención de la rehabilitación arquitectónica y urbana para la promoción del envejecimiento active. In *(Re)Programa. (Re)habilitación+(Re)generación+(Re)programación. El reciclaje y la gestión sostenible del parque edificado andaluz. Gestión de entornos habitables desde criterios de envejecimiento activo, género y habitabilidad urbana*; Fondo Europeo de Desarrollo Regional, Consejería de Fomento y Vivienda, Junta de Andalucía: Andalucía, Españ, 2015; pp. 58–65. ISBN 9788460813825. Available online: https://dialnet.unirioja.es/servlet/libro?codigo=576336 (accessed on 15 December 2019).

25. Di Ruocco, G.; Melella, R. Evaluation of environmental sustainability threshold of "humid" and "dry" building systems, for reduction of embodied carbon (CO_2). *Vitruvio. Int. J. Arch. Technol. Sustain.* **2018**, *3*. [CrossRef]

26. Llatas Oliver, C.; Osmani, M. Development and validation of a building design waste reduction model. Waste Management. *Int. J. Integr. Waste Manag. SCI. Technol.* **2016**, *56*, 318–336. [CrossRef]

27. European Parliament and the Council of the European Union. Directive 2008/98/EC of the European Parliament and of the Council of 19 November 2008 on Waste and Repealing Certain Directives (Waste Framework Directive). *Off. J. Eur. Commun.* 2008. Available online: https://eur-lex.europa.eu/legal-content/EN/TXT/?uri=celex%3A32008L0098 (accessed on 20 February 2020).

28. European Commision. Protocolo de Gestión de Residuos de Construcción y Demolición en la UE. European Commisión and ECORYS. 2016. Ref. Ares (2016) 6914779 - 12/12/2016. Available online: http://aridosrecicladosdercd.es/wp-content/uploads/2018/06/ES-TRA-01.pdf (accessed on 1 October 2019).

29. Gobierno de España. Real Decreto 105/2008, de 1 de febrero, por el que se regula la producción y gestión de los residuos de construcción y demolición; Ministerio de la Presidencia «BOE» núm. 38, de 13 de febrero de 2008; Ref. BOE-A-2008-2486. Available online: https://www.boe.es/buscar/pdf/2008/BOE-A-2008-2486-consolidado.pdf (accessed on 5 November 2019).

30. European Commision. European List of Waste (LER). Commission Decision of 18 December 2014 amending Decision 2000/532/EC on the list of waste pursuant to Directive 2008/98/EC of the European Parliament and of the Council. Official Journal of the European Communities 30.12.2014, No L 370/44. Available online: https://eur-lex.europa.eu/legal-content/EN/TXT/?uri=CELEX%3A32014D0955 (accessed on 20 February 2020).

31. Barroso Domínguez, V.M. *Análisis de la Gestión de Residuos de Construcción y Demolición en la Comunidad Autónoma de Andalucía*; Trabajo Fin de Carrera, Escuela Técnica Superior de Ingenieros, Universidad de Sevilla: Sevilla, Spain, 19 July 2013; Available online: http://bibing.us.es/proyectos/abreproy/30186/fichero/Cap%C3%ADtulo+0.pdf (accessed on 2 April 2020).

32. Florín- Constantin, M. Construction and Demolition Waste in Romania: The Route from Illegal Dumping to Building Materials. *Sustain. J.* **2019**, *11*, 3179. Available online: https://www.mdpi.com/2071-1050/11/11/3179 (accessed on 15 January 2020).

33. Villoria Sáez, P.; Del Rio Merino, M.; Porras-Amores, C.; Astorqui, J.S.C.; Pericot, N.G. Analysis of Best Practices to Prevent and Manage the Waste Generated in Building Rehabilitation Works. *Sustain. J.* **2019**, *11*, 2796. [CrossRef]

34. Centro de Estudios y Experimentación de Obras Públicas (CEDEX), Ministerio de Fomento, Gobierno de España. In *Catálogo de Residuos Utilizables en la Construcción*; Ficha Técnica RCD actualizado, 2014; Monografías of Ministerio de Agricultura, Alimentación y Medio Ambiente; Dirección General de Calidad y Evaluación Ambiental: Madrid, España, 2007; p. 125. Available online: http://www.cedexmateriales.es/ (accessed on 15 April 2020).

35. Rolón Aguilar, J.; Blandón Gonzalez, B.; Huete Fuertes, R. Posibilidades de aplicación del material granular obtenido a partir de residuos de construcción y demolición de hormigón y cerámica. *Res. J.* **2005**, *15*, 32–36. Available online: https://dialnet.unirioja.es/servlet/articulo?codigo=1262253 (accessed on 3 March 2020).

36. Council of the European Communities. Directive 91/156/ EEC of 18 March 1991 amending Directive 75/442/EEC on waste. *Off. J. Eur. Commun.* **1991**. No L 78/32. Available online: https://op.europa.eu/en/publication-detail/-/publication/b2f0989d-59ec-426b-9ace-ef16e8494fc4/ (accessed on 15 May 2020).

Article

Towards a Circular Economy for the City of Seville: The Method for Developing a Guide for a More Sustainable Architecture and Urbanism (GAUS)

Milagrosa Borrallo-Jiménez [1,*], Maria LopezdeAsiain [2], Rafael Herrera-Limones [1] and María Lumbreras Arcos [1]

[1]   Instituto Universitario de Arquitectura y Ciencias de la Construcción, Escuela Técnica Superior de Arquitectura, Universidad de Sevilla, Avd. Reina Mercedes 2, 41012 Seville, Spain; herrera@us.es (R.H.-L.); maria.lumbreras.arcos@gmail.com (M.L.A.)

[2]   Departamento de Arte, Ciudad y Territorio, Escuela de Arquitectura, Universidad de Las Palmas de Gran Canaria, 35017 Las Palmas, Spain; mlasiain@ulpgc.es

*   Correspondence: borrallo@us.es; Tel.: +34-954556591

Received: 27 July 2020; Accepted: 4 September 2020; Published: 9 September 2020

Abstract: The article outlines the creation of a method for the development of tools to incorporate sustainability criteria in the field of architectural design. The aim of the research is to provide society with scientific knowledge related to sustainability, evaluating the environmental impact of their actions within the building sector, in a simple and direct manner through specific and contextualised tools. A specific tool is experimentally developed for the context of Seville, called Guide for a more Sustainable Architecture and Urbanism (GAUS), in its first document, GAUS-D1. Based on national and international documentary references, the method principles are defined, and an approach is adopted that prioritises communicative actions with the aim of reaching citizens, professionals, and researchers in the building sector. The specific experiment is developed with GAUS-D1, and an initial evaluation is made of its suitability and the validity of the proposed method. The approach followed ensures that the experience of developing this type of tool is internationally transferable to any other place. Further statistical verification of the use of the tool (which implies a defined testing strategy) is necessary so that the method can be consolidated as a national and international reference.

Keywords: sustainability; Seville; guide; circular economy; decarbonisation; construction

---

1. Introduction

Nowadays, a large volume of documentary reference is available on sustainability and architecture; many researchers are currently developing multiple aspects within this framework that should be studied in depth and, academically, the discourse is clear, concise and increasingly rigorous. However, the transmission of this approach to professionals who have to implement it and/or to society is still scarce [1]. Multiple current tools and directives for the assessment and certification of architecture, from an energy efficiency point of view, as well as environmental and sustainability ones [2–7], are studied and applied in specialised areas such as universities, but their professional practice and social perception, above all, are very scarce. Environmental and sustainability issues related to architecture and urban planning are still a pending challenge [8] whose doctrinal corpus has been considerably strengthened and rigorously defined, but whose professional practice and social perception are still deficient. Many authors [1,9,10] have been demanding for decades the need to incorporate this knowledge into the teaching of architecture [11], and with it, into the profession itself, but it is not clear that this has been put into practice. It may be due, partly, to the difficult relationship between the conceptualisation of sustainability and the physical architecture practice. In this sense, the affirmation of the academic Albert Cuchí [12], according to which, in the specific field of architecture, sustainability inexorably implies the completion of material cycles in construction [13], could be enlightening.

This approach is directly related to the field of energy efficiency and decarbonisation, which is currently under development and in the process of being transferred from the scientific to the regulatory level and even to professional practice. However, it goes further, as it includes the rationalisation in the use of materials and construction systems according to their complete life cycle. That is to say, considering the energy consumption and the environmental impacts involved in their use (from their extraction process, production, transfer, and installation, as well as their useful life and reuse or recycling as new raw materials), closing the cycle and encouraging a circular economy. On this basis, it is sufficiently clear that methodological access from the environmental point of view, on the one hand, and from the economic, on the other, are essential. This means efficiency from the energy point of view and, at the same time, effectiveness in construction.

As the scale expands, it becomes essential to address architectural design, both on a building and urban level. In this case, while maintaining the implicit requirement of sustainability through the effectiveness and efficiency in the management of materials' life cycle, it is also required in the management of information (understood as one of the three aspects of sustainability, when it is approached from a management perspective: Matter, energy, and information [14]).

On the other hand, the exchange of information reveals the degree of physical and psychological comfort achieved by a certain architecture in relation to its occupants. Therefore, from this approach, it is also necessary to address the effectiveness and efficiency of information management. Thus, the bioclimatic knowledge provides it from the technical environmental area, and the involvement of the user or citizen (in the building or urban level) provides it from the social area. If the scale is again extended, the dimension of urban planning and territorial management appears, but, in addition, other non-architectural disciplinary aspects (from disciplines such as politics, geography, history, sociology, and social work) become more relevant. For this reason, in the context of this research, a contribution is made to the field of urban management, policy, and urban governance by introducing, as far as possible, the connection between resource management in architecture [15] and the concept of circular economy [16] applied at the city scale. This relationship is not only beneficial in terms of environmental sustainability, but also brings economic advantages in terms of revaluation of resources, along with new business opportunities [17] and job creation.

Seville is a city committed to sustainable development and climate change, as demonstrated by its adherence to all actions related to Agenda 21, Agenda 2030, and the achievement of the Sustainable Development Goals (SDG), which makes it a suitable case study for the experiment in question. In order to fulfil this commitment, the interrelationship between research, business, and local government [18] is key, as well as the need to disseminate, educate, and train society in this field, generating a framework of "science-informed decision environment" [17]. Therefore, Academic institutions must be put at the service of institutions to facilitate and guarantee the transfer of knowledge to society in areas of major concern at present, such as circular economy and low-carbon economy [19], among others. Scientific knowledge and experience are particularly important for the development of European environmental policies [20], which, at the same time, have an impact on the social level and transcend it. For this purpose, the development of tools such as the Guide for a more Sustainable Architecture and Urbanism (GAUS) are of great importance, since they can represent a key reference in processes of citizen governance that involve the transmission of knowledge in an adapted form, correctly translated for society. Therefore, the aim of this research is to define a method that will allow the development of reference and transferable tools in terms of structure, approach, systematisation of information, content type, and communication capacity, for the introduction of sustainability requirements in architecture design at local level.

The transmission of specialised knowledge to society requires a significant, complex, and meticulous effort to simplify and clarify the means of dissemination so that it is useful and comprehensible. Based on significant references of working with society in the transfer of architectural and urban concepts [21], not only the applied methodologies are outlined [22], but also the communication tools are generated for this purpose and the need to connect and involve society in general, as well as the productive sector and public administration [18].

Therefore, this research focuses on the necessary transmission of knowledge to society, starting from the broad existing doctrinal framework, with the aim of summarising and producing an approach to the knowledge required, addressing architectural aspects from sustainability on a local scale, taking Seville as an experimental reference. Thus, the aim is to bring technical knowledge closer to citizens, emphasising, on many occasions, the tacit knowledge that they possess in the management of domestic and urban architecture, transforming one into the other [23].

In this sense, the GAUS guide, in its first document, GAUS-D1, used as an experiment, represents a mediating tool in this process, and its development involves extensive research in both communicative and technical terms. For this purpose, it undertakes an extensive analysis of the state of the art in terms of methodology and then proposes a sustainability approach at the scale of materials and construction systems (with a significant impact on circular economy), proposing other different scales of architecture such as: Bioclimatic design, public space design, and urban design for further experiments.

The doctrinal corpus on the subject under study is very extensive, so an exhaustive reference to it would not be meaningful in this article. Subsequently, documents of all types (articles, theses, research, online platforms, and publications in general) that have been of special interest to this research in terms of content, review of literature itself, documentary organization, and/or transference capacity, given their graphic or textual expressive characteristics, are referenced.

The potential references analysed for the design and development of the guide are local, as well as national and international. Among the local relevant documents analysed, the following may be mentioned: The report 'Estimation of the Ecological Footprint of Andalusia and its application to the Seville Urban Area' [24], the 'White Paper—guide for the application of energy efficiency criteria in urban planning and local public construction, province of Seville' [25], or the guides published by SODEAN, the forerunner of the Andalusian Energy Agency, on the integration of solar collectors in buildings [26] or in the urban environment [27]. It is also necessary to refer to the research work developed by the project 'System of Sustainability Indicators in Residential Buildings for Andalusia' [28], a precursor of this research in terms of contents, and a fundamental technical reference complemented with international reference proposals [29–32], as well as the recent guide published earlier this year on life cycle analysis (Renovation LCA), for renovation [13]. These works, despite being of great interest at a technical level, are nevertheless extremely complex for the general public, which is why, from research, the type of communication this proposal was intended to address has to be redirected.

In addition to the experiences, tools [33], and specific scientific–technical documentation from Seville, other national or international cases have been considered as references [34–36], which could complement what has been locally referred in terms of strategies, tools [37], and methodologies. There are several systems of indicators [38–49], which both at building and urban level have been of significant interest as references. They have been objectively analysed to establish the priority issues, scales, and fundamental references. Some international guides [50] have also served as a reference in terms of content, but above all as a reference for structuring the transmission of knowledge. Additionally, several educational strategies that allow working on sustainability issues in architecture have been studied and experienced by many European universities [1,51]. Finally, other related publications are also considered: Guides on sustainability in other cities and Autonomous Communities [52], doctoral theses [53], national and international conference proceedings on sustainability, and even reports on the adaptation of future European regulations [54] to the situation in Spain.

Finally, it must be admitted that sustainability in absolute terms for buildings, and even on an urban scale, is a utopia: It is impossible to meet the specific requirement of closing all material cycles at the local level. For this reason, this proposal is framed within the objective of improving architecture in terms of sustainability and seeking to move towards promoting a greater circular economy for any city, with Seville initially as a case study. The method proposed for the development of transference tools such as the present GAUS tool in general, and GAUS-D1 tool in particular, also aims to make it a reference in terms of structure, approach, systematisation of information, content type, and communication capacity. Thus, the method constitutes an opportunity of development of this type of tool that can be transferred—as already indicated—to other local situations at both national and international level.

The full document that will constitute the GAUS guide will cover all scales of action, from the urban environment to the construction area, including architectural design. However, this article only describes the experiment of the development of "Document 1: Guide for a more sustainable construction in Seville". Starting from this document and checking the results obtained in terms of use and management, the subsequent documents will be developed, and the method will be finally statistically confirmed.

2. Materials and Methods

The specific objective of the research presented in this article is to define a method for the development of tools that are capable of correctly and effectively transmitting scientific knowledge to society in technical, appropriate, and practical terms for its understanding and application in the field of sustainability in architecture. To this end, an experimental case study is developed, GAUS-D1, in the local area of Seville, which, after verification, will consolidate the method for its subsequent use in future complementary documents, both for the experimental case study under consideration and for other similar scenarios that can be extrapolated nationally and/or internationally.

The developed GAUS-D1 tool is therefore part of a set of tools that comprises the potential practical application of scientific knowledge in the field of sustainability to architecture at all scales. This set of tools is defined as GAUS and represents the framework of the experiment.

2.1. Research Methodology

The methodology of the conducted research includes the following phases:

- Definition of the theoretical conceptual framework linking the concept of sustainability in architecture with its technical implications in terms of circular economy and its practical application in terms of scientific transfer to the professional field.
- Study of the scientific literature related to this framework as well as previous national and international experiences in transferring such information to society.
- Study and definition of agents involved in professional technical processes as generators of knowledge and information as well as potential users of the tool to be developed following the method defined by this research.
- Establishment of possible connections between agents, information, scientific knowledge, and knowledge transfer tools.
- Determine what type of tool is more appropriate for the transfer of knowledge on sustainability in architecture to society, and what features should define it.
- Proposal of method for the development of such a tool.
- Experimental development of the tool for a local, specific, and defined case study within the framework of action.
- Identification of the features that define this tool as useful, concise, and rigorous in scientific terms.
- Verification of results from the use of the tool.
- Review of the proposed method based on the results of using the experimental tool and consolidation of the outcomes.

2.2. Justification of the Case Study for the Experiment

The city of Seville, as a region that is highly committed to sustainable development, climate change, and circular economy, demands such tools to improve governance, bringing together citizens, government institutions, and private companies, using the academic and scientific network as a mediator between these areas. The case study for the experiment, constituted by GAUS-D1, is developed within the framework of the "Local Regulation for Energy Management, Climate Change and Sustainability" [55], issued in 2012, and reinforced in 2018 by the "Urban Agenda for Andalusia 2030" [56]. This first document is based on the need to achieve the commitments adopted for the city of Seville in three specific areas: Sustainable development (as a result of signing the Aalborg Charter in 1994 and joining the European Sustainable Cities and Towns Campaign [57]), climate emergency (having signed the Covenant of Mayors against Climate Change [58]) and, finally, continuous improvement in energy management at local level, a task that had already been carried out since 1997, but which requires adaptation of standards and regulations [56]. In addition, it is worth noting that Seville is closely committed to Agenda 21, the global programme of action in all areas related to sustainable development of the planet, approved at the United Nations Conference on Environment and Development, held in Rio de Janeiro in June 1992. The Agenda calls for changes in economic development activities, based on a new understanding of the impact of human actions on the environment. Nowadays, these commitments acquire even greater relevance following the appearance of Agenda 2030 (New Urban Agenda of the International Habitat Conference III [59]), approved in September 2015, together with the SDG [60], a roadmap to fight poverty and inequality [61] with a centred approach on people, the planet, prosperity, peace, and partnership [62]. For all this, the city must begin to develop tools for its management and urban transformation.

Similarly, the Local Administration of Seville has signed the "Declaration of Principles on Circular Economy" (Paris 2015) [63], which stresses the importance of Local Governments for its implementation. According to this commitment, the involvement of the scientific community in the awareness and transfer of programmes that promote this cyclical economy and the development of local strategies is of considerable importance. Therefore, this guide has been designed and conceived as a fundamental tool to achieve, among others, the SDG and the principles of circular economy in the local area of Seville.

3. Results

The conceptual theoretical framework has been developed as described in the introduction, connecting the concept of sustainability in its broadest conception with its implications in the field of architecture and the promotion of a circular economy for our cities. Similarly, the suitability of certain tools for the transmission of scientific knowledge to society has been studied, and those that represent a more valuable reference for this research are specified, although the documentary analysis has been much broader than what is specifically mentioned in this article.

A research of agents involved in the processes of construction and architectural and urban design has been undertaken, defining the connections between them and the contributions that they can make in terms of knowledge. The requirements in terms of knowledge transfer have been determined through consultation with managers and representatives of the different groups, including businesspersons, technicians and government officials, gathering and determining the practical focus of the required tool to be developed. Afterwards, the type of tool needed has been defined, which meets the expectations and requirements specified, named in the case study developed as GAUS. The results are explained in the following graph (Figure 1).

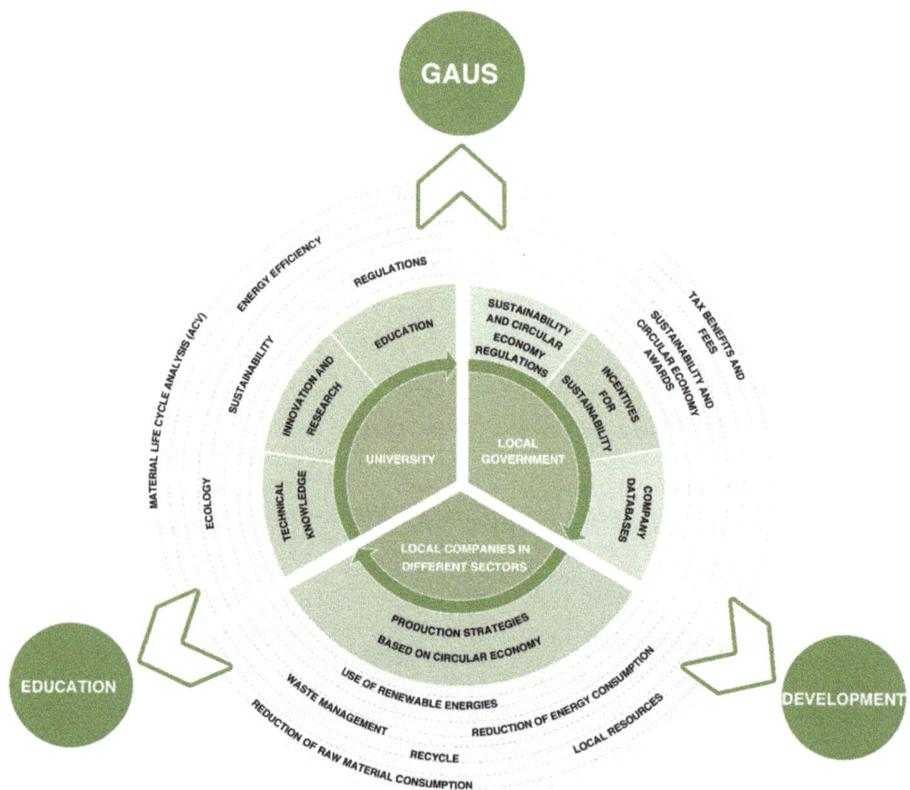

Figure 1. Diagram representing the relationship between agents, knowledge, and technical information in architectural processes. Defining Guide for a more Sustainable Architecture and Urbanism (GAUS) potential contribution in terms of circular economy and sustainability at a local level. Source: The authors.

3.1. Proposed Method

- The proposed method for the development of transference tools is based on the following actions:
- Definition of the approach, scope, reach and objectives of the tool based on the criteria and needs defined in collaboration and agreement with the different local agents involved in the process of consultation and development of the tool. This phase is key and decisive since it is essential to properly define the scope of the tool in order to guarantee its conciseness as well as its scientific rigor.
- Exhaustive documentary research based on the search for tools, methodologies, and strategies in the field of sustainability in architecture as well as strategies and tools for dissemination and/or transfer to society in different contexts: International, national, and local, applicable to the case study. This will enable well-founded decisions to be made, both in terms of form and substance, for the tool being developed, meaning the suitable format it should have, as well as the simplified but rigorous content required.
- Development of local specific climate characterisation determining the potential bioclimatic strategies to be implemented. Understanding local climate conditions is essential to provide adequate passive solutions for comfort, as well as energy efficiency improvements within the framework of sustainability. This allows us to decide the type of actions in the field of

architectural design, construction, and management to be included in the tool, based on the potential development of bioclimatic strategies in specific passive design systems.

- Historical, cultural, and anthropological analysis, from the field of architecture and limited to the scope of work previously defined according to this method, to determine which architectural cultural references should be considered. In order to do this, a sample of these references is compiled and prioritised according to their traditional use over time, excluding solutions that are unlikely or unusual due to the context, thus simplifying the number of specific proposals to be included in the tool.

- Once the context has been fully defined, the specific architectural references would be identified and their properties studied in terms of sustainability, taking into account their bioclimatic performance, their potential contribution to decarbonisation through their use and the potential improvement of the local circular economy that they could represent. This phase requires an important in-depth study of the architectural references involved and the scale considered. In the experimental case being developed, these would be the traditional materials and construction methods of Seville. Regarding architectural design, it would be required to study the bioclimatic performance of the most common building typologies for Seville and their defining characteristics, as well as the management in their design and construction that would involve the use of local resources and therefore, an improvement in terms of circular economy.

- Selection of cases that are justified and documented as proposals to be promoted by the tool, clearly defining the qualities that they should have, and that guarantee both their best bioclimatic performance and their contribution to the decarbonisation and/or improvement of the circular economy for the city.

- Definition of the most appropriate features and graphic format of the tool for the transmission of information and its dissemination. Finally, all this information must be transferred to the dissemination tool and its content with a well-studied graphic design that is capable of reaching users, in this case both citizens and technicians.

- Development and publication of the tool proposing the appropriate social and technical dissemination of the document with the aim of testing results, developing a specific planning for this purpose that enables it to be addressed to the appropriate and required agents for its verification.

- Adjustment and improvement of the tool after its final verification, and development of a management plan for its update and maintenance that consolidates it as a local reference tool in the long term.

This method has been experimentally developed using GAUS-D1 for Seville, and the results and features of the tool are described in detail in the next section.

The GAUS guide is a tool under development, created by different groups of professionals, each group responsible for the experiment of the method for diverse scales of action in Seville (D1–D5). We have initially applied the method to one of them, document D1, at the scale of building construction, and it will be progressively applied to every scale through the rest of documents once the first one has been completely tested. This article defines the method used for this first document and sets out the results obtained from the experiment.

3.2. Development of GAUS-D1 for Seville

Once the need for a specific tool has been determined for the case of Seville, thanks to the contribution of all parties involved, the approach, scope, and features of this tool are defined in order to meet the goal of being a concise, clear, precise, and objective tool for technicians and citizens when dealing with sustainable construction in our cities.

3.2.1. GAUS Scope

As previously mentioned, the full document that will constitute GAUS will cover a broader scope than the merely material and constructive one and, therefore, will extend to other areas and scales prior to the actual construction of buildings. However, the experiment developed uses this first document "Document 1: Guide for a more sustainable construction in Seville" (GAUS-D1) to experiment the method proposed from the research in order to control accurately the multiple parameters involved.

Therefore, the guide GAUS, once completed by replicating the method proposed by this research, will be comprised of the following documents:

- D1. Guide for a more sustainable construction in Seville (scale at materials and construction methods level): Analysis of the predominant construction techniques in Seville and a protocol for selecting and designing them from a sustainable approach.
- D2. Guide for a more sustainable building design in Seville (scale at place and building level): Protocol for the design of bioclimatic architecture, including water cycle and materials management.
- D3. Guide for a more sustainable urban design in Seville (scale at urban space level): Protocol for the design of urban spaces according to the context of Seville and ensuring their habitability.
- D4. Guide for a more sustainable urban planning in Seville (scale at urban planning level): Urban planning procedures design in terms of sustainability.
- D5. Guidelines for a more sustainable territorial planning of Seville and its surroundings (territorial scale): Guidelines for future territorial plans.

3.2.2. GAUS-D1 Objectives

The tool developed for the experiment presents the following objectives, which have been previously defined together with all the agents involved:

- Promote the culture of sustainable construction, which respects the environment and the ecosystems' energy cycles, applied at the local level, in the experimental case, Seville, according to its morphology, historical and cultural background, geographical location, and particular climatic parameters; offering a common long-term vision of Seville as a Sustainable City.
- Involve and engage governments, technicians (designers, architects, and builders), and citizens (energy consumers), contributing to the development of awareness and transfer strategies to society in order to "support the generation of more responsible users/consumers/citizens, who can make daily decisions focused on preserving resources throughout the production, consumption and waste process and who are better informed about their own consumption patterns", in the words of Jordi Segalás [64].
- Be a reference document for technicians, organisations, and citizens, with scientific and technical endorsement, and equivalent to similar actions under development in other European Union countries.
- Contribute to the development of architectural strategies that generate more efficient products and services and promote the reuse of existing ones, providing a scope of implementation for both new build and renovation projects, and contribute to the goal of reducing energy consumption in buildings, including embodied energy in materials, in order to achieve the highest possible decarbonisation.
- Become a recommendation protocol in all stages of the architectural project (design, choice of materials, and construction techniques), construction (waste), and subsequent maintenance of the building. The legislative development, in terms of requirements, should progressively lead to the implementation of these measures, and could even become a subsequent legislative regulation. It must be continuously reviewed in order to be consistent.
- Do not have a prescriptive or regulatory nature, although it does have an incentive value for citizens to be qualified and certified by the competent government agencies (in the case of the

local study, the Energy and Sustainability Agency of Seville [65]). These incentives, such as tax benefits on urban licenses or property-related taxes, are a real claim for their application.

- Promote the achievement of SDGs in the local area of Seville, specifically concerning the development of sustainability indicators in relation to circular economy based on those proposed by the European Union, with the capacity to transfer technical knowledge to society.

3.2.3. Documentary Research

An exhaustive study of both local and national or international tools potentially applicable has been conducted: From research documents and reports, decrees, manuals, guides, etc., to relevant projects and experiences. The selection procedure for the analysis of these documents has been based on the following criteria:

- Be a recognised, rigorous, and scientifically referenced document.
- Constitute an administrative document with a legislative character, whether it is mandatory or not.
- Respond to the previously mentioned sustainability aspects from an architectural and urban approach, either from an environmental, economic, or social point of view.

From all these local, national, and international documents analysed, the following contents and/or methodological references of interest for the design of GAUS-D1 are extracted as the main references finally used:

- Contents and local characterisation of architectural references from a cultural, climatic, constructive, and urban point of view [26,27,66].
- Referencing relevant applicable regulations [25,55,67].
- Specific contents related to the framework of sustainability in architecture and decarbonisation [1,13,25,29–32,35,36,39,50,53,67–69].
- Specific contents related to the framework of sustainability for territories, cities, and the circular economy [24,27,35,38,70,71].
- Definition of scales of action for decision making in each construction process [68].
- Methodological references of appropriate graphic representation and written expression for the transmission of knowledge to society [1,26,50].
- References regarding information structure and its summarising, expressive, and communicative capacity [50,52,68].

3.2.4. Climate Characterisation and Potential Bioclimatic Strategies

Understanding local climate conditions, particularly in Seville for this case, is essential to provide adequate passive solutions for comfort, as well as energy efficiency improvements and/or refurbishment within the framework of sustainability.

Passive architecture, defined as an architecture that adapts to surrounding climatic conditions, has existed since antiquity [72]. Based on the experience of traditional architecture, it aims to ensure hygrothermal comfort in buildings based on their own architectural configuration. It is also a close definition related to bioclimatic architecture, which takes into account the environment, health, and well-being of people [73].

Seville is located in southern Spain and Europe, at latitude 37.3881, belonging to the region of Andalusia, with an inland climate classified by Strahler (1951) as "Mediterranean climate", within the group of "Mid-latitude climates controlled by tropical and polar air masses". Moderate temperatures and a rainy regime with warm and dry summer periods, presenting significant seasonal variations, define this climate, between the parallels 30°–45° N and 30°–45° S. The annual fluctuation of temperatures is moderate [73]. The "Energy Saving Regulations" applicable in Spain [74] classifies it as climate category B4 (with B being the second in order of increasing severity up to E, corresponding to winter climate severity, and 4, the highest summer climate severity).

It is clear that for centuries, each region has developed unique local passive strategies for comfort and conditioning, which are difficult to standardise, as they depend on many climatic and functional factors [72]. More specifically, in the warm or tempered climate of southern Europe, the use of vegetation and water as shading and cooling strategies for outdoor spaces, seasonal window protection systems against solar gains, natural ventilation at night, or thermal mass as a construction strategy to regulate heat flow, are common local strategies (Figure 2).

Figure 2. Typical interior courtyard building in Seville. Reference for passive strategies to achieve thermal comfort in hot weather considered in GAUS-D1: Awnings, vegetation, and solar control on facades.

Consequently, all of them have been conveniently identified and considered when developing proposals for construction solutions or improvements.

3.2.5. Specification of Local Construction Techniques Related to Building Envelopes

"Local architecture is the architecture of a specific location, defined by the particular use of forms, construction methods and materials" [75], and it is a value in the field of sustainability. Therefore, it is important to identify local construction solutions, of a cultural nature [76], related to the scale of intervention.

In the construction field, technicians can decide which constructive systems to use in order to minimise environmental impact or maximise comfort conditions. However, in the renovation field, it is essential to identify the original state of the building in order to propose improvement solutions based on the existing conditions, with the introduction of new materials and systems. This identification, in the case of major construction works, is guaranteed through compliance with technical requirements imposed by regulations. Furthermore, in the case of minor construction works, without a legal requirement for a competent technician to design the project, the developer lacks objective technical information (not from commercial companies) to adopt effective measures in order to improve energy efficiency within the framework of sustainability. This means improvements such as use of materials with a low environmental and economic impact, produced from local industrial and material resources, and committed to a circular economy.

Some details on the study of the traditional and commonly used construction techniques in Seville are presented below, with the aim of clarifying the level of deepening of the subject undertaken by GAUS-D1.

Concerning the definition of local construction methods in Seville and, more specifically, regarding roof design, there are both flat roofs that can be walked on (typical flat roofs, which recover an elevated free space within a complex urban network in some areas such as the city centre), and sloping roofs with an Arabic tile finish, mixing tradition and innovation (Figure 3).

Figure 3. Seville rooftop landscape. Reference for the characterisation of roofs in terms of local construction methods (flat and sloping roofs with Arabic tiles). Source: The authors.

In terms of predominant façades in the city of Seville, there are several typologies, with one or more layers, and made up of diverse local materials, such as ceramic bricks, wall tiles (or rammed earth walls), stone, or lime plaster. Usually, they lack insulation in their original state, based mainly on thermal mass as a resource of thermal control. In most cases, façades work as supporting structures as well as envelopes (Figure 4).

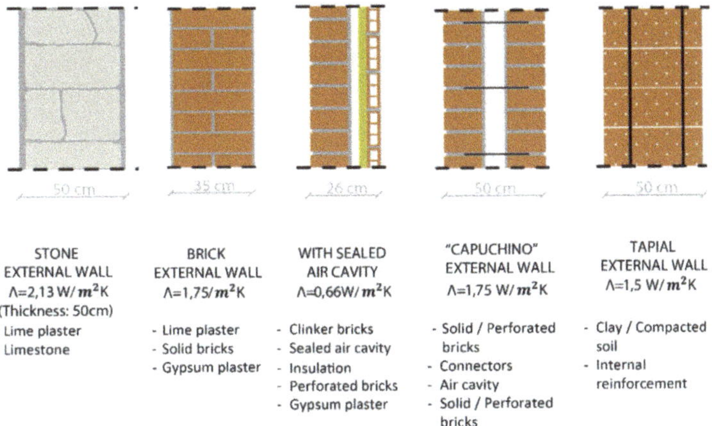

Figure 4. Traditional external wall build-ups in Seville. Example of façade characterisation in terms of local construction methods (GAUS-D1). Source: The authors.

GAUS-D1 includes the most common types of roofs (Figure 5) and façades in the local area of Seville, with all their layers and approximate thermal transmittance values, in order to propose solutions for improving energy efficiency, mainly by incorporating insulation, through new façade cladding, waterproofing, etc. Hence, initial construction solutions of the properties to be renovated are recognised by users and developers as their own, and their improvement proposals are also recognised as adjusted to a technical reality.

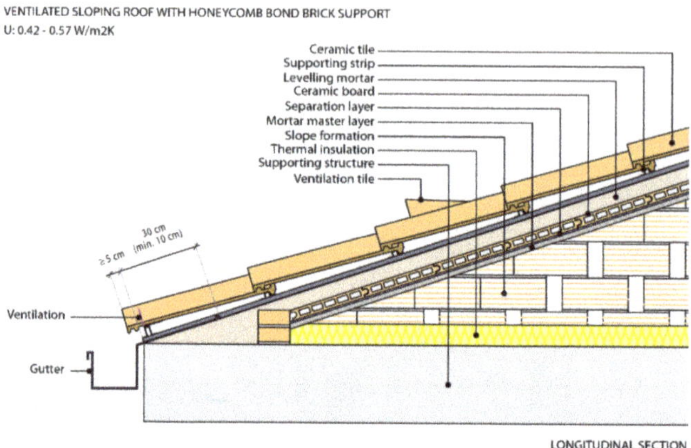

Figure 5. Traditional sloping roof build-up in Seville. Example of roof characterisation in terms of local construction methods (GAUS-D1). Source: The authors and www.hispalyt.es.

3.2.6. Proposal of Actions for Improvement

Given that the building envelope is a key element in the passive conditioning and energy saving of buildings, the proposed actions are mainly focused on façades, with special emphasis on windows, roofs, natural lighting, ventilation, and air quality. It also proposes solutions for water and waste management and the integration of renewable energy systems into architecture (e.g., thermal and photovoltaic solar collectors).

It is assumed that buildings that require renovation are of a certain age, especially in historical cities such as Seville, where useful life of buildings ranges from 50 to 100 years. Some of them were built according to proper construction standards, but lacking the protection of any recognised legal regulation. As a result of this, and the lack of maintenance, the state of conservation is often deficient, with problems of water filtration, air, humidity, deterioration, etc. In this sense, the choice of new materials and products to implement solutions requires specific advice in the field of sustainability. We are referring to ecological and low environmental impact values through Environmental Product Declarations (EPD) in all phases of their life cycle, fundamentally, which can be obtained from prestigious databases such as those of the International EPD® System [33] or the Institut Bauen und Umwelt (IBU) [77], among others.

The most sustainable construction solutions are offered based on quantifiable data obtained from specific bibliographical sources. This facilitates decisions on the choice of a particular paint, for example, justifying in this case the advantages of a mineral or natural product over a synthetic one, for its application on façades. Another example can be related to the advantages of certain insulation materials from renewable sources (cork, hemp, or cellulose), which may be less commercially advertised compared to others of a polymeric nature or based on foams, which are responsible for damaging the ozone layer and global warming. It also offers access to local business databases committed to circular economy that may provide some of these resources.

Therefore, GAUS-D1 tool offers a database of construction solutions and proposals for improvement adapted to local climate, materials, and construction methods. It makes it easily recognisable and useful for local users and building developers who are willing to contribute to environmental preservation and sustainability through their construction activities.

3.2.7. Graphic Design and Structure

GAUS-D1 is structured with "Consultation Cards" focused on both the non-specialised user and the trained technician. It is a tool that, at the same time, can be consulted by the owner of a property to be renovated, as well as by the technician who is going to implement such improvements. That is, it can be used both by the citizen or developer who wants to invest in a new property construction, and by the designer, not an expert in sustainability, who wants to implement it in their project, beyond the exclusive energy efficiency.

Each card responds to an action type, easily located through an exhaustive index. Thus, a clear division into fundamental parts of the building's envelope and into aspects of usability and energy, provides an easy location for the card that describes the intended improvement action.

This method ensures, on one hand, that scientific knowledge, research, and innovation benefit the entire scientific and professional community and, on the other, that citizens are nourished by its results and can use them to improve their quality of life [41]. This means improving the impact on the environment in a simple and direct way and causing a clear action of knowledge transfer.

It is particularly relevant to be concise in order to be accessible to the local citizen, who is committed to the environment, climate change, and sustainability, compared to the conception of an excessively technical document, which is only aimed at professionals in the sector.

The intention of the design chosen for this guide is to approach the user, both technical and non-technical, through explicit, clear, and structured graphics that are intuitive and easy to understand. It is intended that both the graphics and the language used are close, although precise and rigorous; the structure, clear and organised; and the design, chromatic, attractive, and amusing.

This double user-technician approach is directly reflected in the structure and design of each card, where a first more visual part, with adequately defined initial concepts and the fundamental support of the use of representative iconography, is complemented by a second part of a more technical and specific language where the given prescriptions are completed. The guide has a double qualitative and formative reading of concepts related to sustainability. In addition, the local approach allows for specific solutions for particular strategies determined by the climate.

The guide provides referenced data and results of Life Cycle Assessment (LCA) from Environmental Product Declarations (EPD) within the more technical part, which enable professionals in the construction sector as well as citizens and users committed to sustainability, to make decisions related to environmental impact of products. It also incorporates a database of local companies involved in the circular economy and related to products and materials recommended in the construction solutions provided.

Each card is identified by an acronym and a colour corresponding to the block of the table of contents to which it belongs. A simple 3D scheme allows us to locate, within the building, the specified construction element/system; creating a visual explanation that helps us to understand the action that is going to be described. This 3D model is developed using the BIM (Building Information Modelling) methodology, applied to relevant existing examples of sustainable architecture in Seville, functioning as a laboratory or a virtual model of the aspects studied and developed for this guide.

The following approach to the content of each Action Card is developed using simple yet meaningful graphics that highlight the impact on sustainability, in terms of energy, materials, health, or pollution (Figure 6).

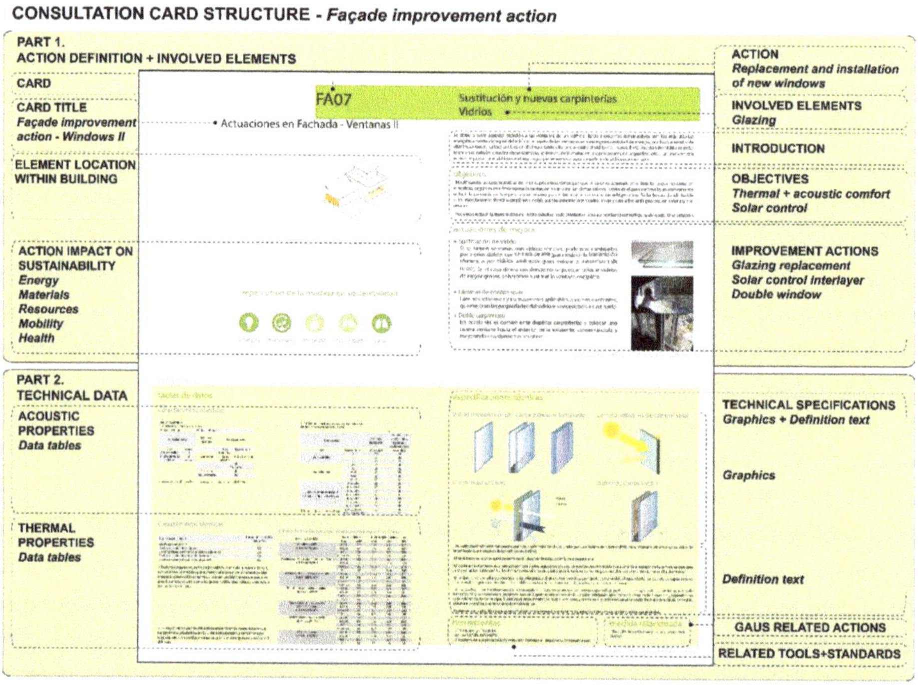

CONSULTATION CARD STRUCTURE - *Façade improvement action*

PART 1.
ACTION DEFINITION + INVOLVED ELEMENTS

CARD

CARD TITLE
Façade improvement action - Windows II

ELEMENT LOCATION WITHIN BUILDING

ACTION IMPACT ON SUSTAINABILITY
Energy
Materials
Resources
Mobility
Health

PART 2.
TECHNICAL DATA

ACOUSTIC PROPERTIES
Data tables

THERMAL PROPERTIES
Data tables

ACTION
Replacement and installation of new windows

INVOLVED ELEMENTS
Glazing

INTRODUCTION

OBJECTIVES
Thermal + acoustic comfort
Solar control

IMPROVEMENT ACTIONS
Glazing replacement
Solar control interlayer
Double window

TECHNICAL SPECIFICATIONS
Graphics + Definition text

Graphics

Definition text

GAUS RELATED ACTIONS

RELATED TOOLS+STANDARDS

FA07 — Sustitución y nuevas carpinterías · Vidrios — Actuaciones en Fachada - Ventanas II

Figure 6. Example of Action Card contained in GAUS-D1. Definition of its graphic design and content structure. In this case, the card explains potential actions to improve building façades. Source: The authors.

The main content of the first part of each Action Card is focused on its definition and objectives. These objectives directly propose a series of improvement actions that are briefly described together with photographs to easily identify the element or system on which a construction, design, or usability proposal is presented.

3.2.8. Social and Technical Dissemination of the Tool

GAUS-D1 is conceived as an open database on sustainability (construction scale) and within the reach of society (tool). Although the design originally proposed is graphic in paper format, the need for continuous updating of databases, regulations, and other sources raises a complementary required development in web or digital format.

4. Discussion

The partial conclusions reached following the development of GAUS-D1, and in accordance with the objectives set for it, are as follows:

- The need for this type of document is currently unquestionable in Spain. The multiplicity of tools compiled in continuous development by numerous professional and educational entities of international scope corroborates this and also shows the need for transference to society.
- It is a relevant contribution that promotes the perception and commitment of Seville as a sustainable city.
- It becomes a reference document and tool for technicians, organisations, and citizens, with scientific and technical endorsement, comparable to similar initiatives being developed in other European Union countries.

- Given the specific measures it provides to reduce environmental impacts and energy consumption, including the energy embedded in materials, it represents an important contribution to the goal of near-zero-energy buildings to achieve the highest possible decarbonisation.
- It is a practical and concise guide that allows both organisations and users to evaluate and quantify renovation and construction activities within the framework of sustainability, in order to apply incentives and subsidies.
- It represents an important progress in terms of transfer to society from the field of architecture to sustainable development and the circular economy. The methodology and development criteria allow the tool to be transferred to another local context of similar scope.
- GAUS-D1 can be considered a basic and solid tool that brings together the technical knowledge needed to achieve the SDG and the transmission of that knowledge to society, getting involved in the daily actions defined by lifestyles.

4.1. Defining Successful GAUS-D1 Features

A subsequent analysis of GAUS-D1 allows us to identify the features that make it a useful, direct, and concise tool, as well as rigorous in scientific terms, which is the objective of this research. These characteristics have been defined during the methodological process of development of the proposed method and will allow us to corroborate its initial validity. They can be summarised as follows:

- Definition and contextualisation of a large part of the parameters due to the selection of a specific local case study, which makes it possible to limit the range of possible construction situations and avoid unjustified simplifications that would make the tool less rigorous.
- Detailed definition of the local context under consideration and all its physical (climate, orography, topography, urban form, etc.) and cultural (traditional building systems, traditional and/or common materials, historical and patrimonial chromaticism, etc.) conditions from the architectural point of view, in order to develop an accurate local analysis.
- Adaptation to the requirements in form and content, either practical, educational, or technical, defined by the local government, social, or business agents according to the analysis of the parties involved.
- Adaptation to the European developments regarding qualification of the construction industry in terms of sustainability, gathering updated regulations, recommendations, and/or methodological proposals, being a reference recommended as an action protocol by the local authorities. Its monitoring can be used to objectively justify the achievement of incentives.
- Contribution and promotion of building improvement strategies that use more efficient methods from an energy and material point of view.
- Conceived as a tool that allows the approach and practical translation of some of the SDG to specific construction design decisions that improve the environmental impact of cities, their habitability, and their local management; contributing to the improvement, at the same time, of the circular economy.
- The exhaustive documentary research on tools, methodologies, and strategies in the field of sustainability and their transfer to society in different contexts provides a database and important background for other studies to be developed along the same path.

All these features, which achieve the objectives initially set by the tool itself, demonstrate the initial validity of the method for the development of practical, precise, and rigorous tools that allow the promotion and incorporation of sustainability principles in architectural construction design for specific local contexts.

4.2. Results Verification and Method Review

The initial theoretical–conceptual verification performed is not sufficient to establish the validity of the proposed method. It is therefore necessary to conduct practical checks on the results of the actual use of the tool by its potential users, both citizens in general, and government technicians and/or professionals in the construction sector. Currently, a campaign is being undertaken to disseminate the tool, from its presentation at congresses and technical meetings to the promotion of its use among professional technical associations and local public authorities. However, relevant statistical data on its use and results will not be available for several years so, despite its initial validity, it will be necessary to perform a subsequent study of its practical relevance in terms of use and results obtained. The terms and indicators of this future study are currently being developed in collaboration with the agents and groups involved.

Once the proposed future study has been completed, it will be required to review the method in depth in order to determine potential changes and adjustments that will guarantee the development of tools that are even better suited to local needs regarding the incorporation of sustainability aspects in architectural construction. However, the method, theoretically proven, although not statistically, can be used for the development of parallel documents related to different architectural scales. This would allow a greater possibility of later statistical verification of the method and thus, its potential transfer to other local contexts that require it for the development of their own specific tools.

The current need for this type of document is unquestionable. The variety of certification and regulation tools [2–6,15,16,29–36,50,54,68], in continuous development by many professional and educational entities corroborates this, and also highlights the need for transfer to society [18,19] in terms of improving the empowerment of citizens and ultimately local resilience and global social sustainability.

5. Conclusions

The presented tool-guide, developed as an experiment by this research, is an important advance in this sense, since it addresses all the required conceptual and technical issues, and it adapts to the user profile and constitutes a major step in terms of knowledge transfer from the field of architecture and circular economy to society.

On the other hand, and here perhaps lies its relevance, it constitutes a validation guarantee of the method proposed by this research and its objective, promoting the use of the proposed method in other similar studies, both for the Seville case study and for other local contexts, while its rigorous statistical validation is being developed.

In this sense, the method defined and developed by this research is useful for the development of reference and transferable tools in terms of structure, approach, systematisation of information, content type, and communication capacity; for the introduction of sustainability requirements in architecture design at local level. Moreover, this method is valid for its extrapolation to other situations, although it can be improved and adjusted, as previously indicated, after a subsequent detailed study that is desirable but not currently scientifically feasible.

The fundamental contribution of the proposed method implies its capacity to develop precise tools useful for citizens and/or technicians that are at the same time scientifically rigorous for the incorporation of sustainability aspects in architectural design, becoming a tool itself for the development and transfer of scientific knowledge to citizens, in a transparent, simple, direct, and rigorous manner.

This contribution is key to the specific development of urban improvement actions based on architectural sustainability principles that result in the improvement of the circular economy in our cities and their resilience.

Author Contributions: All authors have contributed equally to the investigation and final manuscript. All authors have read and agreed to the published version of the manuscript.

Funding: This research received no external funding.

Acknowledgments: Our gratitude to Seville City Council for promoting this research by signing the Framework Agreement for Collaboration between the University and the Seville City Council's Energy and Sustainability Agency.

Conflicts of Interest: The authors declare no conflict of interest.

References

1. EDUCATE Project Partners. *Education for Sustainable Environmental Design. The EDUCATE Project. Summary of Results*; EDUCATE Press University of Nottingham: Nottingham, UK, 2012.
2. Green Building Council Spain LEED v4. Available online: http://spaingbc.org/web/index.php (accessed on 10 July 2020).
3. Building Research Establishment Environmental Assessment Methodology BREEM Spain. Available online: http://www.breeam.es/ (accessed on 10 July 2020).
4. Passivhaus Institut Passivhaus. Available online: https://passivehouse.com/ (accessed on 10 July 2020).
5. Living Building Challenge Certification | Living-Future.org. Available online: https://living-future.org/lbc-3_1/certification/ (accessed on 5 June 2020).
6. European Commission Nearly Zero-Energy Buildings | Energy. Available online: https://ec.europa.eu/energy/topics/energy-efficiency/energy-efficient-buildings/nearly-zero-energy-buildings_en#information-from-individual-countries (accessed on 10 July 2020).
7. Agencia de Ecología Urbana de Barcelona. *Certificación del Urbanismo Ecosistémico*; Rueda, S., Cárdenas, F., Eds.; Ministerio de Fomento, Gobierno de España: Barcelona, Spain, 2015.
8. Higueras García, E. *El reto de la Ciudad Habitable y Sostenible*; Distribución y Asesoramiento de Publicaciones Jurídicas (DAPP): Pamplona, Spain, 2009; ISBN 978-84-92507-19-1.
9. López de Asiain Alberich, M.; Serra Florensa, R.; Coch Roura, H. Reflections on the Meaning of Environmental Architecture in Teaching. In Proceedings of the 21th Conference on Passive and Low Energy Architecture, Eindhoven, The Netherlands, 19–21 September 2004; de Wit, M.H., Ed.; Technische Universiteit Eindhoven: Eindhoven, The Netherlands, 2004; pp. 163–168.
10. López de Asiain, J. La habitabilidad de la arquitectura. El caso de la vivienda. *Dearq* **2010**, *6*, 100–107. [CrossRef]
11. Janssens, B.; Knapen, E.; Winkels, P.; Verbeeck, G. Outcomes of a Student Research Project on Circular Building Systems-Focus on the Educational Aspect. *IOP Conf. Ser. Earth Environ. Sci.* **2019**, *323*. [CrossRef]
12. Cuchí i Burgos, A. *Arquitectura i Sostenibilitat*; Edicions UPC: Barcelona, Spain, 2006; ISBN 84-8301-839-X.
13. Alliance HQE-GBC. *To Assess the Environmental Performance of Renovated Buildings and Compare Their Levels With E + C*; Alliance HQE-GBC: París, France, 2020.
14. López de Asiain Alberich, M. *"La formación Medioambiental del Arquitecto". Hacia un Programa de Docencia basado en la Arquitectura y el Medioambiente*; Universidad Politécnica de Cataluña: Barcelona, Spain, 2005.
15. ARUP. *The Circular Economy in the Built Environment*; ARUP: San Francisco, CA, USA, 2016.
16. United Kingdom Green Building Challenge. *Circular Economy Guidance for Construction Clients. How to Practically Apply Circular Economy*; UK GBC: London, UK, 2019.
17. Remøy, H.; Wandl, A.; Ceric, D.; van Timmeren, A. Facilitating circular economy in urban planning. *Urban Plan.* **2019**, *4*, 1–4. [CrossRef]
18. Masseck, T. Living Labs in Architecture as Innovation Areas within Higher Education Institutions. *Energy Procedia* **2017**, *115*, 383–389. [CrossRef]
19. Losasso, M. Ricerca, progetto architettonico e trasferimento delle conoscenze. *Techne* **2014**, *8*, 8–12.
20. Carmichael, L.; Lambert, C. Governance, knowledge and sustainability: The implementation of EU directives on air quality in Southampton. *Local Environ.* **2011**, *16*, 181–191. [CrossRef]
21. Kuchenbuch, D. Architecture and Urban Planning as Social Engineering: Selective Transfers between Germany and Sweden in the 1930s and 1940s. *J. Contemp. Hist.* **2016**, *51*, 22–39. [CrossRef]
22. Garzón, L.E. Transferir tecnología vs. transferir conocimientos en la arquitectura y construcción con tierra. *Rev. Nodo* **2017**, *12*, 62–72.
23. Evers, H.D.; Gerke, S.; Menkhoff, T. Knowledge clusters and knowledge hubs: Designing epistemic landscapes for development. *J. Knowl. Manag.* **2010**, *14*, 678–689. [CrossRef]

24. Calvo Salazar, M.; Sancho Royo, F. *Estimación de la Huella Ecológica en Andalucía y Aplicación a la Aglomeración Urbana de Sevilla*; Consejería de Obras Públicas y Transportes. Junta de Andalucía: Sevilla, Spain, 2001; ISBN 84-8095-270-9.
25. Herrera-Limones, R.; Parra Boyero, C.; Rodríguez Rodríguez, V. *Libro Blanco-Guía Para la Aplicación de Criterios de Eficiencia Energética en la Planificación Urbanística y la Construcción Pública Municipal, Provincia de Sevilla*; Diputación Provincial de Sevilla, Ed.; Junta de Andalucía: Sevilla, Spain, 2011; ISBN 978-84-932804-9-9.
26. Sodean, S.A.; Sama, S.C. *Integración Arquitectónica de Instalaciones de Energía Solar Térmica*; Junta de Andalucía: Sevilla, Spain, 2000.
27. Sodean, S.A. *Integración de la Energía Solar en el Urbanismo*; Junta de Andalucía: Sevilla, Spain, 2001.
28. López de Asiain Alberich, M. *Sistema de Indicadores de Sostenibilidad en Arquitectura y Urbanismo Para Andalucía*; Junta de Andalucía: Sevilla, Spain, 2010.
29. EFFINERGIE Association Effinergie-Label for Energy Efficient Buildings. Available online: https://www.effinergie.org/web/ (accessed on 4 June 2020).
30. German Sustainable Building Council DGNB System–Sustainable and Green Building. Available online: https://www.dgnb-system.de/en/ (accessed on 4 June 2020).
31. Minergie Association The MINERGIE-Standard for Buildings. Available online: https://www.minergie.ch/ (accessed on 4 June 2020).
32. Agenzia per l'Energia Alto Adige-CasaClima CasaClima KlimaHaus. Available online: https://www.agenziacasaclima.it/it/home-1.html (accessed on 4 June 2020).
33. EPD International AB The International EPD®System. EPD International AB: Stockholm, Sweden. Available online: https://www.environdec.com/es/ (accessed on 5 June 2020).
34. Pelsmakers, S. *The Environmental Design Pocketbook*, 2nd ed.; RIBA Publishing: London, UK, 2012; ISBN 9781859465486.
35. Royal Institute of British Architects. *Sustainable Outcomes Guide*; Royal Institute of British Architects: London, UK, 2019.
36. German Federal Ministry of the Interior Building and Community. *Guideline for Sustainable Building. Future-proof Design, Construction and Operation of Buildings*; German Federal Ministry of the Interior Building and Community: Berlin, Germany, 2019.
37. Georgiadou, M.C. Future-Proofed Energy Design Approaches for Achieving Low-Energy Homes: Enhancing the Code for Sustainable Homes. *Buildings* **2014**, *4*, 488–519. [CrossRef]
38. López de Asiain Alberich, M. Indicadores de sustentabilidad en urbanismo. In *Diálogos Entre Ciudad, Medio Ambiente y Patrimonio*; Valladares Anguiano, R., Ed.; Universidad de Colima: Colima, Mexico, 2014; pp. 100–106. ISBN 978-607-8356-08-9.
39. López de Asiain Alberich, M. Autoevaluación Ambiental versus Certificación Ambiental. Nuevos procesos y herramientas educativas. In *Procesos de Certificación Ambiental de Edificios Sustentables*; Ávila Ramírez, D.C., Arias Orozco, S., Córdoba Canela, F., Eds.; Universidad de Guadalajara: Guadalajara, Mexico, 2012; pp. 65–83. ISBN 978 607 8193 46 2.
40. Vargas Yáñez, A. Sustainability Indicators of the Spanish Municipalities: A methodological proposal to view of its evolution between 2002–2015. In Proceedings of the II International and IV National Congress on Sustainable Construction and EcoEfficient Solutions, Seville, Spain, 25–27 May 2015; Universidad de Sevilla, Escuela Técnica Superior de Arquitectura: Seville, Spain, 2015; pp. 1250–1285, ISBN 978-84-617-3964-6.
41. Leva, G. *Indicadores de Calidad de Vida Urbana*; Politike: Buenos Aires, Argentina, 2005.
42. Hiremath, R.B.; Balachandra, P.; Kumar, B.; Bansode, S.S.; Murali, J. Indicator-based urban sustainability-A review. *Energy Sustain. Dev.* **2013**, *17*, 555–563. [CrossRef]
43. Braulio-Gonzalo, M.; Bovea, M.D.; Ruá, M.J. Sustainability on the urban scale: Proposal of a structure of indicators for the Spanish context. *Environ. Impact Assess. Rev.* **2015**, *53*, 16–30. [CrossRef]
44. Kim, J.T.; Todorovic, M.S. Towards sustainability index for healthy buildings-Via intrinsic thermodynamics, green accounting and harmony. *Energy Build.* **2013**, *62*, 627–637. [CrossRef]
45. Siew, R.Y.J.; Balatbat, M.C.A.; Carmichael, D.G. A review of building/infrastructure sustainability reporting tools (SRTs). *Smart Sustain. Built Environ.* **2013**, *2*, 106–139. [CrossRef]
46. Tupenaite, L.; Lill, I.; Geipele, I.; Naimaviciene, J. Ranking of sustainability indicators for assessment of the new housing development projects: Case of the Baltic States. *Resources* **2017**, *6*, 55. [CrossRef]

47. Huedo Dorda, P.; Lopez-Mesa, B.; Mulet, E. Analysis of sustainable building rating systems in relation to CEN/TC 350 standards. *Inf. Constr.* **2019**, *71*, 1–19. [CrossRef]

48. Kamali, M.; Hewage, K.; Milani, A.S. Life cycle sustainability performance assessment framework for residential modular buildings: Aggregated sustainability indices. *Build. Environ.* **2018**, *138*, 21–41. [CrossRef]

49. Cordero, A.S.; Melgar, S.G.; Márquez, J.M.A. Green building rating systems and the new framework level(s): A critical review of sustainability certification within Europe. *Energies* **2019**, *13*, 66. [CrossRef]

50. London Energy Transformation Initiative (LETI). *LETI Climate Emergency Design Guide. How New Buildings can Meet UK Climate Change*; LETI: London, UK, 2020.

51. Herrera-Limones, R.; Rey-Pérez, J.; Hernández-Valencia, M.; Roa-Fernández, J. Student competitions as a learning method with a sustainable focus in higher education: The University of Seville "Aura Projects" in the "Solar Decathlon 2019". *Sustainability* **2020**, *12*, 1634. [CrossRef]

52. Administración de la Comunidad Autónoma del País Vasco. *Departamento de Vivienda Obras Públicas y Transportes e IHOBE Sociedad Pública de Gestión Ambiental. Guía de Edificación y Rehabilitación Sostenible Para la Vivienda*; Servicio Central de Publicaciones del Gobierno Vasco: Donostia-San Sebastián, Spain, 2011; ISBN 9788445729069.

53. Mercader-Moyano, P. Cuantificación de los Recursos Consumidos y Emisiones de CO_2 Producidas en las Construcciones de Andalucía y sus Implicaciones en el Protocolo de Kioto. Ph.D. Thesis, Universidad de Sevilla, Sevilla, Spain, 2010.

54. Green Building Council España. *PAS-E _ Pasaporte del edificio. Instrumento Para la Rehabilitación Profunda por Pasos*; Green Building Council España: Madrid, Spain, 2020.

55. Agencia local de la energía de Sevilla. *Ordenanza Para la Gestión de la Energía, el Cambio Climático y la Sostenibilidad de Sevilla*; Agencia local de la energía de Sevilla: Sevilla, Spain, 2012.

56. Consejería de Medioambiente y Ordenación del Territorio. *Agenda Urbana de Andalucía 2030*; Junta de Andalucía: Sevilla, Spain, 2018.

57. *European Sustainable Cities Platform | Sustainable Cities Platform. City of Aalborg, Denmark; The Basque Country, and ICLEI—European Secretariat GmbH: Bilbao, Spain*. 2016. Available online: https://sustainablecities.eu/sustainable-cities-platform/ (accessed on 11 July 2020).

58. Ayuntamiento de Sevilla. *Pacto de los Alcaldes para el Clima y la Energía. Sevilla, 2016*; Ayuntamiento de Sevilla: Sevilla, Spain, 2016.

59. Secretaría de Habitat III. *Nueva Agenda Urbana*; Secretaría de Habitat III: Quito, Ecuador, 2017.

60. United Nations. *Transforming our World: The 2030 Agenda for Sustainable Development*; United Nations: New York, NY, USA, 2015.

61. United Nations Development Group for Latin America & Caribbean (UNSDG). *Desarrollo Sostenible en América Latina y el Caribe: Desafíos y Ejes de Política Pública*; UNSDG: Panamá City, Panama, 2018.

62. UNESCO. *Education for All 2000–2015: Achievements and Challenges*; UNESCO: Place de Fontenoy, France, 2015.

63. Municipios y Economía Circular. *Declaración de Sevilla: El Compromiso de las Ciudades por la Economía Circular*; Federación Española de Municipios y Provincias: Sevilla, Spain, 2017.

64. el Clima, R.C.P.; de Gavá, A.; de Barcelona, Á.M.; Provincias, F.E.D.Y.; Circular, S.D.A.L.V.Y. *Gavá Naturalment Por una Economía Circular y Competitiva. 3ª Jornada de Debate y Experiencias*; Ajuntament de Gavà: Barcelona, Spain, 2020; Volume 1, p. 32.

65. Ayuntamiento de Sevilla Agencia de la Energía y Para la Sostenibilidad—Planificación Estratégica. Available online: https://www.sevilla.org/servicios/planificacion-estrategica/agencia-energia-sostenibilidad (accessed on 13 July 2020).

66. Calvo Salazar, M.; Castro, J.M.; Ruíz Hernández, V.; del Moral Ituarte, L.; Díaz Quidiello, J.; González de Molina Navarro, M.; Guzmán, G.I.; Alonso Mielgo, A.M.; García Trujillo, R.; Cano Orellana, A.; et al. *Introducción a la Sostenibilidad en Andalucía*; Junta de Andalucía: Sevilla, Spain, 2005; ISBN 8496329542.

67. Ayuntamiento de Sevilla. *Ordenanza Para la Gestión Local de la Energía de Sevilla*; Ayuntamiento de Sevilla: Sevilla, Spain, 2016.

68. Architecture 2030 2030 Palette–A Database of Sustainable Design Strategies and Resources. Available online: http://2030palette.org/ (accessed on 4 June 2020).

69. Luna-Tintos, J.F.; Cobreros, C.; Herrera-Limones, R.; López-Escamilla, A. Methodology comparative analysis in the solar decathlon competition: A proposed housing model based on a prefabricated structural system. *Sustainability* **2020**, *12*, 1882. [CrossRef]

70. Herrera-Limones, R.; Pineda, P.; Roa, J.; Cordero, S.; López-Escamilla, A. Project AURA: Sustainable social housing. In *Sustainable development and Renovation in Architecture, Urbanism and Engineering*; Mercader Moyano, P., Ed.; Springer International Publishing: New York, NY, USA, 2017; ISBN 978-3-319-51442-0.

71. Calvo Salazar, M. Sostenibilidad en el urbanismo: Una propuesta. *Ciudad. y Territ. Estud. Territ.* **2006**, *XXXVIII*, 61–84.

72. Wassouf, M. *De la Casa Pasiva al Estándar Passivhaus. La Arquitectura Pasiva en Climas Cálidos*; Gustavo Gili: Barcelona, Spain, 2014; ISBN 9788425224522.

73. Neila González, J. *Arquitectura Bioclimática en un Entorno Sostenible*; Munilla-Lería: Madrid, Spain, 2004; ISBN 9788489150645.

74. Instituto de Ciencias de la Construcción Eduardo Torroja. *Código Técnico de la Edificación*; Instituto de Ciencias de la Construcción Eduardo Torroja: Madrid, Spain, 2015.

75. Diíez Reyes, M. *Glosario de Sostenibilidad en la Construcción*; AENOR, Asociacion Española de Normalizacion y Certificacion: Madrid, Spain, 2007; ISBN 9788481435009.

76. Desogus, G.; Felice Cannas, L.G.; Sanna, A. Bioclimatic lessons from Mediterranean vernacular architecture: The Sardinian case study. *Energy Build.* **2016**, *129*, 574–588. [CrossRef]

77. Über den Verein | IBU-Institut Bauen und Umwelt e.V. Available online: https://ibu-epd.com/ibu/ (accessed on 25 June 2020).

Article

The Importance of the Participatory Dimension in Urban Resilience Improvement Processes

Maria LopezDeAsiain * and Vicente Díaz-García

Art, City and Territory Department, School of Architecture, University of Las Palmas de Gran Canaria, 35017 Las Palmas, Spain; vicente.diaz@ulpgc.es
* Correspondence: mlasiain@ulpgc.es

Received: 31 July 2020; Accepted: 31 August 2020; Published: 6 September 2020

Abstract: This article discusses the approach adopted by the researchers into citizen participation in urban regeneration actions and projects. It describes the concepts of sustainability and habitability in relation to the urban environment and architecture within the framework of improving the resilience of our cities through the circular economy and decarbonisation processes in architecture. The authors review the participatory dimension of different urban regeneration actions carried out in Spain and the impact of this dimension on the results obtained by environmental, economic and social urban improvements. They then define possible strategies and methodological tools for integrating this dimension into traditional urban regeneration processes. The article presents case studies and their specific characteristics, and draws conclusions about their effectiveness and relevance. It also compares citizen-led interventions with interventions led by public administrations. Lastly, the authors analyse the potential reasons for success in these processes and projects, identifying weaknesses and proposing possible strategies for future development by researchers.

Keywords: citizen participation; resilience; urban regeneration; bioclimatic refurbishment; sustainable city

---

1. Introduction

The concept of integrated urban regeneration is defined in the Toledo Charter as "a planned process that must transcend the partial areas and approaches that have been commonly adopted until now, instead addressing the city as a functional whole and its parts as components of the urban organism, with the aim of fully developing and balancing the complexity and diversity of the social, productive and urban structures while simultaneously driving greater eco-efficiency" [1] (p. 7). As an umbrella concept bringing together the numerous city refurbishment aspects and approaches developed during the last half-century, including both urban spaces and buildings, it prioritises vulnerable neighbourhoods over the traditional classification of historic centres defined in numerous European urban plans in the last 20 years [2]. Adopting a European approach and based on specific programmes to strengthen and support the improvement of cities, it aims to address the environmental and economic crisis triggered at the beginning of this century and greatly exacerbated since 2008 [3]. Integrated urban regeneration prioritises a coordinated approach to the three dimensions of sustainability (environmental, economic and social) in a territorial framework with a global vision from the local perspective and bearing in mind the future determinants imposed by climate change. It places the emphasis on achieving the objectives of the Europe 2020 strategy [4] by improving the economic performance, eco-efficiency and social cohesion of cities to improve quality of life for citizens, underlining the need for their involvement in urban development through citizen participation [1]. In short, it recognises and highlights the importance of participatory processes as a tool for ensuring that citizens are involved in city improvement processes, from urban, territorial and building perspectives as well as from the perspective of the neighbourhood as a unit of identification and identity.

The urban regeneration processes promoted by Europe, which under the last edition of the Integrated Sustainable Urban Development strategy (ISUD, 2014–2020) allocated 730.917 billion Euros to integrated urban development in Spain [2], present a significant shift in the priorities of urban interventions, placing a greater focus on objectives that no longer apply solely to aspects of physical development but also include the social and economic improvement of towns and cities. These urban regeneration processes highlight the need to adopt new methodological approaches to solve some of the difficulties encountered in integrated projects, such as "obtaining results in the long term, changes in budget allocations, governance and coordination between the different actors (public, private, social groups, etc.), and managing different public scales (state, regional in the case of Spain, municipal)", as pointed out by Fernández-Valderrama et al. [2] (p. 37).

At best, these projects, which often involve both building refurbishments and the redesign of urban spaces or modernisation of urban infrastructures, include poor or incipient participatory processes led by institutions. These weak processes threaten to undermine the collaborative capacity of citizens to take part in urban environment improvements.

1.1. The Sustainability Concept

From a sustainability point of view, the urban regeneration processes carried out in our cities should include an environmental dimension, related to the bioclimatic design of the urban space and buildings; an economic dimension, related to the efficient and effective management of material, energy and information resources; and social management, related to satisfying the needs of citizens and improving their quality of life. That is, it is necessary in terms of urban regeneration "to find a balance among preservation instances, economic development, urban quality and the well-being of the population" [5] (p. 1). Effective social management demands the definition and development of the participatory dimension of the urban regeneration processes and its inclusion in specific actions, consolidating citizen involvement as a fact as well as a need, and developing the necessary tools to implement these processes. This way, citizen participation can be a positive turning point between the economic interests of business entities around urban development and governmental objectives related to urban space qualification and the improvement of citizens' life quality, which may tip the balance in the conflict between these two forces [6] (p. 47). It is also necessary to make the most of these conflict of interests situations [5], promoting their role of articulating and catalysing improvement in the city.

To be able to talk about urban regeneration, it is necessary to first contextualise the concept in the field of sustainable development and the future sustainability of our cities, based on the European approach included in the Toledo Charter [1]. Without this wider perspective, there would be no point to urban regeneration because it would be presented as a change, perhaps a development, but not necessarily as forward-looking regeneration capable of improving local and global resilience, both of the city itself and the territory in which it is situated. Our premise is therefore a concept of urban regeneration that is directly linked to and developed on the basis of the concept of sustainability, fulfilling and implementing its requirements.

We therefore take the initial reference of the concept of sustainable development that includes the economic, environmental and social dimensions. When we address the environmental dimension in architecture, we are referring to aspects related to bioclimatic architecture [7], to understanding the place, harnessing the benefits of its characteristics and designing in response to them, controlling both the flow of the materials and the flow of energy, using nature as a reference, attempting to close all material and energy cycles through the architectural processes. Numerous authors have recently examined this area, directly linked to decarbonisation processes in architecture related to the energy distribution and generation system [8–12], energy efficiency, life cycle analysis and, in general, the management of the design, use and construction of buildings from the perspective of energy and materials.

The concept of decarbonisation initially applied by the European directive [13] to the global energy generation and distribution system of a country also has applications for architecture since energy is required in every architectural process, during construction, use and even, it if comes to that, demolition. Nearly zero-energy buildings [14] are a reference in this aspect. They are based on minimal energy consumption and a level of self-produced energy that permits a zero or nearly zero balance. Some studies highlight the importance of understanding the energy requirements of each building, not only during its use [11] but also related to the energy employed in its construction [8]. Other studies prioritise the global energy approach related to generation and distribution systems at the territorial level [9,11,12], defining the degree of decarbonisation assumed and developed by each country in relation to the use of energy in construction. On occasion, strategies based on polycentric and decentralised governance policies have proven to be very effective in these types of processes [10], which clearly demonstrates the necessary link between greater sustainability and political approaches based on improving governance at every level of society.

When we speak about sustainability in terms of the economic aspects, in the field of architecture and urban planning it makes sense to view it as understanding the need for a certain economy of resources: material, energy, human capital, information, et cetera. This enables us to focus on what is meant by efficient architecture and urban planning, which use materials chosen for their specific characteristics to cover the comfort and operational needs of buildings and urban spaces. In other words, we are referring to the construction logics but also the local logics, the use of the appropriate local materials that will last over time according to the needs or requirements of both the building and the urban space. It is also necessary to refer to the correct and efficient management of information as a resource, to the need to maximise the use of existing architectural knowledge in order to propose better and more appropriate designs, designs that work better and are better suited to the needs of future users. This implies efficient information management [15–17] of the information the architect needs to propose an appropriate response, and the management of information to genuinely understand citizens' needs, whether of those who will live in our buildings or those who will use the urban space that we are designing and characterising.

Meanwhile, the concept of the circular economy, developed by pioneers of ecological economics like Kenneth Boulding [18], can be directly adapted, albeit with a certain complexity [19], to architectural processes related to construction as well as the design of urban space [20]. This concept broadens the approach related to efficiency and effectiveness in resource management and applies it in greater detail to the field of environmental sustainability, insofar as it is intrinsically related to the closure of material cycles. In terms of the social field studied here, the aspects that may bear a direct relationship to the concept of the circular economy are those that are closely linked to citizen management and community [21] as a space for mutual cooperation and the sharing of goods and services. For example, some networks based on the circular economy [22] have developed their own type of currency that is used in very controlled and relatively small local areas and has had fairly positive results, although very few scientific data are available.

When we look more closely at the social dimension, the matter becomes more complicated because in recent years we have seen a paradigm shift in how architecture addresses the social aspect. Earlier studies have focused somewhat timidly on this matter, introducing certain aspects or relevant indicators [23]. But this has been insufficient, and in practical terms, there are only a handful of experiences that confirm its genuine application in urban regeneration projects. Barely analysed, in the urban regeneration encouraged by the institutions (mainly, European Urban Plans), this social dimension has occasionally generated major weaknesses in terms of social sustainability, sometimes leading to gentrification processes, accentuated by the actual regeneration strategies implemented [24].

Furthermore, the social dimension is not only referenced to the concept of sustainability but to understanding and defining the concept of quality of life, which has also been defined, analysed and developed in some studies through the use of indicators [25]. This article is based on the premise that the environmental and economic aspects related to sustainability have been fully defined by researchers, leading to a large body of knowledge on the subject, and we therefore focus mainly on the social aspects. Participatory processes are an appropriate dimension for this purpose.

We must also define the last objective of urban regeneration. The Toledo Charter [1] emphasises the need to improve the quality of life of citizens, but after analysing the possible tools [25] for defining this concept, and in view of its complexity, we believe that it is more accurate to use the term habitability as a desirable objective of urban regeneration actions. Accordingly, " . . . the spaces necessary for daily human life must be habitable, offering the necessary characteristics to ensure the appropriate level of physical, psychological and physiological comfort and allow activities to be carried out" [26] (p. 72). This statement is key to our proposition that, if urban space and architecture in general must comply with the habitability condition regarding the ability to carry out activities, this habitability is therefore also necessary for addressing the real and specific needs of citizens as users of these architectural urban spaces [27].

In this respect, we must ask ourselves how we can possibly resolve urban issues without understanding citizens' real needs. As María Álvarez Sainz [28] (p. 5) says, "In the world of architecture and urban design, the key question resides in the ability to listen, to understand the needs and lifestyles of those who are going to live in the space designed." This demands an understanding between the technicians behind the proposal [29] and the ultimate users and their needs [30]; between the more technical urban design and the field of citizen participation to gain a genuine insight into the context of the issues in the urban space subject to improvement. However, the aim is not only to understand, but to build and check whether the regeneration proposed through specific actions, will have a high probability of success [31].

Accordingly, the proposal must provide a real response to the user's needs as expressed by the user; in other words, decisions regarding urban architectural design must take into account the user's experience and judgement. For this purpose, the development and adaptation to participatory practices of certain multi-criteria analysis [5] and multi-criteria assessment tools [32] is of great interest, although rarely explored.

From the point of view of social sustainability, improving urban resilience [33] is partly directly related to actions or processes associated with the development of citizen governance [34] and participatory processes in neighbourhoods, towns and cities, in line with a governance model which Pierre [35] first described as "corporate" and which Tomás Fornés, M., and Cegarra Dueñas, B. [36] subsequently developed and renamed "participatory". In addition to being beneficial for decision-making and the development of improvement strategies through urban regeneration, these inclusive processes are highly instrumental in community building. As such, they are essential for encouraging collaboration between citizens who identify as neighbours [37], establishing a positive bond between them. Insofar as it generates citizen exchange and support networks, this collaboration guarantees a better global response as a community to situations of scarcity or calamity.

We can best understand the relationship or enrichment of governance through participatory processes when they involve relations of power and the ability of non-governmental agents to influence the redistribution of resources, as well as in government processes. This prompts debates about the rights and obligations of all social groups, and about their responsibilities and demands, encouraging access to institutions where they can negotiate and regulate social conflicts [38]. There are different participatory models, some of them one-dimensional, based on a technocratic vision that imposes on citizens a specialist's analysis of their needs and evades citizen participation initiatives. Other multi-dimensional models are obtained from studies of certain urban movements [38] and, according to Castells [39], from demands for urban services and spaces more rooted in the local identity and therefore with a greater capacity for local self-governance. These are developed through contentious interaction between urban actors, a necessary characteristic for developing participation as a meeting between different positions in search of a consensus. Based on the focus of this article, it is these multi-dimensional models that are starting to lead to more successful experiences in Spain, and therefore provide the reference for a more interesting participatory approach.

1.2. The Place of Participatory Processes

The aforementioned arguments clearly demonstrate the need to rethink all actions in the city from the perspective of citizen participation. This refers to the need to analyse, develop, work and build with social and citizens' groups when it comes to improving the habitability of their cities and neighbourhoods, as advocated by María Álvarez Sainz [28] (p. 18) in relation to her commitment to "user-centred urban design and citizen participation as a collaborative model". The starting point for this will always be the need to improve quality of life and even address it from the subjective and perceptive dimension [31].

This need is examined by authors like Borja [40], who explains how the social movements of the latter decades of the 20th century (mainly referring to the 1960s, 1970s and 1980s) made enormous contributions to urban management and development, with significant achievements such as "the revaluation of public space as a place of gathering and social interaction, the social demand for quality of life in cities, the demand for democracy, dialogue and participation by local citizens, and the recovery of urban politics led by local governments". Borja [40] goes on to say that this transforms the concept of citizens as passive subjects into active subjects from the moment they intervene and participate in the construction and management of their city.

It is plain to see that, nowadays, there are more and more situations in which certain citizens' groups that demand the right to the city, are achieving success and social support [22], to the extent that, government recognition of the participatory dimension has been translated into a specific requirement of the rules for obtaining European grants for integrated sustainable urban development strategies (ISUDS) [2].

In certain cases, these citizen demands have laid the foundations for partnerships with local bodies and have led to specific actions carried out in the city. Although these cases have not always been successful, we are beginning to glimpse a possibility for change in the approach to urban regeneration in the city. They are usually initiatives that adopt new values [41], make local demands and form part of a logic that is perfectly integrated into the concept of improving the sustainability and local resilience of cities, linked to the dynamics proposed through a strategy based on encouraging the circular economy. In general, they tend to be defined from the perspective of citizen empowerment, prioritising micro initiatives and relationships on a human scale to rescue the molecular dimension of the social fabric [42].

This reflection demonstrates the potential of participatory processes for the success of urban regeneration programmes, and yet relatively few authors [43] have defined the specific reasons that guarantee this success in each case. There is clearly a need to identify successful representative cases that can serve as a guide regarding the dynamics of processes, actors, timeframes and methods, and what makes them successful: their management, the tools used, the actors involved or the socio-political context.

In the case of Spain, there are glaring weaknesses in the citizen participatory tradition apart from that related to the improvement in the urban fabric promoted by the state during the Franco period [28]. In fact, citizen participation started to emerge again in the 1990s. Based on a series of representative cases, we analyse the aims of these early projects as well as their characteristics, methods and tools, and the results achieved, in order to define opportunities for action, successful methodologies and useful tools for ensuring their efficient implementation and encouraging the proliferation of successful cases in the future.

The goal of this article is to define which aspects, strategies and tools can help to guarantee success of participatory processes in the Spanish context. For this purpose, an analysis of European and international methodological and strategic references is carried out, from which reference situations and characteristics are drawn to be valued for our case studies. Some Spanish cases known for being considered successful by citizens are taken into account. The criteria for the selection of specific case studies are defined, as is the methodology to be used in their analysis. A comparative analysis of the cases is carried out. Finally, the strategies and tools that can be potentially extrapolated to other situations are pointed out, shaping the conclusions of the research.

2. Material and Methods

Having defined the conceptual framework for our research and justified the importance of participatory processes in urban regeneration actions in cities, we now develop a non-exhaustive, inductive process based on selected case studies that offer clear guidelines, methods and tools that may serve as the basis for successful participatory processes in the future.

We examine a selection of Spanish cases, chosen for their specific innovative characteristics, and analyse the general context, dynamics, agents involved, timeframes, and the method and tools used. Next, we draw conclusions about the explicit opportunities and weaknesses presented by each case, and finally we highlight the lessons that may be extrapolated to other situations according to the terms and dynamics analysed.

2.1. Analysis from the International Perspective

There are public administration initiatives that attempt one way or another to involve social groups and residents of certain neighbourhoods in the development and improvement of areas of their city, but their approach is always partial, failing to recognise the real value and potential scope of participatory processes [38].

Today, the debate is focused on citizen participation cases that have emerged from the proposals of concrete social groups [44] related to specific needs, and that have been developed outside the margins of public institutions and have sometimes even been much more successful. The survey of numerous "bottom-up regeneration processes" carried out in Europe [3,44–46], which heterogeneously focus on the principles of the progressive redistribution of resources, ecological sustainability and social responsibility [44], highlights both the fundamental aim of these types of processes and the characteristics that set them apart from traditional processes promoted by public administrations. According to Squizzato [3], the fundamental aim of these projects is to improve the urban environment through non-governmental private projects and for a non-speculative purpose. Developed spontaneously by citizens, they are innovative because they not only include physical improvements of the urban space but social improvements as well, using different models of social interaction, appropriation and possession of the space as well as alternative financing [47].

The conclusion drawn from the analysis of European cases is that the success of each individual case depends on numerous local factors related to the specific characteristics of each situation, and the scope of this article, therefore, focuses on an in-depth analysis of some interesting Spanish cases. However, this analysis of European cases also reveals interesting defining circumstances and characteristics [3] that must be analysed in the Spanish cases examined here, in order to determine whether these characteristics influence the degree of success. These circumstances and characteristics are as follows:

1. They incorporate the use of new economies and values, including identity ones, which in turn define a new method of intervention sustained by its own development [45].
2. They use strategies based on very limited financing, promoting realisation through creativity [46,47].
3. They use urban activists, entrepreneurs, or both, as catalysts for the social potential [48].
4. They address the changing role of professionals and technicians [45,48].
5. They involve citizens as the true actors in the city self-organisation process [45].
6. They include the active participation of the community in the design process [45].
7. They include the innovative use of graphic, technological and digital tools for management and active communication during multiple phases of the process [45,48].
8. They benefit the community involved in several ways, not just by improving physical aspects of the space [44].

2.2. Situation in Spain

Our investigation involved numerous cases at the national level in which the participatory dimension has played a prominent role. Examples of projects led specifically by citizens' groups without any input from the local authorities include the following: "Playa Luna" in Madrid, where Ecosistema Urbano [49] and local residents carried out a protest action concerned with the construction of an artificial pseudo beach in a disused square, Plaza Luna, in the centre of Madrid; the occupation of "Can Batllo" [50] in Barcelona, with assistance from the Lacol group, and its subsequent partial refurbishment as a civic space for local residents; the famous "Campo de la Cebada" [51] project in Madrid, with input from a number of citizens' groups, including Basurama y Zuloark; and "Oasis" [52] at the Ruedo de Moratalaz housing development, in partnership with GSA Madrid and Asociación Caminar. All of these cases represent landmark demands by certain social groups that have defined a modus operandi based on self-organisation and self-management. They have mainly arisen out of neglect by the public administration and actions that were "extralegal" a priori but were subsequently agreed with by the administration following social and media pressure.

There are also interesting cases in which the public administration has been involved and has worked closely with citizen groups. These examples include "Vamos a hacer la calle" (Let us build the street) by eP espacio elevado al público group [53], carried out in the 3000 Viviendas and Martínez Montañés area of Seville's Polígono Sur district. The aim of this action is to highlight the importance of local residents' abilities to improve the public space, enabling them to participate in and own the project. Another example is "El Ejido Elige" (El Ejido Chooses) in Málaga by Fundación Rizoma [54], Paisaje Transversal [55] and the Omau (Urban Environment Observatory), a participatory process in which the financial resources are used to build designs agreed with the local residents. In these cases, it is the public administration that activates the participatory aspect with the aim of generating preliminary processes that can then be continued through self-development.

As a particularly interesting complementary aspect of our research, we have been able to glimpse the possible links between citizen-led strategies and projects and their formal institutional support through different local, national and international programmes, although in practice the technical and administrative requirements sometimes prevent their implementation and even access [3] to European funds, as in the case of the Community Led Local Development instrument proposed by the European Commission [56].

After a preliminary analysis of the cases, we can affirm that, in general, a high percentage of the actions undertaken through citizen initiatives achieve a considerable degree of success [31]. However, their subsequent maintenance and development demands committed involvement from a local group of residents or citizens to ensure that they evolve and also to fill them with activities and uses. Otherwise, many of these improvement actions constitute a fleeting moment in time with no real continuity, and we therefore cannot conclude that they improve the local resilience of the community, although they do improve the short-term habitability of the space.

Meanwhile, the actions undertaken through public administration initiatives tend to develop the participatory process on a partial level (with specific exceptions such as "El Ejido Elige"). They achieve a notable success in the initial stages but if they are not redirected properly, they may not fulfil citizen expectations and may even undermine citizen trust in these types of processes. Sometimes, this leads social groups to deliberately exclude public administrations in their processes and the change and improvement actions they undertake.

All the cases studied offer enormous scope for extracting lessons but we have limited our in-depth analysis to a small selection to determine which of their characteristics are potential reasons for success, according to the European cases previously analysed, and to be able to draw more specific, substantiated conclusions. The cases chosen for our study fulfil the following selection criteria:

1. They represent landmark successes, corroborated by the awards received, express academic recognition or express acceptance by citizens.

2. They represent landmark cases as pioneers in citizen participation in specific areas of Spain.

3. They represent landmark cases in terms of research and academic education as reference models for social service-learning and transfer strategies.

4. They represent landmark cases that challenge the role of architects as professionals and trigger debate about their responsibilities and abilities.

5. They represent landmark cases that exemplify the different roles that a local administration and a group of citizens or professionals can play as the drivers of a participatory action, whether self-managed or collaboratively managed.

2.3. Case Analysis Methodology

The case analysis methodology developed is based on the following points:

1. Analysis of the project technical information, plans, timeline, regulations, academic publications and dissemination in the media, awards and distinctions. This information is decisive in order to carry out an in depth study of the case, and in case it did not exist, the case study would be dismissed in this research work.

2. Mapping of actors involved, citizens' and government groups, public administrations, and technical and professional teams. This is a key tool for understanding the process and it must include, not only the actors involved, but also the relationships among them and the collaboration dynamics developed, if any.

3. Interviews with technicians and mediators, as well as with the accessible groups, citizens and administration technicians. Deep understanding of the processes can be clarified by documenting opinions, preferably from different types of stakeholders to have a broad and complete perspective.

4. In depth analysis of the development of the process, focusing on timescales and schedules, actions, strategies and tools used. This analysis has been done in great depth, using all the data initially extracted. In the first stage, the analysis is carried out individually, and is later used in a comparative way with other case studies, providing a wider perspective for the drawing of conclusions.

5. Extraction of conclusions about the characteristics and circumstances predefined as possible catalysts for the success of these projects. They will respond initially to the circumstances and characteristics observed in the international cases as potentially successful. In a later stage, conclusions will be drawn from the case studies and their circumstances and specific development.

6. Identification of interesting tools used. Beyond those referred to by international cases such as the map of actors or the graphic, technological and digital tools for management and communication, another type of interesting tools is found.

7. Identification of possible guidelines or management methods used in the process that may provide a reference for future cases. The possible use of specific methodologies with a previously established scientific basis is considered, against the construction of new methods.

The four cases analysed in depth are briefly described below.

2.4. Case Studies

2.4.1. Moret Park, Huelva

This case is a pioneer in Spain insofar as the citizens' demands were addressed by the public administration and there was a participatory process for the design of a public space. The case began around 2000 and the initial phase was completed in 2005 with an official inauguration. Huelva city council was interested in recovering the so-called Moret Park in the centre of the city, a 50-hectare area that had been neglected for decades and was in an advanced state of decay.

Local residents had already been demanding use of this public space and had set up the Moret Park Platform to campaign for the recovery of the park as the "green lung of Huelva". The platform, which brought together numerous groups (see Figure 1), succeeded in persuading the public administration to invite tenders for the project based on a set of specifications that both parties (platform and administration) had previously agreed during a series of workshops.

Figure 1. Moret Park citizen group. Source: Own elaboration.

A set of activities were carried out throughout the process (Figure 2) with the ultimate aim of developing an agreed design with the community platform:

1. Regular meetings to establish objectives and contents.
2. Contact with groups, political parties and organisations to involve them in the meetings.
3. Talks, exhibitions, educational, artistic and environmental routes, and a panel of experts to provide a greater understanding of the Moret Park Complex.
4. Briefing on the history of the area to contextualise the Moret Park phenomenon.
5. Proposals for the design of Moret Park.

(a)   (b)   (c)

Figure 2. (a–c) Different activities carried out to design Moret Park. Source: Own elaboration.

The design of the park was finally undertaken by the architecture studio Seminario de Arquitectura y Medioambiente, based on bioclimatic and environmental criteria and with the innovative inclusion of a professional mediator between the public administration and citizens. Thanks to the participatory process, the park was officially opened (Figure 3) to great success and with mass attendance by citizens. In spite of being an incipient urban regeneration action, the initial participatory dimension led by citizens was key to the successful execution of the project and its subsequent acceptance by the public.

The design process for the park has won several awards and distinctions: first prize national competition for the Moret Park, Huelva; first prize of the Andalusian Federation of Municipalities as an example of citizen participation with the Moret Park Platform, 2005; and being selected at the Dubai International UN-Habitat Best Practices Awards, 2008, earning the classification "Good Practice". The dynamics involved in the project have been analysed and disseminated extensively in research articles, at academic meetings and through postgraduate programmes.

Figure 3. Official opening of Moret Park. Source: Own elaboration.

2.4.2. Arraijanal Park, Málaga

Arraijanal Park is a vast urban space on the edge of Málaga with a history of enormous social pressure as the subject of numerous demands. The only remaining virgin piece of coastline belonging to the city of Málaga, it is extremely picturesque as well as boasting great ecological, historical and archaeological value. Numerous projects have been proposed for the space but to date none have been approved. Within the scope of our research, we examined the preliminary proposal commissioned by the Regional Environment, Climate Change and Land Management Ministry in 2015. In spite of the approach envisaged by the regional ministry, this proposal does not constitute a design project as such, but rather the design of a participatory process. During the development and execution, it would allow a preliminary in-depth analysis of the characteristics of the context subject to intervention, such as the participatory design of the future park with residents' and citizens' groups, and the collective construction of the parts of the project that are physically feasible and buildable with the groups and citizens. This is the first time that a public administration has ever allowed the transformation of a project into a participatory design process, although in light of the recent political changes, the project currently under way, has been awarded by tender under the traditional terms of architectural design.

The specific interest of this proposal as a case study is the analysis of the systematisation and design of a concrete process, which may provide a reference for the development of a specific method or tool for designing participatory processes (Figure 4). The key aspects have been analysed, classified and characterised to serve as a basis for the proposal included in the scope of this research.

2.4.3. Pepe Dámaso Cultural Centre, Las Palmas de Gran Canaria

This is a case of the encouragement of citizen participation without the existence of a prior explicit demand by local residents. The initiative emerged in La Isleta, a neighbourhood of Las Palmas de Gran Canaria, as part of the "La Isleta Participa" research project carried out between 2008 and 2010 by the University of Las Palmas. At the time, the neighbourhood did not have any adequate spaces for community building and partnerships. In the context of a severe economic crisis, during which Las Palmas City Council made concerted efforts to encourage citizen participation, the project furthered the demands for greater participation in the fields of architecture and urban planning.

One of the most resounding successes of the project was undoubtedly the fact that it capitalised on the existence of different national investment plans (known at the time as "Zapatero plans") to secure a budget for the construction of a cultural centre as a reference point for participation and cultural creation in the neighbourhood.

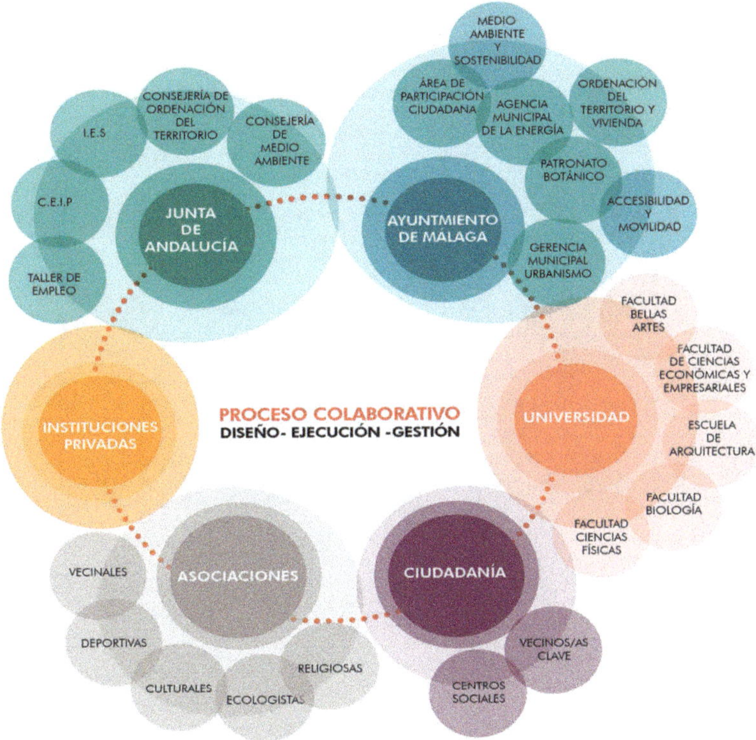

Figure 4. Collaborative design process for Arraijanal Park. Actor tree. Source: Own elaboration.

The Pepe Dámaso Cultural Centre occupies a plot in a public space designated for educational use, bounded to one side by a residential building and to the other by an infant and primary school. This was once the site of an old fighting ground, and in the 1970s it was used for student assemblies. The building has three storeys and a basement. The ground floor contains the entrance hall, a triple-height courtyard forming a visual connection with the adjacent building, the toilets and a large hall. The basement comprises an exhibition hall, a courtyard and various storage areas. The debating hall or Tagoror, which is also used as an exhibition space, occupies the first floor. Lastly, the second floor contains the quadrangle that forms the rooftop space, with access to the adult education centre next door.

In 2018 this was selected as one of the top five new builds of the last 10 years in Gran Canaria, but its merits also include the variety of excellent activities that are conducted inside the building. It is used by the children from the adjacent school, by residents' groups and by the city council for district meetings. Theatre rehearsals, concerts, exhibitions, workshops, courses, social gatherings, et cetera, all take place in the cultural centre. It has been used as a polling station for a referendum, commercials have been filmed here, and it has been the venue for miscellaneous events, political rallies and meetings. No programme of uses could ever have anticipated the enormous variety of events (Figure 5) that have been hosted at this centre. In fact, in his book "Architecture Depends", the theorist Jeremy Till [57] discusses how the dependent nature of architecture often makes it difficult to envisage the uses that a space should provide. In recent years we have witnessed a procession of technical, professional and artistic profiles that have discovered these spaces as places in which to carry out a whole range of extraordinary activities. This clearly demonstrates that citizen participation does not only concern social or residents' groups but impacts society as a whole.

Figure 5. Example of the various social and cultural activities currently being carried out in the Pepe Dámaso Cultural Centre. Source: Own elaboration.

What was originally planned as a building for citizen participation, more related to debate and discussion, has become a place for the participation of the entire neighbourhood, on every level. This invites us to reflect on the need to recover the ambiguous nature of spaces that architecture undoubtedly offered until the 19th century. It also prompts the demand for participation spaces in every neighbourhood, whether explicitly proposed by citizens or not. This proposal highlights the importance of the role that other sectors of society can play, in this case university researchers, teachers and students, as mediators between citizens and technicians to enrich citizen participation.

2.4.4. Majanicho Citizen Participation Programme, Fuerteventura

Our last case concerns the application of a citizen participation tool in an action led by the public administration, but precisely at a time when the European Commission is questioning interventions in the ecosystem that lack an impact assessment and citizen participation.

The case is centred on the town of La Oliva, in the north of the island of Fuerteventura, specifically a space that has been impacted by the construction of a housing development that contained multiple irregularities from the outset. Between 2000 and 2002, the town council of La Oliva approved the Partial Plan ("SAU 12 Houses Majanicho") and the urban development project and granted permission for the construction of 748 homes and a commercial area. That same year, 2002, the approval was refuted by the Canary Islands government for the first time and an environmental association lodged the first complaint. In 2006, during the completion stage of the project, two important events occurred: the Canary Islands High Court of Justice (TSJC, after its Spanish initials) declared the Majanicho Partial Plan null and void (ratified by the Supreme Court in 2011), and the Canary Islands government established a new series of special bird protection areas (SPBAs) affecting part of the area comprising the housing development. It was precisely the invasion of part of an SPBA, among other factors, that prompted the European Commission to send a letter to the Spanish government in 2018 calling for the adoption of measures to mitigate the effects of the Majanicho Partial Plan. The measures finally proposed include the execution of an ex-post environmental impact assessment (i.e., conducted after completion of the housing development) and the preparation of a citizen participation programme (Figures 6 and 7).

Figure 6. Image of citizen protests due to the development of Cotillo and Majanicho urban plans, in La Oliva Town, Fuerteventura. Source: Own elaboration.

This case therefore provides the opportunity to turn a participatory process, as part of an ex-post environmental impact assessment (encouraged partly by citizens and partly by the public administrations involved), into a programme: "Citizen participation programme in the framework of the Environmental Impact Assessment for the Majanicho housing development project—SAU 12, La Oliva (Fuerteventura)". In the short term, this could contribute to the design and adoption of the appropriate compensatory measures, and in the medium to long term, it could result in the coordinated management of the present and future of the island's northern coastline. In this case, citizen participation should help to raise the level of debate and restore the trust of the residents of La Oliva, and by extension those of Fuerteventura as a whole, in their institutions, particularly regarding sensitive issues like interventions in the natural heritage and landscape.

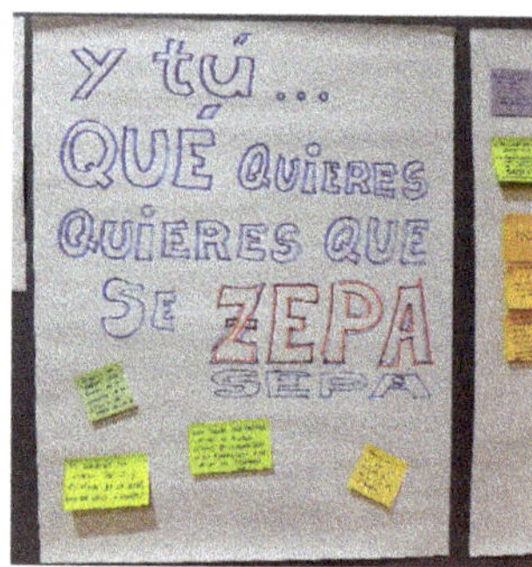

Figure 7. Image of the citizen participation workshop carried out during the development of the Citizen Participation Programme within the frame of the environmental impact assessment. Source: Own elaboration.

Scale has been an important dimension in this process. It was necessary to incorporate global issues like the climate emergency and the UN Sustainable Development Goals [58], and reconcile general laws with the various plans affected (SPBA, island plan, general plan, land management plans, special plans, etc.), while at the same time, taking into account local and even individual interests. The clearest example of the importance of scale is that one of the reasons that has led us to this point is the Kentish plover (*Charadrius alexandrinus*), a small bird that visits this area and whose defence and protection has led us to rethink the use and enjoyment of this place from a more sustainable and resilient perspective.

On a technical level, the circumstances surrounding this citizen participation project have led to the sharing of space and time with a very broad and experienced technical team, mainly with expertise in the environmental field. Lastly, technicians have sought to reconcile the analysis and diagnosis tasks with the desire to draw up a series of compensatory measures for the environmental impacts detected.

In view of the unique nature of this entire process, the institutions and the technicians involved in the work and the discussions, as well as citizens, firmly believe that the project should culminate in decisions that effectively change the way in which these environments are used and enjoyed. The SAU 12 Houses Majanicho project has been halted for technical and legal reasons for years, and today there is a certain degree of consensus about the importance of taking action. The priority is to implement compensatory measures as a means of starting to regain the trust of citizens.

Lastly, this case has demonstrated the importance of participation education, insofar as it should not only reach the local citizens, but professionals, technicians and politicians as well. Each actor must climb up the "ladder of citizen participation" that Sherry Arnstein [59] proposed more than 50 years ago. Getting past the non-participatory rungs (manipulation and therapy) and involving more people on the intermediate rungs (informing, consultation, placation) or even on the top rungs (partnership, delegation, citizen control) are tasks for which we all share responsibility.

3. Results

In this section, we present the data resulting from the analysis of the characteristics of the participatory processes studied, from the approach and layout of the study to the results obtained. A synthesis of such results is presented in Table 1.

Table 1. Analysis of the characteristics, approach, considerations and results of the case studies.

	Moret Park	Arraijanal Park	Pepe Dámaso Cultural Centre	Majanicho Citizen Participation Programme
New economies and values	Highlights the identity of the neighbouring residential areas linked to the history of the park's use and its collective memory. This memory is linked to a natural design of the park demanded by the citizens. Attention paid to ecological dynamics. Approach based on flexible use.	Highlights the identity of the natural and archaeological space following the demands of local associations. It rejects economic and speculative performance and it promotes previous interventions from local associations. The land is understood as a common good (right to the town and space justice). Attention paid to ecological dynamics. Inclusive participatory approach, also cross-generation and gender approach. Design approach based on flexibility and reversibility. Extension of time frame.	Highlights the cultural identity of a neighbourhood following proposals emerging from the academic sphere. Experimental emphasis on the concept of citizen participation in Las Palmas. Public funding through traditional means. Chance of highlighting social, cultural, political, commercial and educational activities previously non-existent. Approach based on the flexibility of the built spaces.	Highlights the natural ecosystem and its species. Calls into question urban use, claiming a series of compensatory measures.
Financial resources	In this case, the collaboration agreement with the city council provides the necessary resources.	Scarce financial resources coming from the regional government. Boosting of alternative social resources in kind	Governmental financial resources.	Financial resources from the local government are used as a compensation for a wrong legislative procedure.
Urban activists and entrepreneurs as catalysts for the social potential	Citizen associations themselves are able to start and manage the process without the need of external activists.	The catalysts in the process are the technicians appointed by the public administration themselves, despite the fact that their proposal is eventually dismissed.	University researchers as initial catalysts of the process.	The group Ben Magwc-Ecologistas en Acción as catalysts of the process thanks to their claims.
Role of professionals and technicians.	Technicians act as mediators between the city council and the Moret Park platform of associations, providing answers to the demands through debates and workshops.	Proposal of technicians as mediators, speakers and process managers.	Technicians (in this case researchers) take a mediation role between public administrations and citizens. Technicians are also architecture professionals. The new potential technical role for architects is specifically studied.	Four of the five technical teams involved are professionals coming from different fields linked to environmental science (ornithology, oceanography, etc.). The fifth team carries out a mediation role between public administrations, technicians and citizens.
Role of citizens in the process	Citizens are the triggers of the process, although they are involved mainly in the initial stages of it.	Citizens' proposals as true actors of the process, always working hand-in-hand with public administrations and technicians. Self-pedagogical process.	In this case citizens have a passive role until the building is finished and activities started. The building is used in a variety of ways that go beyond the cultural sphere. Process led by the institutions.	Environmental associations are the ones that initially claim the process. At a later stage, citizens are involved in the process of "fixing the problem".
Community's participation in the design process	Only in the initial phases.	Participation in all phases of the process, from planning to design and construction.	No involvement in the design process.	There is no involvement in the design process, but it does occur in the later critical process and potential solutions.
Innovative use of graphic, technological and digital tools.	Not known.	Use of graphic and technological means of communication between actors, from social networks to purpose-built platforms.	From the starting phase of the research project graphic means are used to inform the public administration of the importance of spaces for participation.	Not known.
Provide benefits to the community	All the associations and citizens' groups are represented, providing specific spatial answers to their needs.	The demands of the social associations are met, involving them in the development of the project.	Technical and academic deficiencies found in the social, political and cultural sphere are taken care of. All sorts of unplanned opportunities arose, providing tangible or intangible benefits to the community.	The right of citizens to decide is acknowledged in the case of a poor management process by the local government.
Results of the work	Adequate development of the project and later social acceptance.	Proposal accepted by the administration but turned in a later stage into a traditional management model.	Project developed and finished with the built centre; current management by the local government.	Participatory project developed, pending the implementation of the agreed actions.

All four projects highlight relevant aspects for citizens, both cultural (historical, anthropological, artistic, etc.), natural and ecological. They enhance both group identity and citizens' memories on the one hand, and attention paid to ecological and systemic dynamics on the other, both approaches being key issues for the improvement of resilience in our communities and towns. It should be pointed out how environmental approaches are changing chronologically and are more and more ambitious. Whereas in the case of Moret Park the focus is centred on maintaining the existing ecosystem adequately, in Majanicho the inappropriate use of a natural protected area is claimed. In Arraijanal, on the other hand, the focus is on the potential values of the ecosystem that they want to restore and maintain, taking into account the effects that climatic change may bring to it on the basis of a specific study on the matter.

Whereas three of the projects emphasise the need to make use of architectural and urban spaces flexible (Pepe Dámaso, Moret and Arraijanal), as well as the benefit coming from a certain reversibility of some projects; the fourth project (Majanicho) provides answers to the deficiencies resulting from the lack of flexibility of one-sided urban projects. The introduction of inclusive participatory dynamics strategically develops into a need for flexible architectural and urban spaces and land use, as well as for a flexible management of such spaces, as they are heritage of all citizens.

The four projects follow an inclusive approach from a gender and cross-generation perspective, at least as far as the development of the participatory process is concerned. However, the actual dynamics of every process, their characteristics, timing and circumstances, cause the results to be distinctive. In the cases of Moret and Pepe Dámaso (already completed) the benefits for the whole community resulting from the actions are clearly seen. Regarding Arraijanal and Majancho, they are still to be studied when the works are finally carried out. In any case, the greatly important gender aspects related to participatory processes could not be studied in further depth in these cases, due to the lack of objective data available. It is clear that the approach followed in the processes has aimed to be inclusive in gender terms, although it is not possible to provide clear evidence of it, at least in the frame of the this research.

Two of the projects (Pepe Dámaso and Arraijanal) have been developed in a context of poor financial resources, without this implying a cause or an excuse for its abandonment. The other two projects (Moret and Majanicho) were developed according to traditional economic dynamics, noting that there are sufficient public resources for well-managed participatory projects, despite the fact that in the case of Majanicho the context was that of a solid economic crisis. Beyond what they may seem at first sight, the participatory processes are economically affordable, if compared to the building processes involved in any urban qualification project. Besides, they can use social resources, which make them even more affordable if well managed. However, it is also true that it should be taken into account that the participatory processes take longer to develop than traditional top-down processes.

The actors acting as catalysts in each case are very different, from the citizens themselves in the case of Moret Park to the technicians appointed by the public administration in Arraijanal, as well as environmental groups and researchers of the public university. This reality implies that organised groups of citizens are no longer the only potential activists of participatory processes; such processes are starting to be seen as necessary and convenient by the government and their professionals and technicians themselves. The next step would be to transfer this need to the political agenda.

In all four cases, the changing role of architecture professionals is posed and developed, more linked to mediation rather than to the design of architectural spaces in the participatory processes. Nonetheless, their technical abilities (graphic, communicative and in media terms, as well as in constructive terms) continue to be necessary for a correct development of such processes, regardless of the fact that they may or may not lead to the design and construction of specific architectural spaces. In case where they do, architects can also develop their traditional technical roles and associated responsibilities.

The role of citizens in the different cases is very diverse; in three of them (Moret, Arraijanal and Majanicho) citizens directly or indirectly trigger the participatory process itself. In the case of Pepe Dámaso, they are collateral beneficiaries of it. In the case of Arraijanal, the citizens are called to empower themselves and be the true actors of the process as ultimate historical claimants, the process understood as self-pedagogical. That is, the process is seen as an opportunity for social learning linked to the different actions and professions involved in the design and construction of the urban space. The process can thus be also defined as a deep educational, emotional, environmental and even technical practice.

In general terms, the participation of the different social actors in the design process is scarce. It was initially proposed in Arraijanal, but was dismissed when the approach of the project changed and the participatory process was cancelled. Numerous cases exist (some previously mentioned yet not selected as case studies) in which the involvement of citizens as actors in the design and construction of the architectural project is a reality. This leads us to think that such participatory dimension is an opportunity for the progress of this kind of process, despite the fact that it has hardly been addressed in the reference cases.

The use of innovative graphic, technological or digital tools is not known in any of the projects except for Arraijanal, where they were proposed but not used. In the case of the Pepe Dámaso Centre, some graphic instruments were used at the beginning of the process to inform the authorities, but they were traditional ones. This circumstance is despite the fact that in many references previously analysed as possible case studies such tools were seen as an interesting asset for the success of certain participatory processes. We can deduce that they are desirable tools still little developed in this sort of process and which are not decisive for its success.

It is clearly seen that all four cases provide different kinds of benefits to the community, not only in terms of space but also in ecological, social, educational, cultural, political and even economic terms. In environmental terms, this sort of process implies a step forward as they follow a serious and rigorous approach, demanding conclusive results regarding the respect for the ecosystem, a good bioclimatic performance of both indoor and outdoor spaces and, ultimately, an answer to climatic change through the improvement of resilience of environmental communities. They also imply an educational opportunity from the perspective of social inclusion and following a cultural approach, which is attractive as a communication strategy to involve different types of actors from the social sphere. In the political and economic spheres, the benefits emerging from these sorts of processes include a sense of belonging to a political community capable of making decisions in terms of citizen empowerment and improvement of urban governance; an improved management of economic resources; and the involvement of citizens, which become responsible for their correct management.

Two of the participatory projects have been developed and are in the process of later management and the other two are currently being developed. One, the Arraijanal case study, can be considered a failure in participatory terms due to the government's rejection of the continuous action taken by the citizens to defend their rights and because no technical solution was achieved. The other, the Majanicho case study, is being developed at the moment and we may expect that it will be carried out successfully.

The strategies used for the correct development of the process are diverse and have been defined according to the characteristics of each case study, as can be observed in Table 2. The importance of multidisciplinary teams and the mediation work among administrations, technicians and citizens is noteworthy. The latter, has sometimes been developed by the architects themselves, exploring this new role of their profession, and other times directly by researchers of the participation field, therefore being more flexible in their functions and methodological skills.

Table 2. Analysis of strategies, tools and methods developed by the case studies.

	Parque Moret	Parque Arraijanal	Pepe Dámaso Cultural Centre	Majanicho Citizen Participation Programme
Process strategies	Multidisciplinary teams. Citizen-government mediation. Fostering trust between the parts	Multidisciplinary teams, multidisciplinary process. Inclusion of all agents involved throughout the whole process. Adapting time span and strategies to social context. Citizen-government mediation. Information adapted to users, transparency. Community building and fostering trust between the parts. Flexible and adaptable methodology [60]. Time and space for creative innovation	Technical and academic teams in collaboration. Methodology based on the involvement of research project both in the neighbourhood resources desk (promoted by the administration) and in different communal activities. Technicians as drivers of the participatory process and citizen involvement. Short times adapted to traditional local administrative procedures.	Multidisciplinary team with a strong representation of environmental issues. Inclusion of all agents involved during participatory process. Recovery of trust on the part of citizens. Involvement of academic actors that provide credibility and prestige to the process.
Tools	Workshops and meetings	Map of actors, workshops, mediation strategies, surveys, interviews, cultural actions. Definition of "motor team" as manager of the participatory process.	Fieldwork is carried out, analysing the neighbourhood's situation and obtaining a diagnosis of deficiencies in terms of participation.	Map of actors, workshops, mediation strategies, interviews, provision of alternatives.
Method participatory phases	1. Citizens' claims. 2. Definition of demands and needs. 3. Project acceptance. 4. Successful inauguration.	1. Citizens' claims. 2. Public intervention proposal. 3. Analysis of opportunities in the project's approach and management. 4. Comprehensive participatory process proposal (from design to construction). 5. Withdrawal of public administration and traditional tender. 6. Citizens' struggles continue.	1. Academic proposal. 2. Acceptance and local government initiative. 3. Project design. 4. Citizens' consultation. 5. Citizens' acceptance and building works. 6. Local government management with great involvement of citizen groups. 7. Very successful outcome of the project in social terms.	1. Citizens' claims through environmental organisation. 2. Legal requirement. 3. Proposal of compensatory participatory process by the government and university. 4. Successful participatory process. 5. Waiting for building works of project agreed through participatory process.

In the four cases under study, generating a climate of trust among the actors involved has proved to be greatly important, allowing the development of actions with a constant involvement of all parts, and at the same time narrowing down the expectations linked to the process in real terms.

On the other hand, the processes ought to be flexible in terms of adaptation to administrative times, which are radically different to those of social management in participatory processes.

With regards to the tools used, they are numerous, depending on each case and the specific needs of citizens' involvement, communication and technical and administrative management of the process. It should also be noted that in those processes led by the government, cultural strategies aimed at calling people's attention are used, whereas in those processes led by citizens peer communication attains greater significance, using graphic, audiovisual and any other means which are adapted to the common language of citizens. The proposal of defining a "motor team" is highlighted, with the involvement of actors from all the spheres (public administration, technicians, citizens, university). This team will be responsible for the management of the participatory process and embraces the following functions: coordination; communication and dissemination; design and production of the necessary material; stimulation of activity; and legal-economic management. This tool, although only proposed in one of the cases, should be further investigated in the light of its potential relevance as an opportunity to improve management within the processes.

As far as the participatory phases of the method are concerned, it is not possible to define an organisational tendency beyond the existence of three main distinct phases: a phase of proposal (by citizens, university and institutions); a phase of design and construction of the project (with different degrees of participation in each case); and a final phase of use and maintenance of the project (sometimes managed by the administrations and other times self-managed by the citizens). Every case is unique and its circumstances mark the way it is used through time.

4. Discussion

Transferable Opportunities and Learning

Following the case studies carried out, some learning can emerge from them which can later be useful for future situations:

In this sort of participatory process, the figure of the technician as mediator and translator is greatly important; an agent of citizen participation " … creating a dialogue between divergent and often opposing perspectives, of transforming citizens' claims into proposing strategies, and of implementing new tools and communication channels that speed up the processes and prevent lack of coordination between different governmental areas, as well as between them and the citizens." [61]. This ultimately poses the need of new professionals and new working opportunities around citizen participation.

We can confirm, as commented before that, in general terms, a high percentage of actions developed by citizens' initiatives achieve considerable success [31]. However, for them to continue and evolve through time they need some additional factors that favour them, such as involvement of local social or neighbourhood organisations; their usefulness and link to existing dynamics that provide activities and use. Otherwise, many of these actions will represent a specific moment in time and improved habitability of a space in the short term, but will not last in time and will not improve local community resilience.

On the other hand, the actions proposed by the administration usually obtain partial participation. They are successful initially in their approach, but without a good management method, they may not fulfil citizens' expectations and could even undermine citizen trust in this sort of process. In some cases, this provokes rejection from certain citizen groups that deliberately exclude public administrations from their processes and actions for change and improvement.

As well as the instruments that nowadays provide a greater transparency in public management and bring institutions and people closer to one another, the rise of citizen participation and its quality, without a doubt, leads to a greater degree of involvement on the part of citizens in common issues and in public management, at least at three different levels: at the urban-land level, at the technical-environmental level and at the social-political level. In the urban-land level we can point out, firstly, the centrality of participation, that is, the importance of including citizen participation in the early stages of any urban planning instrument, achieving a highly qualified degree of participation, defined by Pretty as active participation [62] and closer to the two upper rungs of Pretty's ladder. The coming together of citizen participation and urban planning is providing both concepts with new dimensions. Architecture and urban planning are proving they have a lot to offer to participatory processes, given their ability to anticipate through projects or forward thinking, on the one hand, and their approach to complex thinking through a skilful use of scaling, on the other. The field of urban planning conveys an opportunity to all kinds of participatory processes, while at the same time such participatory processes are an opportunity for improvement in urban regeneration processes that aim at being successful in the medium or long term. In this sense, it would be interesting to apply the new paradigm proposed by Mussinelli et al. [63] (p. 66) that "combines two levels: the long-term strategic vision and the short-medium term experimental vision".

In the political and social sphere, the map of actors becomes a key tool to determine the place of citizen participation. Given the complexity of our societies, sometimes it is difficult for the different actors involved in a territory or in a given issue to be aware of the multiple dimensions at play. One of the goals of the citizen participation project and particularly the map of actors (or sociogram) is to increase the numbers of variables involved. In the case of Majanicho, we have found that from the macro scale of the European Union to the micro scale of Majanicho's neighbours, all parties have something to say.

The dialogic process starts when the higher number of possible actors has been identified and the chance to participate in each case is balanced. Not all actors can or wish to participate in every case, as sometimes not all actors should or need to participate. It is from the moment when the map of actors is established that we can start planning the fieldwork, implying a higher or lower degree of participation in the participatory project designed. In this sense, the definition of a "motor team" for the management of the participatory process can also be a useful tool as long as it implies a previous selection of those actors and agents capable of involvement, management and communication within the participatory process. It is necessary to go deeper into the research of the potential of this tool.

Similarly, the importance of teaching the mechanics of participation has also been stated. That is, the pedagogy of citizen participation should reach the citizens, the professionals, the technicians and the public representatives. Every actor should in their own way move up in the ladder of participation as proposed by Sherry Arnstein [59] half a century ago. Overcoming the non-participatory rungs (manipulation and therapy) [62] and getting as many people as possible involved in the intermediate rungs (informing, consultation, placation) and even in the higher rungs (partnership, delegation, citizen control) [3] is a task that we are all responsible for.

In this research, the existence of appropriate tools which are common to different fields and help participatory processes, adapting to different scales, is clearly stated. These tools should be further studied to determine in which cases they can be used and their scope in each kind of process. This way, as in the case of Majanicho's Citizen Participation Programme, they can become a minimal agreed cell for citizen participation to be applied in all aspects related to the town and its safeguarding. It should also be pointed out that the complex nature of public policies that take social and cultural values into account, as well as their participatory dimension, call for the development of adequate tools capable of dealing with large databases and decision-making systems. Some research works [64] have developed a work frame for the use of analytics in supporting the policy cycle. This progress is a key reference for the further study of the potential of certain tools.

In particular, in the legal administrative sphere, the existence of Bologna Regulation for the Care of Urban Commons is a very relevant tool to support these participatory processes [65]. This regulation regulates partnership between citizens and governments for the care and regeneration of the urban commons. It sets off from an organisational model based on partnership instead of the struggles between citizens and the government. Here, citizens are not considered mere users and recipients of the State, but are finally acknowledged as key actors in the treatment and management of the common goods. It regulates the terms of collaboration for the safeguarding and management of urban commons through "Collaboration Pacts" based on mutual trust and a relationship of equality. This approach, developed by laws, is a very important step forward and a reference for the development of local and national legislation based on the same terms.

To sum up, successful participatory processes are characterised in general terms by the following: Understand the territory as a common good; work from the understanding of residents as active citizens to improve governance through shared management; work with proposals of inclusive, cross-generation and gender-sensitive processes, including all phases of analysis, design and construction. This process should at the same time respond to the environmental dynamics of the land and provide an environmental, flexible and reversible project (fostering conservation rather than building) that is also social, self-pedagogic and that stands for the right to the city and spatial justice for all citizens.

As a consequence, this sort of process is expected to provide proposals that are built on an increased level of creativity thanks to collective work, that enhance the territory safeguarding habitability of urban spaces, increasing investment on green areas and, with an adequate design, increasing its biodiversity.

We could also expect that these proposals work with a certain critical mass of population and enough activity in order to be representative in real terms; that they generate a certain degree of self-sufficiency in terms of energy, water and material resources, as well as in the provision of self-managed facilities, fostering improved alternative means of transport such as public or non-motorised transport. And it could also be expected that this dynamics builds on urban resilience in terms of social cohesion and adaptability to climate change.

There are still no predefined methods to guarantee the success of this kind of process and subsequent positive consequences, only some strategies and opportunities such as the ones elucidated by this research work.

5. Conclusions

It is clear that the processes of citizens' involvement linked to urban regeneration are still at an early developmental stage. We can also prove their important roles for improvement in terms of urban resilience from a social point of view. A number of tools exist [66] that can help the development of such processes (coming from different fields), but the processes are currently so diverse and unique that there is no specific pattern or methodology to follow for guarantying their success.

However, taking as a starting point the unique character of each of these processes and of the partial generic tools from other fields (and gradually also from the urban planning field) it is indeed possible to establish a series of criteria, strategies and opportunities that can guide such processes.

The precise definition of these opportunities, strategies and processes has been outlined in this research work (as has been described in the partial conclusions of the different chapters) and should be further developed in future investigations. In any case, the opportunities more clearly defined by the present research are listed below:

5.1. Citizen Participation and Environmental Sustainability

The participatory processes imply an opportunity to include aspects of environmental sustainability linked to the preservation of ecosystems, to establish social-systemic synergies [67] and ultimately to mitigate climatic change. For this purpose, an approach based on flexibility as a strategy and action criteria is necessary, adapting every action to include the needs, actors and specific local situation. The advantages of developing reversible actions from an environmental and social point of view also imply a decisive advantage for the correct development of these processes whose context and conditions often change through time.

5.2. Citizen Participation and Economic Sustainability

In the participation processes studied, we have clearly seen that its low economic cost in relation to the number of actions developed or the use of human resources could well compensate the economic cost derived from longer building times. This issue must be subject to further scientific study in any case.

5.3. Citizen Participation and Social Sustainability

Concerning the complex map of participation processes, we have stated how the different dimensions involved in citizen participation demand the use of useful tools to deal with them. Thus, creating a map of actors becomes one of the first necessary actions to be taken in participatory processes. This map should be able to show the social, economic or political complex picture that we are going to face in any participatory process. Trying to introduce to the discussion "the interests, goals and logics of action that the various categories of actors have, but also the dimension, from a local to global scale, of the interests themselves, since it has important consequences on the solution of collective problems" [68] (p. 53).

The catalysts of participatory processes are more and more diverse and closer to political management, potentially improving governance processes with the inclusion of citizens. This could imply a chance to come closer to political power through a meaningful and solid approach, allowing citizens and government to work hand in hand naturally and on the basis of agreement in urban planning and improvement processes.

The involvement of citizens in the design and even more so, the building stage of urban improvement projects, is still at an early stage and lacks a specific legal framework. This involvement means a great opportunity for the development of participatory processes that needs to be studied in further detail. Similarly, the potential of a rigorous gender approach to these processes has still not been sufficiently studied and needs further attention.

5.4. The Spanish Case

This article is only a starting point for the scientific study of such processes in the Spanish context. The results of the research carried out are of great interest, although they need to be further developed due to the complex nature of the issue, as has been stated throughout this article.

Author Contributions: All authors have contributed equally to the investigation and final manuscript. All authors have read and agreed to the published version of the manuscript.

Funding: This research received no external funding.

Conflicts of Interest: The authors declare no conflict of interest.

References

1. Unión Europea. *Declaración de Toledo*; Unión Europea: Toledo, Spain, 2010.
2. Fernández-Valderrama, L.; Martin-Mariscal, A.; Ureta-Muñoz, C. Rehabilitación urbana integrada: Un proceso complejo pero ineludible. *CIC- Centro Informativo de la Construcción* **2017**, *540*, 34–37.
3. Squizzato, A. Urban Regeneration: Understanding and Evaluating Bottom-up Projects. *Urbanities* **2019**, *9*, 19–35.
4. Comisión Europea. *Estrategia Europa 2020. Una Estrategia Europea para un Crecimiento Inteligente, Sostenible e Integrador*; Comisión Europea: Bruselas, Belgium, 2010.
5. Capolongo, S.; Sdino, L.; Dell'Ovo, M.; Moioli, R.; Della Torre, S. How to assess urban regeneration proposals by considering conflicting values. *Sustainability* **2019**, *11*, 3877. [CrossRef]
6. Sdino, L.; Rosasco, P.; Lombardini, G. The Evaluation of Urban Regeneration Processes. In *Regeneration of the Built Environment from a Circular Economy Perspective*; Springer: Berlin/Heidelberg, Germany, 2020; pp. 47–57, ISBN 9783030332563.

7. Ló pez de Asiain Alberich, M.; Serra Florensa, R.; Coch Roura, H. Reflections on the Meaning of Environmental Architecture in Teaching. In Proceedings of the 21th Conference on Passive and Low Energy Architecture; de Wit, M.H., Ed.; Technische Universiteit Eindhoven: Eindhoven, The Netherlands, 19 September 2004; pp. 163–168.

8. Arcas-Abella, J.; Alcaraz, M.; Bas, A.; Bilbao, A.; Catalan, P.; Cunill, L.; Melo, M.; Huerta, D.; Sauer, B. *Impulsa Cíclica [space community ecology] Green Building Council España (GBCe) | Instrumento para la rehabilitación profunda por pasos |*; Green Building Council España: Madrid, Spain, 2020; ISBN 9788409183890.

9. Hidalgo, F.; Arquitecto, A.; Por, T.; Upc, L.; Sevilla, E. *La Termografía y la Eficiencia Energética Aplicada a la Construcción Una Salida Laboral en el Ámbito de la Arquitectura Técnica*; COAAT: Sevilla, Spain, 2019.

10. Sovacool, B.K.; Martiskainen, M. Hot transformations: Governing rapid and deep household heating transitions in China, Denmark, Finland and the United Kingdom. *Energy Policy* **2020**, *139*. [CrossRef]

11. Valor, C.; Escudero, C.; Labajo, V.; Cossent, R. Effective design of domestic energy efficiency displays: A proposed architecture based on empirical evidence. *Renew. Sustain. Energy Rev.* **2019**, *114*. [CrossRef]

12. Scamman, D.; Solano-Rodríguez, B.; Pye, S.; Chiu, L.F.; Smith, A.Z.P.; Gallo Cassarino, T.; Barrett, M.; Lowe, R. Heat Decarbonisation Modelling Approaches in the UK: An Energy System Architecture Perspective. *Energies* **2020**, *13*, 1869. [CrossRef]

13. Unión Europea. *Reglamento (UE) 2018/1999 del Parlamento Europeo y del Consejo de 11 de diciembre de 2018 sobre la gobernanza de la Unión, de la Energía y de la Acción por el Clima*; Unión Europea: Estrasburgo, France, 2018; Volume 2018.

14. Comisión Europea. Recomendación (UE) 2016/1318 de la Comisión de 29 de julio de 2016 sobre las directrices para promover los edificios de consumo de energía casi nulo y las mejores prácticas para garantizar que antes de que finalice 2020 todos los edificios nuevos sean edi. In *Diario Oficial de la Unión Europea*; Unión Europea: Bruselas, Belgium, 2016; pp. 46–57.

15. Cubillos González, R.A.; Trujillo, J.; Cortés Cely, O.A.; Rodríguez Álvarez, C.M.; Villar Lozano, M.R. La habitabilidad como variable de diseño de edificaciones orientadas a la sostenibilidad de Colombia. *Rev. Arquit.* **2014**, 114–125. [CrossRef]

16. Grau, F. Proyectos del realismo crítico en la era de la simultaneidad: Debates sobre la gestión de la información y las actuaciones en la ciudad construida. *Palimpsesto* **2014**. [CrossRef]

17. Triviño Barros, G. The building as a system of management of information. *Inf. Construcción* **2003**, *55*, 53–62. [CrossRef]

18. D'Alisa, G.; Economía circular. *Pluriverse: A Post-Development Dictionary*; Kothari, A., Salleh, A., Escobar, A., Demaria, F., Acosta, A., Eds.; Icaria Editorial: Barcelona, Spain, 2019; Volume 8, p. 55, ISBN 978-84-9888-884-3.

19. Wandl, A.; Balz, V.; Qu, L.; Furlan, C.; Arciniegas, G.; Hackauf, U. The circular economy concept in design education: Enhancing understanding and innovation by means of situated learning. *Urban Plan.* **2019**, *4*, 63–75. [CrossRef]

20. ARUP. *The Circular Economy in the Built Environment*; ARUP: San Francisco, LA, USA, 2016.

21. Remøy, H.; Wandl, A.; Ceric, D.; van Timmeren, A. Facilitating circular economy in urban planning. *Urban Plan.* **2019**, *4*, 1–4. [CrossRef]

22. de Asiain Alberich, M.L.; Latapié Sére, M. Propuestas para el empoderamiento de los ciudadanos; Participación social ante el cambio climático desde un enfoque arquitectónico y urbano. In *Memoria del XXXVI Encuentro de la Red Nacional de Investigación Urbana, AC. Cambio Climático y Expansión Territorial*; Valladares Anguiano, R., Chávez González, M.E., Eds.; Programa Editorial de la Red de Investigación Urbana: México D.F., México, 2014; pp. 281–301, ISBN 978-968-6934-33-5.

23. de Asiain Alberich, M.L. Indicadores de sustentabilidad en urbanismo. In *Diálogos Entre Ciudad, Medio Ambiente y Patrimonio*; Anguiano, R.V., Ed.; Universidad de Colima: Colima, Mexico, 2014; pp. 100–106, ISBN 978-607-8356-08-9.

24. Bottero, M.; Bragaglia, F.; Caruso, N.; Datola, G.; Dell' Anna, F. Experimenting community impact evaluation (CIE) for assessing urban regeneration programmes: The case study of the area 22 @ Barcelona. *Cities* **2020**, *99*, 1–16. [CrossRef]

25. Leva, G. *Indicadores de Calidad de Vida Urbana.Teoría y Metodología*; Politike: Buenos Aires, Argentina, 2005.

26. de Asiain Alberich, M.L.; Valladares Anguiano, R.; Chávez González, M. Habitabilidad y Calidad de Vida como indicadores de la función adaptativa del habitar en el entorno urbano. In *Diversas Visiones de la Habitabilidad*; Anguiano, R.V., Ed.; Programa Editorial de la Red de Investigación Urbana de México: Puebla, México, 2015; pp. 71–89, ISBN 9789686934366.

27. López de Asiain Alberich, M. Participatory Approach to Urban Regeneration Processes. In Proceedings of the 3 rd International Congress on Sustainable Construction and Eco-Efficient Solutions; Universidad de Sevilla. Escuela Técnica Superior de Arquitectura: Seville, Spain, 27–29 March 2017; pp. 77–89.

28. Alvarez Sainz, M. La ciudad en la economía de la experiencia y el rol de los ciudadanos. Necesidad de participación ciudadana en Bilbao. In Proceedings of the XIII Coloquio Internacional de Geocrítica. El control del espacio y los espacios de control; Universitat de Barcelona: Barcelona, Spain, 5–10 May 2014; Volume 18, p. 24.

29. Carmichael, L.; Lambert, C. Governance, knowledge and sustainability: The implementation of EU directives on air quality in Southampton. *Local Environ.* **2011**, *16*, 181–191. [CrossRef]

30. Longo, D.; Boeri, A.; Gianfrate, V.; Palumbo, E.; Boulanger, S.O.M. Resilient cities: Mitigation measures for urban districts. A feasibility study. *Int. J. Sustain. Dev. Plan.* **2018**, *13*, 734–745. [CrossRef]

31. de Asiain Alberich, M.L.; Cano Ruano, B.; Mendoza, S. Proyecto EUOBs. Mejorando la calidad de vida de los ciudadanos desde la sostenibilidad EUOBs Project. Trying to Improve the Quality of Life of Citizens by Working in terms of Sustainability. *WPS Rev. Int. Sustain. Hous. Urban Renew.* **2015**, *1*, 1768–2387.

32. Gamboa, G.; Munda, G. The problem of windfarm location: A social multi-criteria evaluation framework. *Energy Policy* **2007**, *35*, 1564–1583. [CrossRef]

33. Berkes, F.; Ross, H. Community Resilience: Toward an Integrated Approach. *Soc. Nat. Resour.* **2013**, *26*, 5–20. [CrossRef]

34. Campos-Sánchez, F.S.; Abarca-Álvarez, F.J.; Domingues, Á. Sostenibilidad, planificación y desarrollo urbano. En busca de una integración crítica mediante el estudio de casos recientes. *Archit. City Environ.* **2018**, *12*, 39–72. [CrossRef]

35. Pierre, J. Models of urban governance: The institutional dimension of urban politics. *Urban Aff. Rev.* **1999**, *34*, 372–396. [CrossRef]

36. Tomás Fornés, M.; Cegarra Dueñas, B. Actores y modelos de gobernanza en las Smart cities. *URBS. Rev. Estud. Urbanos Cienc. Soc.* **2016**, *6*, 47–62.

37. Margot Breton, M. Neighborhood Resiliency. *J. Community Pract.* **2001**, *9*, 21–36. [CrossRef]

38. Martínez López, M. Dimensiones múltiples de la participación ciudadana en la planificación espacial. *Reis* **2011**, *133*, 21–42. [CrossRef]

39. Castells, M. *La Ciudad y las Masas: Sociología de los Movimientos Sociales Urbanos*; Alianza Universidad. Textos 98; Alianza: Madrid, Spain, 1986; ISBN 84-206-8098-2.

40. Borja, J. Ciudadanía y espacio público. *Ambiente Desarro.* **1998**, *XIV*, 13–22.

41. Alguacil Gómez, J. La Calidad de Vida y el Tercer Sector: Nuevas Dimensiones de la Complejidad; Ciudades para un Futuro más Sostenible. *Boletín* **1997**, *3*, 35–47.

42. Max-Neef, M.; Elizalde, A.; Hopenhayn, M. *Desarrollo a Escala Humana. Opciones Para el Futuro*; Centro de Alternativas al Desarrollo CEPAUR: Santiago, Chile, 1986.

43. Webb, R.; Bai, X.; Smith, M.S.; Costanza, R.; Griggs, D.; Moglia, M.; Neuman, M.; Newman, P.; Newton, P.; Norman, B.; et al. Sustainable urban systems: Co-design and framing for transformation. *Ambio* **2018**, *47*, 57–77. [CrossRef]

44. Rabbiosi, C. Urban regeneration 'from the bottom up'. *City* **2017**, *4813*. [CrossRef]

45. Venturini, G.; Riva, R.; Architettura, D.; Costruito, A.; Milano, P. Regie e processi innovativi nel progetto di riattivazione sociale e rigenerazione ambientale degli spazi pubblici residuali. *TECHNE* **2017**, *14*, 343–351. [CrossRef]

46. Müller, M.; Schmid, J.F.; Schönherr, U.; Weiß, F. *Null Euro Urbanismus*; Universität Hamburg: Hamburgo, Germany, 2008.

47. Bialski, P.; Derwanz, H.; Otto, B.; Vollmer, H. 'Saving' the city: Collective low—Budget organising and urban practice. *Ephemera* **2015**, *15*, 1–19.

48. Manzini, E. *Design, When Everybody Designs: An Introduction to Design for Social Innovation*; Design thinking, design theory; The MIT Press: Cambridge, MA, USA, 2015; ISBN 0-262-32864-X.

49. Ecosistema Urbano Ecosistema Urbano. Available online: https://ecosistemaurbano.com/es/ (accessed on 5 September 2020).

50. La Col LaCol. Available online: http://www.lacol.coop/proj/rehabitar-can-batllo (accessed on 5 September 2020).

51. Chen, Y. *El Campo de la Cebada*; YouTube: Madrid, Spain, 2012.

52. GSA Madrid. *Oasis El Ruedo*; Vimeo; GSA: Madrid, Spain, 2011.

53. [Espacio Elevado al Público]—Plataforma de Acción + Reflexión en Búsqueda de Mejoras en Las Lógicas y Dinámicas Urbanas Que Eleven Al Público Como Protagonista. Available online: https://espacioelevadoalpublico.wordpress.com/ (accessed on 5 September 2020).

54. Rizoma Fundación. Available online: http://www.rizoma.org/rizoma_fundacion.html (accessed on 5 September 2020).

55. Paisaje Transversal—Oficina de Planificación Urbana Integral. Available online: https://paisajetransversal.com/ (accessed on 5 September 2020).

56. European Commission—Directorate-General for Agriculture and Rural Development—Unit, C. 4 *Guidelines, Evaluation of Leader/Clld*; European Commission: Brussels, Belgium, 2017.

57. Till, J. *Architecture Depends*; MIT Press: Cambridge, MA, USA, 2009; ISBN 9780262012539.

58. United Nations. *Transforming our World: The 2030 Agenda for Sustainable Development*; United Nations: Quito, Ecuador, 2015.

59. Arnstein, S.R. A Ladder of Citizen Participation. *J. Am. Plan. Assoc.* **1969**, *35*, 216–224. [CrossRef]

60. Ministerio de Vivienda y Urbanismo de Chile. *Inventario de Metodologías de Participación Ciudadana en el Desarrollo Urbano*; Ministerio de Vivienda y Urbanismo de Chile: Santiago, Chile, 2010; ISBN 9789567674145.

61. Paisaje Transversal La Regeneración Urbana Integral: El Paso Hacia la Sostenibilidad de Las Ciudades. Available online: https://paisajetransversal.org/2019/08/regeneracion-urbana-integral-participativa-el-paso-sostenibilidad-ciudadades-ruip-barrio/ (accessed on 31 July 2020).

62. Pretty, J.N. Participatory learning for sustainable agriculture. *World Dev.* **1995**, *23*, 1247–1263. [CrossRef]

63. Mussinelli, E.; Tartaglia, A.; Fanzini, D.; Riva, R.; Cerati, D.; Castaldo, G. New Paradigms for the Urban Regeneration Project Between Green Economy and Resilience. In *Regeneration of the Built Environment from a Circular Economy Perspective*; Springer: Berlin/Heidelberg, Germany, 2020; p. 380, ISBN 978-3-030-33255-6.

64. Tsoukias, A.; Montibeller, G.; Lucertini, G.; Belton, V. Policy analytics: An agenda for research and practice. *EURO J. Decis. Process.* **2013**, *1*, 115–134. [CrossRef]

65. Silli, F.; Cuidado de Los Comunes: Reglamento de Bolonia. Un Comentario: Prototyping. Available online: http://www.prototyping.es/procomun/cuidado-de-los-comunes-reglamento-de-bolonia-un-comentario (accessed on 31 July 2020).

66. Johnson, L. The community planning handbook: How people can shape their cities, towns and villages in any part of the world. *Aust. Plan.* **2016**, *53*, 263–264. [CrossRef]

67. Borrallo-Jimenez, M.; LopezDeAsiain, M. Confort y sostenibilidad en la arquitectura habitada. Aplicación del conocimiento a la sociedad para la toma de decisiones. In *De Forma et vita La Arquitectura en la Relación de lo Vivo Con lo no Vivo*; Athenaica: Sevilla, Spain, 2020; ISBN 978-84-17325-94-7.

68. Dente, B. *Understanding Policy Decisions*; Springer: Berlin/Heidelberg, Germany, 2014; ISBN 9783319025193.

Article

Implications of BREEAM Sustainability Assessment on the Design of Hotels

María M. Serrano-Baena [1], Paula Triviño-Tarradas [1], Carlos Ruiz-Díaz [2] and Rafael E. Hidalgo Fernández [1,*]

1   Department of Graphic Engineering and Geomatics, University of Cordoba, Campus de Rabanales, 14071 Cordoba, Spain; ep2sebam@uco.es (M.M.S.-B.); ig2trtap@uco.es (P.T.-T.)
2   Independent Researcher, Manchester M3 6GE, UK; cruizarchitect@outlook.com
*   Correspondence: ig1hifer@uco.es

Received: 23 July 2020; Accepted: 12 August 2020; Published: 13 August 2020

Abstract: This original research paper analyses the actual and important topic of the implications of BREEAM sustainability assessment on the design of hotels and it is a personal response to "The Agenda 2030 for Sustainable Development" and its influence on the Tourism and Hospitality Industry. The paper aims to examine the influence of the sustainable assessment method BREEAM on the design of hotels by using seven case studies and studying the changes that were implemented in order to achieve their targets. Qualitative data were obtained by conducting in-depth interviews and analyzing the supplied documentation. The authors notice that the results revealed that a BREEAM approach might limit the design of the hotels but, including the right measures at the early design stage of the project, the target can be easily achieved.

Keywords: hospitality; hotels; sustainable assessment; BREEAM methodology; green hotel

---

1. Introduction

Over the past decades, sustainable development has become very important in every economic sector but especially in the Construction and Tourism sectors [1]. Tourism is one of the most powerful drivers of development for the world economy and in particular, international tourism has been considered as a tool for development in emerging economies. Developing countries are leaders in including sustainability procedures for building and hotels are becoming more eco-friendly by including environmental standards in their design, architecture and management [2,3].

Hotels are key in this process of sustainability; in places where tourism is the main contributor to GDP (Gross Domestic Product), the addition of a sustainable design and sustainable certifications for new buildings and renovations is crucial [3]. For a decade, the Hospitality sector has been pressured to include environmental strategies and reduce the carbon footprint by government legislation, in order to minimize its impact on the environment [4,5].

Despite its involvement in sustainability, there is not much research on what is the best way to approach a "green hotel" and how the design can be affected by including sustainable measures [6]. This article is organized into three main sections. The first offers a brief overview of the impact of the Agenda 2030 on Tourism. The second exposes the existing literature about the importance of sustainable certificates on the Hospitality Industry and the background of one of the first sustainable European building assessment methods applicable to residential and commercial buildings, the Building Research Establishment Environmental Method (BREEAM) [7]. The last part aims to analyse by a qualitative method the impact of BREEAM on the design of hotels.

Impact of 2030 Agenda on the Tourism

In 2015, the Heads of State and Government of 193 countries met at the 70th General Assembly of United Nations and approved the resolution of "The 2030 Agenda for Sustainable Development". This plan is call to action to protect our planet, end poverty and improve the lives and prospects of everyone. The 2030 Agenda sets out 17 Sustainable Development Goals (SDGs) and 169 targets to be achieved by 2030 that are considered at the moment a global emergency to mitigate and balance the three dimensions of sustainable development, the economic, social and environmental [8,9]. This plan must be implemented entirely and as a whole, as all the different goals and targets are related to each other. The United Nations Resolution states that in order to achieve the completion of the SGDs, all the nations must take responsibility. The European Union has played an active role and will implement the 2030 Agenda internally and globally in cooperation with partner countries [10].

The UNWTO (World Tourism Organization) establishes that tourism contributes directly or indirectly to the achievement of all SDGs and in particular, it is included as a target in the Goals 8, 12 and 14 [8,11].

- SDG 8. Promote sustained, inclusive and sustainable economic growth, full and productive employment and decent work for all. According to a study from WTTC (World Travel and Tourism Council), Travel and Tourism was the sector with the fastest growth in 2018 (3.1%), ahead of the Construction (2,8%) and Banking (2.6%) sectors. In addition, Travel and Tourism sustained a total of 319 million jobs across the world and induced 10% of all jobs, exceeding the impacts of the Financial, Health and Banking sectors, among others [2,3]. The contribution of the Tourism sector is specified in Target 8.9 "By 2030, devise and implement policies to promote sustainable tourism that creates jobs and promotes local culture and products" [8].
- SDG 12. Ensure sustainable consumption and production patterns. Tourism can play an important role in the evolution of a green economy and inclusive growth [2]. If the tourism sector adopts sustainable consumption and production practices (SCP), it can significantly accelerate the shift to a more sustainable planet [11]. The One Planet Sustainable Tourism Programme has the objective to improve the sustainable development impacts of tourism by promoting SCP that use natural resources and produce less waste [12]. The inclusion of "green hotels" is key for this goal. Clients are experiencing an awareness of environmental damage and the addition of sustainable measures in hotels is becoming a very important factor for their design [13].
- SDG 14. Conserve and sustainably use the oceans, seas and marine resources for sustainable development.

Coastal and maritime tourism are tourism's biggest segments. It is specified in target 14.7 that "by 2030, increase the economic benefits to small island developing States and least developed countries from the sustainable use of marine resources, including through sustainable management of fisheries, aquaculture and tourism" [8]. Small Island Developing States (SIDS) rely on healthy marine ecosystems. Integrated Coastal Zone Management must include tourism development to preserve fragile marine ecosystems and promote a blue economy [11].

2. Literature Review

2.1. The Importance of Sustainability Certifications for the Hospitality Industry

In 1993, the World Tourism Organisation (UNWTO) put forward the concept of Sustainable Tourism Development. In 1995, United Nations Educational, Scientific and Cultural Organization (UNESCO), United Nations Environment Programme (UNEP) and UNWTO held in Spain the first World Conference on Sustainable Tourism and the Charter for Sustainable Tourism was adopted in this meeting. Since then, Sustainable Tourism has occupied a dominant position in the Tourism industry [14,15]. Although the UNWTO has established that tourism contributes to the completion of the SDGs [10], the Hospitality industry faces the challenge of determining which of the 17 SDGs and associated targets are its priority [16].

Widely, the building sector accounts for one third of energy-related carbon emissions [17,18]. In the UK, buildings are responsible for half of the carbon emissions, half of the water consumption, one third of the landfill waste and one quarter of all raw materials used in the economy [19]. Hotels' users consume a great amount of water and energy at the same time that generates an important amount of solid waste and effluents on a daily basis [20], but hotels may have a significant and positive impact on the environment by changing some of their management and design aspects. Customers have radically changed their attitude towards adopting environment-friendly practices and as a result, hotels cannot ignore their environmental and social responsibilities [13,21]. For this, many hotels are now in the process of becoming "green hotels", but it has been demonstrated that most of them only focus on small changes such as reusing linen and towels [22,23]. A green hotel must operate according to the principles of green hospitality; this does not mean planting some trees in front of it, but being environmentally friendly, implementing waste management systems, recycling, and saving in water and energy, among other procedures [24].

The guests expect the implementation of more sustainable and eco-friendly practices such as water conservation, energy efficiency and proper waste division [25,26] and this approach should be incorporated from the planning stage to the demolition phase of the building. A green building must be designed as a whole and covers all the phases such as design, construction and operation [27].

As it has been studied before, the term smart tourism is closely related to the application of new technologies [28,29]. They propose to rethink the traditional approach to tourism and include the latest technology such as smartphones and tablets in its planning and programming. New technologies can enhance tourism in different ways; several studies demonstrate that hotels display, on the internet, their certificates and environmental practices to show their customers their awareness of the environment [30,31]. Studies indicate that the application of environmental procedures and their consequent certifications can improve the image of the company and its operating performance [32]. As a consequence, environmental assessments of buildings have become one of the most important steps in the sustainable built environment [5,7,33]. The European Union Energy Performance of Buildings Directive (EPBD) [34] and the UK Climate Change Act of 2008 set sustainable buildings on the UK policy agenda; since then, a wide variety of tools have been developed to assess and help construction projects, and BREEAM is one of the most successful of these tools [35].

2.2. Background of BREEAM Assessment

At first in the UK and now worldwide, BREEAM is leading the list of sustainability assessments; between 2013 and 2017, over 10,800 certified assessments were issued at both the Design and Post-construction stage [36]. The method was launched in the UK in 1990 by the Building Research Establishment (BRE); it was initially designed to focus predominantly on environmental aspects [33,37], but in the past decade it has also highlighted economic and social aspects. It has been applied in 77 countries [7,38].

BREEAM considers ten categories to measure sustainable value, including management, health and wellbeing, energy, transport, water, materials, waste, land use and ecology, pollution and innovation [38]. Each of these categories is divided into a range of assessment issues with its own target, aim and benchmarks.

A BREEAM assessor will determine when a target or benchmark is reached and will award with score points or credits as per Table 1, then these are weighted and aggregated on a scale of outstanding, excellent, very good, good, pass and unclassified [38], as shown in Table 2.

Table 1. BREEAM Environmental section weightings.

Environmental Section	Weighting (%)			
	Fully Fitted Out	**Simple Building**	**Shell and Core Only**	**Shell Only**
Management	11	7.5	11	12
Health and Wellbeing	14	16.5	8	7
Energy	16	11.5	14	9.5
Transport	10	11.5	11.5	14.5
Water	7	7.5	7	2
Materials	15	17.5	17.5	22
Waste	6	7	7	8
Land Use and Ecology	13	15	15	19
Pollution	8	6	9	6
Total	100	100	100	100
Innovation (additional)	10	10	10	10

This weighting system is defined in more detail within the BRE Global Core Process Standard (BES5301).

Table 2. BREEAM rating benchmarks.

BREEAM Rating	% Score
Outstanding	≥85
Excellent	≥70
Very Good	≥55
Good	≥45
Pass	≥30
Unclassified	<30

BREEAM is regularly updated to ensure it meets the requirements for building sustainability and it is used for public and private projects, applicable to residential and commercial buildings [7]. In some cases, it is a mandatory requirement to satisfy certain planning conditions or regulations, particularly for the public sector projects. In other cases, it is used voluntarily to earn recognition due to its international prestige [33], and it is highly valued when it comes to indicate the carbon emissions of commercial buildings.

It has been demonstrated that hotels and tourism benefit from the application of sustainable assessments such as BREEAM [32]. In addition, the method has shown its support to the SDGs and the Agenda 2030 and has demonstrated how and where the BREEAM family of standards and tools support the SDGs [39]. It has highlighted its significant contribution to meeting the following goals [8]:

- SDG 3. Ensure healthy lives and promote well-being for all at all ages.
- SDG 6. Ensure availability and sustainable management of water and sanitation for all.
- SDG 7. Ensure access to affordable, reliable, sustainable and modern energy for all.
- SDG 9. Build resilient infrastructure, promote inclusive and sustainable industrialization and foster innovation.
- SDG 11. Make cities and human settlements inclusive, safe, resilient and sustainable.
- SDG 12. Ensure sustainable consumption and production patterns.
- SDG 13. Take urgent action to combat climate change and its impacts.
- SDG 15. Protect, restore and promote sustainable use of terrestrial ecosystems, sustainably manage forests, combat desertification, and halt and reserve land degradation and halt biodiversity loss.

Despite its importance, little research has been conducted about the effects the application of this sustainability assessment method might have on the hotel's design and management process. Holmes and Hudson [40] examined how BREEAM affects the design of an office building but the method has been through many changes since this study was taken and they focused more on the perceived value of the method than its design effects. Lowe and Watts [41] analysed the benefits of implementing BREEAM on a two storey medical centre development but this research was focussed on the financial implications of the method. Fenner and Ryce [42] compared BREEAM with the LEED method, analysing their limitations in current practice, but this research did not focus on the design aspect. Several studies taken from Haroglu and Schweber [33,35,43] analysed the implication that BREEAM might have on the design process although these studies used different types of buildings as case studies such as educational, commercial, residential and healthcare. Mengxue Lu [7] compared four rating tools: BREEAM, Leadership in Energy and Environmental Assessment (LEED), Green Star and Building Environmental Assessment method (BEAM Plus), and concluded that all of them were similar, covering the same environmental aspects, in terms of energy audit. The carbon audit, however, was not a common investigation focus in any of them.

This paper examines how the implementation of BREAAM impacts on the design of hotels and their management processes and demonstrates the changes that have occurred to the projects in a specific context as a result of the assessment. The intention of the study is not to generalize across specific design features of the hotels but rather to provide a general understanding of the influences and impact that the BREEAM method might have on the hotels' design stage. Due to the lack of research on this field and its importance in Tourism and Hospitality, this study examines the impact of the BREEAM assessment method on the design process of seven hotels in the UK.

3. Methodology

This study uses a qualitative method of analysis. The qualitative approach to research seeks to analyze the effect of BREEAM on hotels' design decisions; this is not confined to a specific phase of the projects but it comprehends the entire process from conception to completion and beyond. Although some researchers have investigated before about sustainable assessments and specifically BREEAM, little research has been focused on the implications that the method might have on the design of buildings [30,32,40] and no prior research has been found were the case study was a hotel. Hence, the study presents this limitation and the need for further development in this area of research. In addition, the qualitative method used in the research is based on human experience and is dependent on the individual skills of the researcher and the people interviewed.

In the present study, a number of specialized Hospitality professionals, both architects and architect technicians, were selected based on their level of expertise in the field and quality of hotels produced during their careers. All of them are Hospitality focused and the main people responsible for the selected projects of this study. They offered a variety of hotel projects assessed under BREEAM. In total, 7 case study hotels were selected, each of them with different characteristics, stages and BREEAM targets. As a context, all of the hotels are based in the UK due to the importance of this method there.

Architects and architectural technicians responsible for the design of the buildings were interviewed. The interviews were undertaken during April and May of 2020 and, due to the worldwide pandemic of COVID 19 and the mandatory lockdown in the UK during these months, it was not possible to do the interviews face to face. Hence, they were done via Yealink Meeting Server with each technician and architect. Each interview lasted an average of 30 min per project and they were recorded with Windows 10 Screen Recorder. The collection of data was also supplemented with documentation provided by the designers including plans and elevations, BREEAM assessments, planning decision notices and relevant information from several professionals that influenced in the design process of the hotels, with all of the information in pdf format. In Table 3, the details of the selected cases are shown such as the stage of the project, number of rooms, number of storeys and BREEAM target and final score. For confidentiality, the projects are named under the headings A, B, C, D, E, F and G.

Table 3. Case study hotel projects.

Case Studies	Stage	N° Rooms	N° Storeys	BREEAM Target	BREEAM Final Score
A	Completed	216	5	Excellent	Excellent
B	Completed	302	7	Very Good	Very Good
C	Under construction	82	9	Very Good	On going
D	Completed	339	42	Excellent	Outstanding
E	Under construction	456	13	Excellent	On going
F	Under construction	153	9	Very Good	On going
G	Under construction	329	19	Very Good	On going

4. Results

An extensive amount of data was collected from the interviews about the influence that the BREEAM assessment method has on various aspects of the hotel design. This section includes two parts, the first is a description and a brief of the information obtained from the architects and technicians about the required changes during the design process of each case study. The second part exposes the key design changes extracted from the interviews that were crucial in all the cases.

4.1. General Description of Case Studies

In case study A, the client was very interested in getting a BREEAM Excellent score and it finally achieved the highest ever BREEAM score for a hotel in the UK. The building is now utilising a range of low energy technologies to reduce its energy demand and generates its own energy supply, delivering an 87% reduction in CO_2. The negotiation process with the assessor involved many changes in the project, mainly in the Mechanical and Electrical designing phase, where the assessor asked for bigger plant rooms and more plant areas in addition to several systems that were needed to achieve the credits. It was required to incorporate a number of sustainable construction techniques including enhanced isolative building elements and a sustainable surface water drainage system. In addition, the hotel harvests rainwater, which is used to serve the toilets in the building and to water the external landscaping areas. Increasing the size of cycle store was also needed; the technician explained that this is very common practice when working on a BREEAM project since it is an easy way to get the number of credits required and usually it does not impact excessively the design of it. Finally, the designers had to increase the size of the refuse store as a result of accommodating more numbers of bins and also provide an external bin store. The materials used for this building were assessed in order to meet the Building Regulations and BREEAM criteria, resulting in a combination of brick, metal and stone cladding.

Case study B was a regeneration and further extension of an existing historical building; as a result, the design process required special attention in this case. The hotel is located in the city centre and had a planning requirement to achieve a BREEAM Very Good score. A pre-assessment was submitted as part of the planning application, which is the standard process for most of the buildings in the UK. The planners reviewed it and conditioned it to make sure that the score was achieved; as a consequence, certain changes were needed in order to get BREEAM Very Good as the final score. The designers had to increase the number of cycle storage spaces in the building in order to get the credits related. Under the ecology requirements, even though it was a regeneration of an existing site, the architects and technicians had to provide an enhancement. They had to include certain numbers of bat boxes housed on the roof levels designated for nesting bats. This requirement was needed under the premises that the demolition of an adjacent building was needed in order to extend the hotel and it was deemed that bats could have nested there if they wanted to, so in order to replace this, bat boxes were included. The number and size of bins were increased and, as a consequence, the refuse store was also slightly expanded to accommodate them. In regard to the building envelope, some of the windows had to be upgraded due to the fact that the acoustic requirements of BREEAM were higher than the hotel operator's requirements. The specification of the glass was changed but the size and position of the windows remained the same, so the design of the facade was not altered. Finally, the material of the new hotel was proposed as rendered to match the existing building which, from a BREEAM perspective, is not ideal due to its characteristics but planning required it to keep the same material in order to minimise the impact of the extension.

Case study C is an existing building refurbishment and extension, it has planning approval and is currently under construction. The hotel is on the boundary on three sides and they are extending it out on the fourth side. There is no space for landscaping and the existing building has a pitched roof, so there were no opportunities for proposing green roofs, bat or bird boxes and planters at ground floor; as a consequence, the ecology credits were difficult to achieve. In order to compensate this, the BREEAM assessor advised that during the design stage, the architects should carry out an assessment to demonstrate that they were not able to provide ecological enhancements. The Mechanical and Electrical engineer presence was crucial in this hotel; although they always have an important role on the completion of BREEAM projects, in this specific case and due to the lack of ecological enhancements, the engineers had to bring new solutions to compensate this lack. The installation of flow restrictors helps to save water by limiting the flow of particularly wasteful taps and showers. The architect explained that the water consumption is always a common factor to take into account for Hospitality projects and its right management can drastically change the running cost of the hotel. Finally, to meet BREEAM acoustic requirements, the technicians had to enhance the existing glazing by adding a secondary glazing. At the moment, they are on target for achieving BREEAM Very Good.

Case study D is a highly sustainable skyscraper building with a small footprint and 42 floors. It had a planning requirement for a BREEAM Excellent but it finally achieved an Outstanding, the highest BREEAM rating available at the time of completion. Ecology had a major impact on the project. Solar panels and a green roof were included in addition to beehives on the 39th floor to produce honey for guests. On the lower levels, bat and bird boxes for nesting were also incorporated. In addition, due to the small footprint of the building, it was impossible to include a cycle store within it, so a separated external cycle store was designed with a green roof and green walls under the recommendations of the ecologists. Finally, it is important to mention that the mechanical system of the hotel has a vital importance in this project. Several measures were included such as a combined heat power system that contributes to a 30 per cent reduction in CO_2 emissions and light regulation system that adjusts the level of light according to the time of day and season. In addition, waste management is crucial in this case; the hotel recycles cooking oil, general waste, soap and bottled bathroom products. Moreover, it uses eco-friendly products from local suppliers.

Case study E has planning approval and is currently under construction. In this project, there were no planning requirements for BREEAM but the client founders had a BREEAM target of Excellent; they requested the BREEAM assessment method to help to reduce the running cost of the hotel. The architect explained that the energy and water consumption in hotels is massive so by including some procedures that will reduce the impact of that, the benefits will be higher for the owners. As well as that, one of the hotel operators also had, as an employee requirement, the achievement of BREEAM Very Good for their projects, so although in this case it is a lower target than the founder´s, it is still important to take into consideration. The pre-assessment of the project is done and the building has a potential target of 79% score for BREEAM Excellent, higher than the 70% minimum score needed so it is highly possible that the project will achieve the targeted score. In this case study, they also had to increase the cycle and bin stores' sizes in order to achieve the credits needed. For the ecology requirements, the designers had to include a total of 65 square metres of planting distributed on three separate roof levels as a consequence of filling the site completely by the building at ground level. The ecologists made a report with their requirements and due to the lack of space for planting, the landscape architect had to include all these planters on different levels of the building to achieve the needed credits. Internally, the architects had to increase the thickness and specifications of the partitions to meet BREEAM acoustic requirements. The technicians had to enhance the glazing specifications for some of the external windows due to the fact that the operators had some requirements for the windows and, in some instances, BREEAM was above them so they had to upgrade the windows in order to achieve the acoustic requirements. The architect explained that one of the hotel operators had very strict thermal requirements, which were well above Building Regulations and BREEAM requirements so the insulation of the building did not need any adjustment to meet the credits. Finally, an adaptability study was needed, which is a document that studies the spatial, structural, and service strategies of the building and analyzes the malleability of it in response to changing operational parameters over time [44]. The study proved that the hotel will be able to change its use if needed, mainly because it is built in lightweight partitions and blade columns.

Case study F is under construction at the moment. This project comprehends two buildings in the same site, a refurbished existing office building and a newly built hotel with some affordable accommodations included. Due to there being three different uses sharing the site, planning required three separated permanent cycle stores and temporary short-term cycle storage. The temporary cycle store had to be increased as the BREEAM assessor recommended. In this case, it was a planning requirement to increase the number of bird and bat boxes that the ecologists advised, so BREEAM did not influence this decision. Since the hotel is a new building, the designers decided to go for blade columns and lightweight partitions to facilitate the adaptability of the building if it changes its use, so BREEAM rated this positively.

Case study G has planning approval and it is under construction; it has planning and operator's requirements for BREEAM Very Good. There is a planning condition that it must get this targeted score, so a pre-assessment was done in the first place to study the best way to meet this requirement. As in the previous case studies, the main changes were the bin and cycle stores and the inclusion of planters at ground floor level around the building. After a few discussions with the ecologists, some of the planters were omitted as they contained trees that are no longer proposed; the architect explained that this aspect is being revised at the moment. Currently, the design and illuminance of external lighting, including the hotel signage, has been highlighted by the assessor to take into consideration in order to reduce future risks of non-conformance. A lighting assessor will revise these and recommend the best solution; as a consequence, it will have a minor impact on the external design.

4.2. Key Design Changes

Several changes were needed during the design process of the described projects. The key changes applied to the buildings explicitly to obtain the required BREEAM credits and under the instructions of the BREEAM assessor have been identified below. In addition to these changes, the presence of the Mechanical and Electrical engineer is crucial for the design of the hotels. As it has been demonstrated, hotels are linked to produce high levels of water consumption; people tend to consume more water when they stay in hotels than in their homes [45,46]. Implementing the right strategies, technologies and innovation measures help the hotels to reduce their water consumption drastically [47]; case studies C and E included water management systems. The key design changes can be classified under three main groups: upgrade through layout amendments, upgrade through performance and upgrade through additions. On the other hand, Haroglu [33] classified the key design changes under three main areas: building features, materials and water services.

4.2.1. Upgrade through Layout Amendments

These changes were related to both internal and external amendments of the layout.

The increase in the cyclist facilities in order to house a greater number of cycles is a common factor in all the case studies. This amendment falls under the Transport section and credits are given with the adequate provision of cyclist facilities to promote exercise and help to reduce congestion and CO_2 emissions [48]. It is a relatively easy way to get BREEAM credits and does not have a major impact on the layout.

Another key change that was required in the case studies A, B, E and G was the enlargement of the bins storage areas due to increase in the number of bins; this amendment falls under the Waste section. With the provision of dedicated storage facilities, BREEAM aims to promote sustainable waste management and divert recyclable waste from landfill or incineration [49]. With an inefficient management programme, hotel owners pay twice for their waste, one is in the form of packaging and the other is for their disposal. It has been demonstrated that around the 30% of a hotel's solid waste can be recycled and reused [50].

4.2.2. Upgrade through Performance

A different category of measures that can be applied to improve the efficiency of the hotels, especially for refurbished and existing buildings, are those related to the performance of their components.

In case studies B, C and E, the architects had to change the type of windows by upgrading the glazing specifications and also, in case study E, they had to increase the thickness of the partitions to meet BREEAM acoustic requirements. These are associated with the Acoustic performance subcategory of the Health and Wellbeing section of BREEAM [51]. In this same section, the team involved in case study G are working on the enhancement of lighting impact to ensure artificial lighting is considered at the design stage to minimize its impact [52].

4.2.3. Upgrade through Additions

Finally, there are specific elements that help to obtain the BREEAM target score and will help with the completion of a sustainable or green hotel. Mostly two main categories can be classified in this group, Energy and Land Use and Ecology.

Under the first category, solar panels are an immediate response when working on a sustainable hotel as happened in our case study D. BREEAM promotes the reduction in atmospheric pollution and carbon emissions and encourages local energy generation taken from renewable sources [53].

Certain actions are adopted to enhance the ecology of the site. These fall under the Land Use and Ecology BREEAM category, specifically under the subcategory Enhancing site ecology [54]. The installation of bird and bat boxes at strategic locations on the site was a measure used in case studies B, D and E. Planters with native species or that are beneficial to local wildlife were distributed on different levels of the hotels in case studies D, E and G. Lastly, case study D has a green roof that improves air and water quality and creates a wildlife habitat.

5. Discussion

The results obtained in this work have important implications on what is the best way to approach the design of a sustainable hotel. The details obtained from the interviews suggest that the application of the sustainability assessment method of BREEAM on the project, should be planned and done since the first stage of its design in order to facilitate the inclusion of the actions or elements recommended by the BREEAM assessor. In addition, it has been found that the inclusion of a BREEAM assessor at the very early stages of the project and the right technicians will accelerate and facilitate the execution of a green hotel. This aligns with the comments by Haroglu [33] about the significance of how the assessment process is handled and how the early involvement of BREEAM assessors can play an important role in the design. Fenner and Ryce [42] also state that the assessor's approach might have a significant effect on the clarity of the BREEAM assessment. It has been noticed that, in addition of the architects, two other professionals are key for the execution of the project under the BREEAM premises; they are the mechanical and electrical engineer (M&E), and the ecologist. These two technicians will help to adequately address the requirements that the assessor imposes on the hotel. This relates to the results of Lowe and Watts [41], which stated that the introduction of BREEAM on a construction development will cause an increment of the M&E workload, with most of the work being in the design of the systems.

The data collected from the interviews shows that the impact that BREEAM might have on the design depends on whether the hotel is existing or newly built. It has been explained by the architects that if the building is existing, both Planning and BREEAM are more flexible in qualifying the project under the BREEAM requirements due to the difficulty of adapting an existing hotel into a green hotel. In addition, it has been extracted that specifically for hotels, the measures to be taken are common in almost all the cases. In the case of existing buildings that require a refurbishment for their new use, the most important categories to consider are Health and Wellbeing and Land Use and Ecology. In this instance, an update of the existing windows will improve the thermal and acoustic insulation conditions. In addition, the introduction of bat and bird boxes and planters will immediately add an enhancement of the ecology of the hotel. In the case of newly built hotels, the Waste and Transport categories have been a common factor in all of them. An adequate design of the project that includes enough area designated to cycle and bins stores, will help in acquiring the necessary credits to classify the project as a sustainable hotel. Again, Health and Wellbeing and Land Use and Ecology categories are very important for newly built hotels; additionally to the elements mentioned above for refurbished hotels, the incorporation of green roofs will contribute positively to acquiring the BREEAM target score. As it has been already highlighted, the inclusion of new technological and sustainable measures, such as solar panels, is also key in new buildings.

6. Conclusions

The present study has shown the results of the investigation of seven case study hotels in order to examine the impact that BREEAM has on their design. It was found that the sustainability assessment has a major impact on the design of these buildings and the approach for designing a BREEAM hotel might seem challenging. The main implications of the sustainability assessment were found to be in the Health and Wellbeing and Land Use and Ecology categories for both refurbished and newly built hotels but some categories, such as Waste and Transport, have a major importance for newly built hotels.

The study utilises BREEAM as the tool for measuring the sustainability of different hotels because it is the most common sustainable method used in the UK, but the results can be extrapolated and used for reference for any other hotel that aims to become more sustainable or for those newly built hotels that need to include sustainable elements in their design to be green hotels. For instance, the key changes shown in this research can be applied for newly built or refurbished hotels.

BREEAM is expected to become even more popular and international in the following years ahead. This will implicate new restrictions and measures taken to incorporate BREEAM requirements and the need for further research on this field.

While this paper makes a contribution to understanding the impact that BREEAM has on the design elements of hotels, the importance of water and waste management in hotels has been noticed. Hence, future research could seek to identify how BREEAM impacts on the water and waste management in hotels and how the right actions can enhance their performance.

Author Contributions: Conceptualization, M.M.S.-B.; Data curation, M.M.S.-B., C.R.-D. and R.E.H.F.; Formal analysis, M.M.S.-B.; Investigation, M.M.S.-B.; Methodology, M.M.S.-B.; Project Administration, P.T.-T. and R.E.H.F.; Resources, M.M.S.-B.; Supervision, P.T.-T. and R.E.H.F.; Validation, M.M.S.-B. and P.T.-T.; Writing—original draft, M.M.S.-B.; Review—original draft, P.T.-T. and C.R.-D.; Final editing, M.M.S.-B. All authors have read and agreed to the published version of the manuscript.

Funding: This research received no external funding.

Conflicts of Interest: The authors declare no conflict of interest.

References

1. Khan, A. Tourism and Construction Have Power to Lead Move to Sustainable Economies | UNWTO. Available online: https://www.unwto.org/global/press-release/2018-07-18/tourism-and-construction-have-power-lead-move-sustainable-economies (accessed on 19 July 2020).
2. Khan, A.; Bibi, S.; Ardito, L.; Lyu, J.; Hayat, H.; Arif, A.M. Revisiting the Dynamics of Tourism, Economic Growth, and Environmental Pollutants in the Emerging Economies—Sustainable Tourism Policy Implications. *Sustainability* **2020**, *12*, 2533. [CrossRef]

3.   Abokhamis Mousavi, S.; Hoşkara, E.; Woosnam, K.M. Developing a Model for Sustainable Hotels in Northern Cyprus. *Sustainability* **2017**, *9*, 2101. [CrossRef]

4.   Berezan, O.; Millar, M.; Raab, C. Sustainable Hotel Practices and Guest Satisfaction Levels. *Int. J. Hosp. Tour. Adm.* **2014**, *15*, 1–18. [CrossRef]

5.   Pham, N.T.; Vo Thanh, T.; Tučková, Z.; Thuy, V.T.N. The role of green human resource management in driving hotel's environmental performance: Interaction and mediation analysis. *Int. J. Hosp. Manag.* **2020**, *88*, 102392. [CrossRef]

6.   Han, H.; Ariza-Montes, A.; Giorgi, G.; Lee, S. Utilizing Green Design as Workplace Innovation to Relieve Service Employee Stress in the Luxury Hotel Sector. *Int. J. Environ. Res. Public Health* **2020**, *17*, 4527. [CrossRef] [PubMed]

7.   Lu, M.; Lai, J. Review on carbon emissions of commercial buildings. *Renew. Sustain. Energy Rev.* **2020**, *119*, 109545. [CrossRef]

8.   Transforming Our World: The 2030 Agenda for Sustainable Development. Available online: https://sustainabledevelopment.un.org/post2015/transformingourworld/publication (accessed on 24 April 2020).

9.   Trivino-Tarradas, P.; Gomez-Ariza, M.R.; Basch, G.; Gonzalez-Sanchez, E.J. Sustainability Assessment of Annual and Permanent Crops: The Inspia Model. *Sustainability* **2019**, *11*, 738. [CrossRef]

10.   Boto-Álvarez, A.; García-Fernández, R. Implementation of the 2030 Agenda Sustainable Development Goals in Spain. *Sustainability* **2020**, *12*, 2546. [CrossRef]

11.   UNWTO. Tourism in the 2030 Agenda. Available online: https://www.unwto.org/tourism-in-2030-agenda (accessed on 10 April 2020).

12.   UNWTO. One Planet. Available online: https://www.unwto.org/sustainable-development/one-planet (accessed on 25 April 2020).

13.   Mbasera, M.; Plessis, E.D.; Saayman, M.; Kruger, M. Environmentally-friendly practices in hotels. *Acta Commer.* **2016**, *16*, 1–8. [CrossRef]

14.   Guo, Y.; Jiang, J.; Li, S. A Sustainable Tourism Policy Research Review. *Sustainability* **2019**, *11*, 3187. [CrossRef]

15.   Qian, J.; Shen, H.; Law, R. Research in Sustainable Tourism: A Longitudinal Study of Articles between 2008 and 2017. *Sustainability* **2018**, *10*, 590. [CrossRef]

16.   Joner, P.; Hillier, D.; Comfort, D. The Sustainable Development Goals and the Tourism and Hospitality Industry. *Athens J. Tour.* **2017**, *4*, 7–18. [CrossRef]

17.   IPCC. Climate Change 2014 Mitigation of Climate Change (Fifth Report). Available online: https://www.ipcc.ch/site/assets/uploads/2018/02/ipcc_wg3_ar5_chapter9.pdf (accessed on 17 July 2020).

18.   United Nations Environmental Program (UNEP) & IEA. Global Status Report 2017. Available online: https://ec.europa.eu/energy/sites/ener/files/documents/020_fatih_birol_seif_paris_11-12-17.pdf (accessed on 17 July 2020).

19.   HM Government Strategy for Sustainable Construction. Available online: https://webarchive.nationalarchives.gov.uk/+/http:/www.bis.gov.uk/files/file46535.pdf (accessed on 2 May 2020).

20.   Ad, A. Green Hotels and Sustainable Hotel Operations in India. *Int. J. Manag. Soc. Sci. Res. (IJMSSR)* **2017**, *6*, 2319–4421.

21.   Verma, V.; Chandra, B. Intention to implement green hotel practices: Evidence from Indian hotel industry. *Int. J. Manag. Pract.* **2018**, *11*, 24–41. [CrossRef]

22.   Wymer, W. Which theory is more effective for predicting hotel guest participation in towel and linen reuse programmes, social influence theory or attribution theory? In *Behaviour Change Models: Theory and Application for Social Marketing*; Brennan, L., Binney, W., Parker, L., Aleti, T., Nguyen, D., Eds.; Edward Elgar Publishing: Cheltenham, UK, 2014; pp. 268–275. ISBN 978-1-78254-814-0.

23.   Shang, J.; Basil, D.Z.; Wymer, W. Using social marketing to enhance hotel reuse programs. *J. Bus. Res.* **2010**, *63*, 166–172. [CrossRef]

24.   Kostić, M.; Ratković, M.; Forlani, F. Eco-hotels as an example of environmental responsibility and innovation in savings in the hotel industry. *Hotel Tour. Manag.* **2019**, *7*, 47–56. [CrossRef]

25.   Batra, A.; Shreshta, P. A Study on Green Practices and Perception of Guests in Selected Green Hotels in Bangkok. In Proceedings of the International Conference of Inclusive Innovation and Innovative Management (ICIII 2014), Pathum Thani, Thailand, 11–12 December 2014.

26.   Ogbeide, G.-C. Perception of Green Hotels in the 21st Century. *J. Tour. Insights* **2012**, *3*, 1. [CrossRef]

27. Ahn, Y.H.; Pearce, A. Green luxury: A case study of two green hotels. *J. Green Build.* **2013**, *8*, 90–119. [CrossRef]

28. Garau, C. Emerging Technologies and Cultural Tourism: Opportunities for a Cultural Urban Tourism Research Agenda. In *Tourism in the City: Towards an Integrative Agenda on Urban Tourism*; Bellini, N., Pasquinelli, C., Eds.; Springer: Cham, Germany, 2017; pp. 67–80. ISBN 978-3-319-26877-4.

29. Mehraliyev, F.; Chan, I.C.C.; Choi, Y.; Koseoglu, M.A.; Law, R. A state-of-the-art review of smart tourism research. *J. Travel Tour. Mark.* **2020**, *37*, 78–91. [CrossRef]

30. Fernández-Robin, C.; Celemín-Pedroche, M.S.; Santander-Astorga, P.; Alonso-Almeida, M.D.M. Green Practices in Hospitality: A Contingency Approach. *Sustainability* **2019**, *11*, 3737. [CrossRef]

31. Karaman, A.; Sayin, K. The Importance of Internet Usage in Hotel Businesses: A Study on Small Hotels. *J. Internet Appl. Manag.* **2017**, *8*, 65–74. [CrossRef]

32. Bagur-Femenias, L.; Celma, D.; Patau, J. The Adoption of Environmental Practices in Small Hotels. Voluntary or Mandatory? An Empirical Approach. *Sustainability* **2016**, *8*, 695. [CrossRef]

33. Haroglu, H. The impact of Breeam on the design of buildings. *Proc. Inst. Civ. Eng. Eng. Sustain.* **2013**, *166*, 11–19. [CrossRef]

34. Energy Performance of Buildings Directive. Available online: https://ec.europa.eu/energy/topics/energy-efficiency/energy-efficient-buildings/energy-performance-buildings-directive_en (accessed on 10 May 2020).

35. Schweber, L.; Haroglu, H. Comparing the fit between BREEAM assessment and design processes. *Build. Res. Inf.* **2014**, *42*, 300–317. [CrossRef]

36. Prior, J.; Holden, M.; Ward, C. *The Digest of BREEAM New Construction and Refurbishment Statistics 2013 to 2017*; BRE Global Ltd: London, UK, 2019; Volume 2.

37. Cooper, I. Which focus for building assessment methods—Environmental performance or sustainability? *Build. Res. Inf. Build. Res Inf.* **1999**, *27*, 321–331. [CrossRef]

38. How BREEAM Certification Works. Available online: https://www.breeam.com/discover/how-breeam-certification-works/ (accessed on 23 May 2020).

39. BREEAM Strategy. Available online: https://www.breeam.com/discover/resources/strategy/ (accessed on 11 May 2020).

40. Holmes, J.; Hudson, G. The application of BREEAM in corporate real estate: A case study in the design of a city centre office development. *J. Corp. Real Estate* **2003**, *5*, 66–77. [CrossRef]

41. Lowe, J.; Watts, N. An evaluation of a Breeam case study project. *Sheff. Hallam Univ. Built Environ. Res. Trans.* **2011**, *3*, 42–53.

42. Fenner, R.; Ryce, T. A comparative analysis of two building rating systems. Part I: Evaluation. *Proc. Inst. Civ. Eng. Eng. Sustain.* **2008**, *161*, 55–63. [CrossRef]

43. Schweber, L. The effect of BREEAM on clients and construction professionals. *Build. Res. Inf.* **2013**, *41*, 129–145. [CrossRef]

44. Pinder, J.; Schmidt, R.; Austin, S.; Gibb, A.; Saker, J. What is meant by adaptability in buildings? *Facilities* **2017**, *35*, 2–20. [CrossRef]

45. The Impact of Climate Change on Water Use in the Tourism Sector of Cyprus. Available online: https://www.researchgate.net/publication/268267953_The_impact_of_climate_change_on_water_use_in_the_tourism_sector_of_Cyprus (accessed on 4 July 2020).

46. Eurostat, European Commission. Eurostat Medstat II: Water and Tourism Pilot Study. Available online: https://ec.europa.eu/eurostat/documents/3888793/5844489/KS-78-09-699-EN.PDF/04c900a4-6243-42e0-969f-fc04f184a8b6 (accessed on 5 July 2020).

47. Kasim, A.; Gursoy, D.; Okumus, F.; Wong, A. The importance of water management in hotels: A framework for sustainability through innovation. *J. Sustain. Tour.* **2014**, *22*, 1090–1107. [CrossRef]

48. Cyclist Facilities. Available online: https://www.breeam.com/BREEAMUK2014SchemeDocument/content/07_transport/tra03.htm (accessed on 28 June 2020).

49. Operational Waste. Available online: https://www.breeam.com/BREEAM2011SchemeDocument/content/10_waste/wst03.htm (accessed on 28 June 2020).

50. Mohan, V.; Deepak, B.; Sharma, D. Reduction and Management of Waste in Hotel Industries. *Int. J. Eng. Res. Appl.* **2017**, *7*, 34–37. [CrossRef]

51. Acoustic Performance. Available online: https://www.breeam.com/BREEAMUK2014SchemeDocument/content/05_health/hea05.htm (accessed on 5 July 2020).

52. Visual Comfort. Available online: https://www.breeam.com/BREEAMUK2014SchemeDocument/#05_health/hea01_nc.htm%3FTocPath%3D5.0%2520Health%2520and%2520Wellbeing%7C_____1 (accessed on 5 July 2020).
53. Low and Zero Carbon Technologies. Available online: https://www.breeam.com/BREEAM2011SchemeDocument/content/06_energy/ene04.htm (accessed on 5 July 2020).
54. Enhancing Site Ecology. Available online: https://www.breeam.com/BREEAMUK2014SchemeDocument/content/11_landuse/le04.htm (accessed on 5 July 2020).

MDPI

St. Alban-Anlage 66

4052 Basel

Switzerland

Tel. +41 61 683 77 34

Fax +41 61 302 89 18

www.mdpi.com

Sustainability Editorial Office

E-mail: sustainability@mdpi.com

www.mdpi.com/journal/sustainability